HUM

PHYSIOLOGY

Jones & Bartlett Learning Titles in Biological Science

AIDS: Science and Society, Seventh Edition
Hung Fan, Ross F. Conner, & Luis P. Villarreal

AIDS: The Biological Basis, Sixth Edition
Benjamin S. Weeks & Teri Shors

Alcamo's Fundamentals of Microbiology, Body Systems Edition, Second Edition
Jeffrey C. Pommerville

Alcamo's Laboratory Fundamentals of Microbiology, Tenth Edition
Jeffrey C. Pommerville

Alcamo's Microbes and Society, Fourth Edition
Jeffrey C. Pommerville & Benjamin S. Weeks

Biochemistry
Raymond S. Ochs

Bioethics: An Introduction to the History, Methods, and Practice, Third Edition
Nancy S. Jecker, Albert R. Jonsen, & Robert A. Pearlman

Bioimaging: Current Concepts in Light and Electron Microscopy
Douglas E. Chandler & Robert W. Roberson

Biomedical Graduate School: A Planning Guide to the Admissions Process
David J. McKean & Ted R. Johnson

Biomedical Informatics: A Data User's Guide
Jules J. Berman

Botany: A Lab Manual
Stacy Pfluger

Case Studies for Understanding the Human Body, Second Edition
Stanton Braude, Deena Goran, & Alexander Miceli

Electron Microscopy, Second Edition
John J. Bozzola & Lonnie D. Russell

Encounters in Microbiology, Volume 1, Second Edition
Jeffrey C. Pommerville

Encounters in Microbiology, Volume 2
Jeffrey C. Pommerville

Encounters in Virology
Teri Shors

Essential Genetics: A Genomics Perspective, Sixth Edition
Daniel L. Hartl

Essentials of Molecular Biology, Fourth Edition
George M. Malacinski

Evolution: Principles and Processes
Brian K. Hall

Exploring Bioinformatics: A Project-Based Approach, Second Edition
Caroline St. Clair & Jonathan E. Visick

Exploring the Way Life Works: The Science of Biology
Mahlon Hoagland, Bert Dodson, & Judy Hauck

Fundamentals of Microbiology, Tenth Edition
Jeffrey C. Pommerville

Genetics: Analysis of Genes and Genomes, Eighth Edition
Daniel L. Hartl & Maryellen Ruvolo

Genetics of Populations, Fourth Edition
Philip W. Hedrick

Guide to Infectious Diseases by Body System, Second Edition
Jeffrey C. Pommerville

Human Biology, Eighth Edition
Daniel D. Chiras

Human Biology Laboratory Manual
Charles Welsh

Human Body Systems: Structure, Function, and Evironment, Second Edition
Daniel D. Chiras

Human Embryonic Stem Cells, Second Edition
Ann A. Kiessling & Scott C. Anderson

Laboratory Investigations in Molecular Biology
Steven A. Williams, Barton E. Slatko, & John R. McCarrey

Lewin's CELLS, Third Edition
George Plopper, David Sharp, & Eric Sikorski

Lewin's Essential GENES, Third Edition
Jocelyn E. Krebs, Elliott S. Goldstein, & Stephen T. Kilpatrick

Lewin's GENES XI
Jocelyn E. Krebs, Elliott S. Goldstein, & Stephen T. Kilpatrick

The Microbial Challenge: A Public Health Prespective, Third Edition
Robert I. Krasner & Teri Shors

Microbial Genetics, Second Edition
Stanley R. Maloy, John E. Cronan, Jr., & David Freifelder

Molecular Biology: Genes to Proteins, Fourth Edition
Burton E. Tropp

Neoplasms: Principles of Development and Diversity
Jules J. Berman

Precancer: The Beginning and the End of Cancer
Jules J. Berman

Principles of Cell Biology
George Plopper

Principles of Modern Microbiology
Mark Wheelis

Principles of Molecular Biology
Burton E. Tropp

Science and Society: Scientific Thought and Education for the 21st Century
Peter Daempfle

Strickberger's Evolution, Fifth Edition
Brian K. Hall

Symbolic Systems Biology: Theory and Methods
M. Sriram Iyengar

20th Century Microbe Hunters
Robert I. Krasner

Understanding Viruses, Second Edition
Teri Shors

HUMAN PHYSIOLOGY

CHERYL WATSON
Central Connecticut State University

JONES & BARTLETT
LEARNING

World Headquarters
Jones & Bartlett Learning
5 Wall Street
Burlington, MA 01803
978-443-5000
info@jblearning.com
www.jblearning.com

Jones & Bartlett Learning books and products are available through most bookstores and online booksellers. To contact Jones & Bartlett Learning directly, call 800-832-0034, fax 978-443-8000, or visit our website, www.jblearning.com.

Substantial discounts on bulk quantities of Jones & Bartlett Learning publications are available to corporations, professional associations, and other qualified organizations. For details and specific discount information, contact the special sales department at Jones & Bartlett Learning via the above contact information or send an email to specialsales@jblearning.com.

Production Credits
Chief Executive Officer: Ty Field
President: James Homer
Chief Product Officer: Eduardo Moura
Executive Publisher: William Brottmiller
Publisher: Cathy L. Esperti
Senior Acquisitions Editor: Erin O'Connor
Editorial Assistant: Raven Heroux
Production Manager: Louis C. Bruno, Jr.
Marketing Manager: Lindsay White
Manufacturing and Inventory Control Supervisor: Amy Bacus
Composition: Laserwords Private Limited, Chennai, India
Illustrations: Elizabeth Morales
Cover Design: Kristin E. Parker
Cover Illustrator: David Forbes
Photo Research and Permissions Coordinator: Lauren Miller
Cover Image: © Nick White/age fotostock
Printing and Binding: Courier Companies
Cover Printing: Courier Companies

To order this product, use ISBN 978-1-284-03517-9

Library of Congress Cataloging-in-Publication Data
Watson, Cheryl L., 1948– author.
Human physiology / Cheryl Watson. — First edition.
pages cm
ISBN 978-1-284-03034-1
1. Human physiology—Textbooks. 2. Human physiology. I. Title.
QP34.5.W378 2015
612—dc23 2013015276

6048

Printed in the United States of America
18 17 16 15 14 10 9 8 7 6 5 4 3 2 1

To my husband, Steve, for all his patience, humor, and support.

Brief Contents

Contents

Preface

This book was written for human physiology students. During 18 years of teaching physiology I've discovered that students learn physiology most quickly and retain it the longest when it is presented as a series of concrete problems. This observation gave birth to my use of case studies and storytelling as tools for understanding human physiology. This text uses case studies involving healthy adults to introduce all the most important concepts in physiology and simultaneously help students to think as physiologists. To accomplish this, students must see the physiological consequences of any change in our physical state, such as exercise, dehydration, or hunger. These consequences involve multiple organ systems and a whole-body response. This text is designed to emphasize these interactions and encourage students to see physiology as a cascade of mechanisms.

Physiology is fascinating in its own right. We all want to know why we get a fever, why a cut finger stops bleeding, or why being frightened makes the heart beat faster. Physiology, however, is the foundation of medicine, pharmacology, and pathophysiology. A good working knowledge of physiology, therefore, is indispensable for those scientific disciplines or any health profession that utilizes them. To emphasize these connections, pharmacology case studies and clinical case studies are included at the end of each chapter.

Brevity

This text is designed to be used in its entirety during a single-semester, undergraduate, human physiology course, so it is shorter than many standard texts. No organ system need be eliminated due to a lack of time, and, therefore, students have the opportunity to learn about the interplay of all the organ systems. The text is not intended to be all-inclusive nor does it attempt to be a comprehensive resource for all of human physiology. It is constructed to give undergraduate students an integrated framework that will allow them to understand the fundamentals of human physiology from protein to organ system to whole body response. With this background, students will be well-prepared for studying advanced physiology, pathophysiology, or pharmacology.

Organization

There are several techniques in this book designed to help students grasp and use essential physiological concepts.

Chapter Sequence

The chapters are in a slightly different order than commonly found in other physiology textbooks. Systems that affect all other organ systems, such as the autonomic nervous system, are introduced early in the text. Because these controller systems are covered at the beginning, they can be applied to other systems later in the text. Thus, immediately following the chapter on cell physiology is the chapter on autonomic physiology, which manages most of our vegetative functions, followed by a chapter on endocrine physiology, which relies heavily on the receptors and signal transduction introduced in the first chapter to coordinate our activity in a long-term fashion. My intention is to continuously reuse concepts throughout the text. In this way, students will be able to integrate concepts introduced earlier, such as receptors, with organ system responses. This will help students see the human being as a coordinated physiological system. Throughout the book, the organismal consequences of protein function or hormonal processes will be introduced, so students can have a real-world example of molecular phenomena. I will always "close the loop," so that relevant connections are illustrated and explained, not just implied. Too often, as instructors, we anticipate that students will make connections that they simply are not scientifically educated enough to make.

End-of-Chapter Pharmacology Case Studies

Pharmacology case studies appear at the end of each chapter. While each pharmacology case is system based, it often refers back to specific receptors or metabolic processes that have been studied in previous chapters. This helps students to see the cumulative nature of physiology and its application to pharmacology. To follow through with tissue receptor distribution and the possible effects when they are targeted by drugs, there are end-of-chapter pharmacology questions. For example, receptors and their signaling pathways, introduced in the first chapter, are utilized again in the final chapter. Because receptors are frequently drug targets, understanding that the same receptor type can exist in many tissues and cause different actions is important.

Case Studies: Normal Physiology and Clinical Case Studies

Case studies at the beginning of a chapter are referenced throughout the chapter, bridging the gulf between academic material and application. This is important because students sometimes compartmentalize what they know, separating academic concepts from the "real" world. The case studies are based on normal physiology, so they are events experienced by healthy young people, allowing students to see physiology at work within themselves. These case studies are frequently cumulative, so new systems are added to those previously considered. All of this is designed to promote thinking about human physiology in an integrated way.

Finally, there are clinical case studies at the end of each chapter. These cases are based on two prevalent diseases: diabetes and heart failure. These two diseases form a clinical thread throughout the text, illustrating both the commonality of cellular mechanisms between tissues and how organ systems interact with one another. Diabetes, for example, is introduced during the cell physiology chapter (1) through a discussion of glucose transport and insulin receptor signaling. It reappears in the endocrine chapter (3) in a section on glucose regulation; in the sensory-motor system chapter (5) regarding loss of sensory ability; in the digestive system chapter (6) in a section on nutrient distribution; in the cardiovascular system chapter (7) during a discussion of vascular inflammation; in the immune system chapter (4), also under inflammation; and finally in the renal system chapter (9) in discussion of glucose uptake. The case studies of these two conditions reintroduced in most of the chapters illustrate the far-reaching consequences of apparently simple physiological malfunctions and help students understand the interdependencies of organ systems.

To the Student

Read each opening case study carefully as it is referenced throughout the chapter and will help you apply physiological principles to your daily life. You will notice quickly that this text does not emphasize memorization of facts but an understanding of process. This is a slightly different way of studying, so don't forget to sit back after your reading and ask "How do I move?" or "How does the sugar in the doughnut I just ate get into cells?" or "How do I distinguish between touch and pain?" and then try to answer your own question. Directed daydreaming is an important component of learning physiology!

To the Instructor

Stories are powerful teaching tools and an important part of human history. Stories help us understand and are easy to remember. In science, storytelling is accomplished through case studies. Normal physiology case studies are incorporated into this text as an application of principles. These will help you bring concepts to life. Additional case studies, both pharmacology cases and clinical cases are available at the end of the chapter and in the instructional materials on-line.

Resources

For the Instructor

Compatible with Windows and Macintosh platforms, the Instructor's Media CD provides instructors with the following:

- The *PowerPoint Image Bank* provides the illustrations, photographs, and tables (to which Jones & Bartlett Learning holds the copyright or has permission to reproduce digitally) inserted into PowerPoint slides. You can quickly and easily copy individual images or tables into your existing lecture presentations.
- The *PowerPoint Lecture Outline* presentation package provides lecture notes and images for each chapter of *Human Physiology*. Instructors with the Microsoft PowerPoint software can customize the outlines, art, and order of presentation.

The *Test Bank*, provided as text files (with LMS-compatible options available), is offered online as a secure download. Please contact your sales representative for more information.

For the Student

Jones & Bartlett Learning has prepared extensive electronic support to further student understanding in a website to accompany *Human Physiology* (go.jblearning.com/HumanPhysCWS). It is designed to help you organize, prioritize, and apply what you learn from the text and in class. The website hosts a variety of review materials, including multiple-choice study quizzes, chapter summaries, an interactive glossary for key-term review, interactive flashcards, crossword puzzles, and a host of topic-specific web links to help you use this vast source of information. In tandem with the case studies found in the text, additional case studies on asthma and chronic obstructive pulmonary disease (COPD) have been developed and will be hosted exclusively on the companion site. Access to this site is free with every new print copy of the text and is available for purchase separately at go.jblearning.com/HumanPhysCWSAccess.

Digital Learning Solutions

To be used in conjunction with the printed textbook or as a stand-alone course resource, Jones & Barlett Learning offers **Navigate Human Physiology**, a completely customizable, comprehensive, and interactive courseware solution for the undergraduate introductory human physiology course. **Navigate Human Physiology** transforms how students learn and instructors teach by bringing together authoritative and interactive content aligned to course objectives with student practice activities and assessments and learning analytics reporting tools. **Navigate Human Physiology** empowers faculty and students with easy-to-use, web-based curriculum solutions that optimize student success, identify retention risks, and improve completion rates.

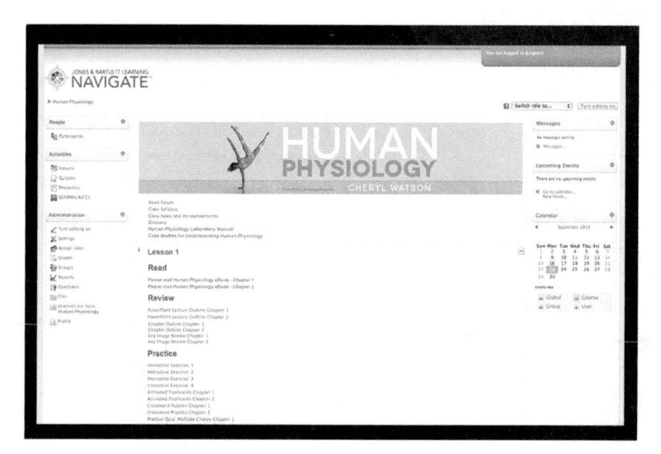

Incorporated into **Navigate Human Physiology** is **Navigate eFolio**, an interactive eBook with enhanced activities and engaging learning tools. **Navigate eFolio** is an exciting new choice for students and instructors looking for a more interactive learning experience. **Navigate eFolio** reinforces important concepts through exercises, enhanced diagrams, activities, videos, and animations—bringing key concepts to life! Students can practice what they read and see immediate feedback on their performance. **Navigate eFolio: Human Physiology** is also available for stand-alone purchase by visiting go.jblearning.com/NavigateEfolioHumanPhys.

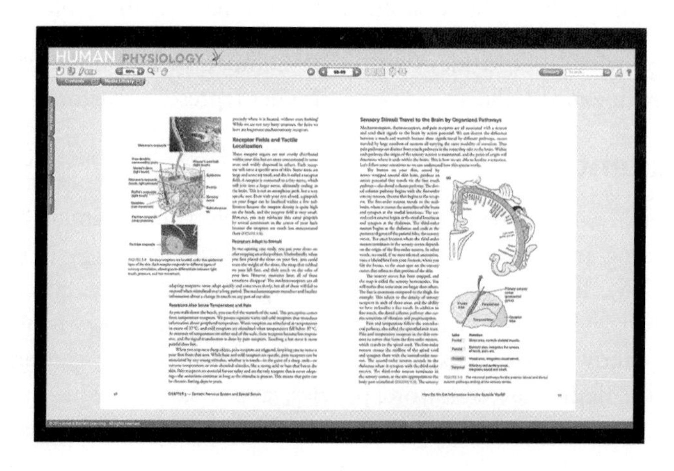

Acknowledgments

I would like to thank the team at Jones & Bartlett Learning, who very graciously and professionally guided me through the production process: Erin O'Connor, Michelle Bradbury, Rachel Isaacs, Louis Bruno, and Lauren Miller have been a joy to work with. Elizabeth Morales, always charming, did an excellent job with the art program.

Most importantly, this text would have been impossible without the kind input from hundreds of human physiology students at Central Connecticut State University. Their questions and thoughtful evaluation of my course materials inspired the organization of the text. Several recent classes used drafts of the text and have generously provided insights for its improvement. Their enthusiasm for this text was heartening and made all revisions easy work.

Cheryl Watson

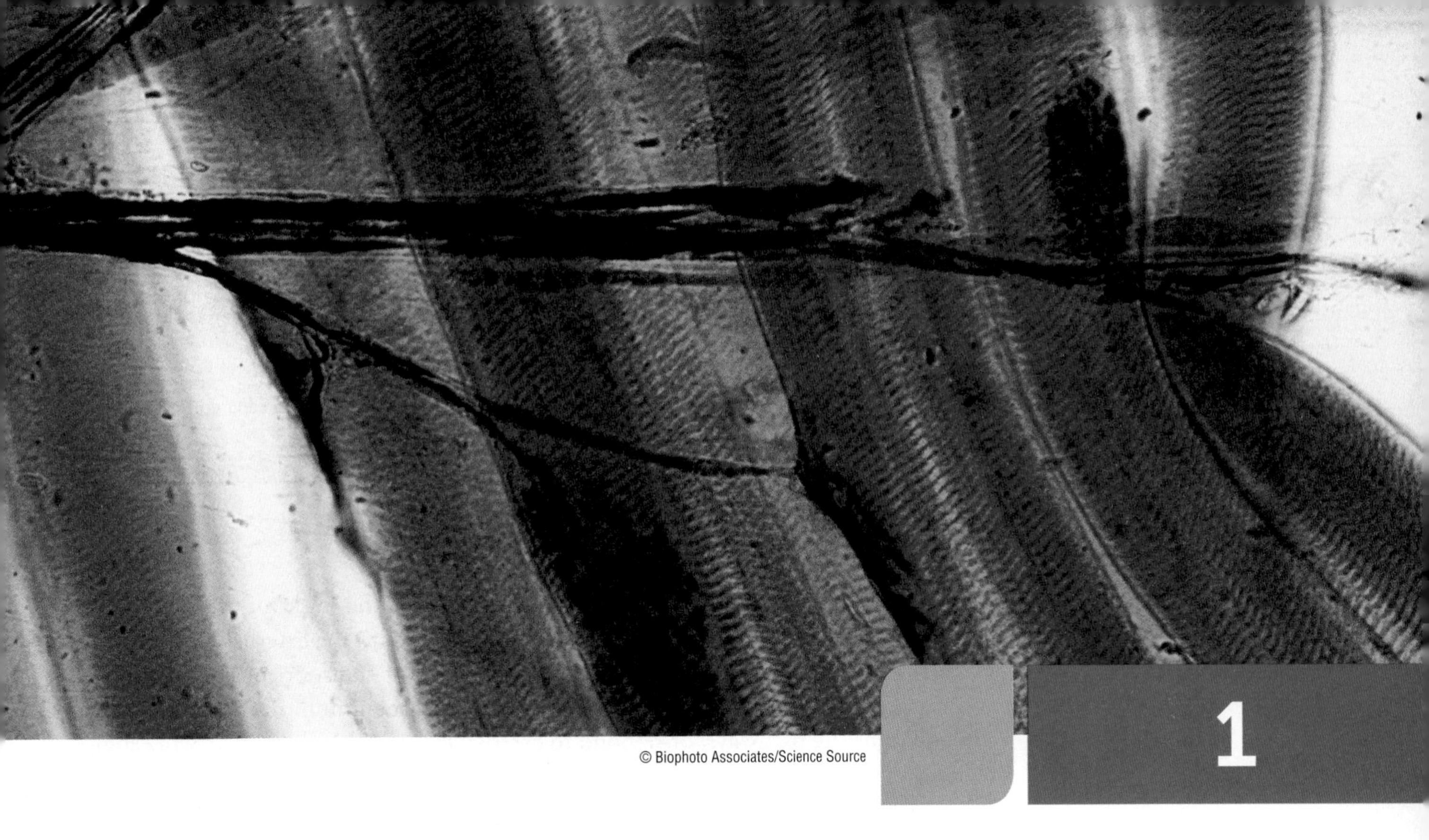

1

Cellular Physiology

Case 1

Last month, you decided to increase your leg muscle strength, which you notice is working nicely. It is late afternoon, and you have just finished your regular three-mile run. The workout has made you both hungry and thirsty, so you gratefully break for a large bottle of water and a candy bar before you take a shower. As you stand in the shower, you begin to wonder about where the water and sugar from the candy bar were going after you swallowed them. How did the candy bar contribute to renewed energy? What is happening to your legs to increase muscle strength? How do muscles contract? How do cells communicate with one another?

Introduction

If the cells within tissues are to work together to perform the "task" of the tissue, there must be communication between cells and responsiveness to the external environment. Cellular physiology is concerned with the mechanism of transport of nutrients, ions, and water into and out of the cell, as well as how cells communicate with each other through signaling pathways, or respond to external cues. In this chapter we will explore (1) how the body is separated into compartments, between which all transport is regulated, (2) how cells communicate with each other, electrically or chemically, (3) how proteins are made, and (4) how we make the energy to support all this activity.

Compartmentalization: Cells Are Separated from Extracellular Fluid by a Plasma Membrane

Cells are enclosed by a lipid bilayer—a double layer of phospholipids that is impermeable to large molecules and charged ions. The basic composition of the lipid bilayer is two layers of phospholipids that arrange spontaneously, with the phosphate heads facing the extracellular fluid (ECF) and intracellular fluid (ICF) (i.e., the water layers), and the fatty acid tails oriented toward the center, the hydrophobic core of the membrane (**FIGURE 1.1**). Cholesterol is an inherent part of the membrane and serves to stiffen it. The plasma membrane is highly fluid, with the consistency of olive oil, yet this oil-like layer is an effective barrier to large, charged, or hydrophilic molecules. Little movement across this membrane would be possible if it were not for the proteins that float in this lipid sea. The proteins, partially mobile within the bilayer, form channels and transporters, which regulate movement of large or charged molecules between the ECF and the inside of the cell. This lipid bilayer with its integral proteins forms a semipermeable membrane through which water, ions, and nutrients can cross in a regulated way.

The ECF on the outside of this cell has a very different composition from the fluid within the cytoplasm. The ECF is high in Na^+, high in Ca^{2+}, and low in K^+, while the ICF is low in Na^+, low in Ca^{2+}, and high in K^+. This inequality of ions is maintained by the lipid bilayer and, as we will see later, is a source of potential energy. Movement of these ions from the ECF to the ICF or from the ICF to ECF can occur only through membrane proteins, such as ion channels or pumps. Larger molecules, such as glucose or amino acids, must also be moved across the plasma membrane via

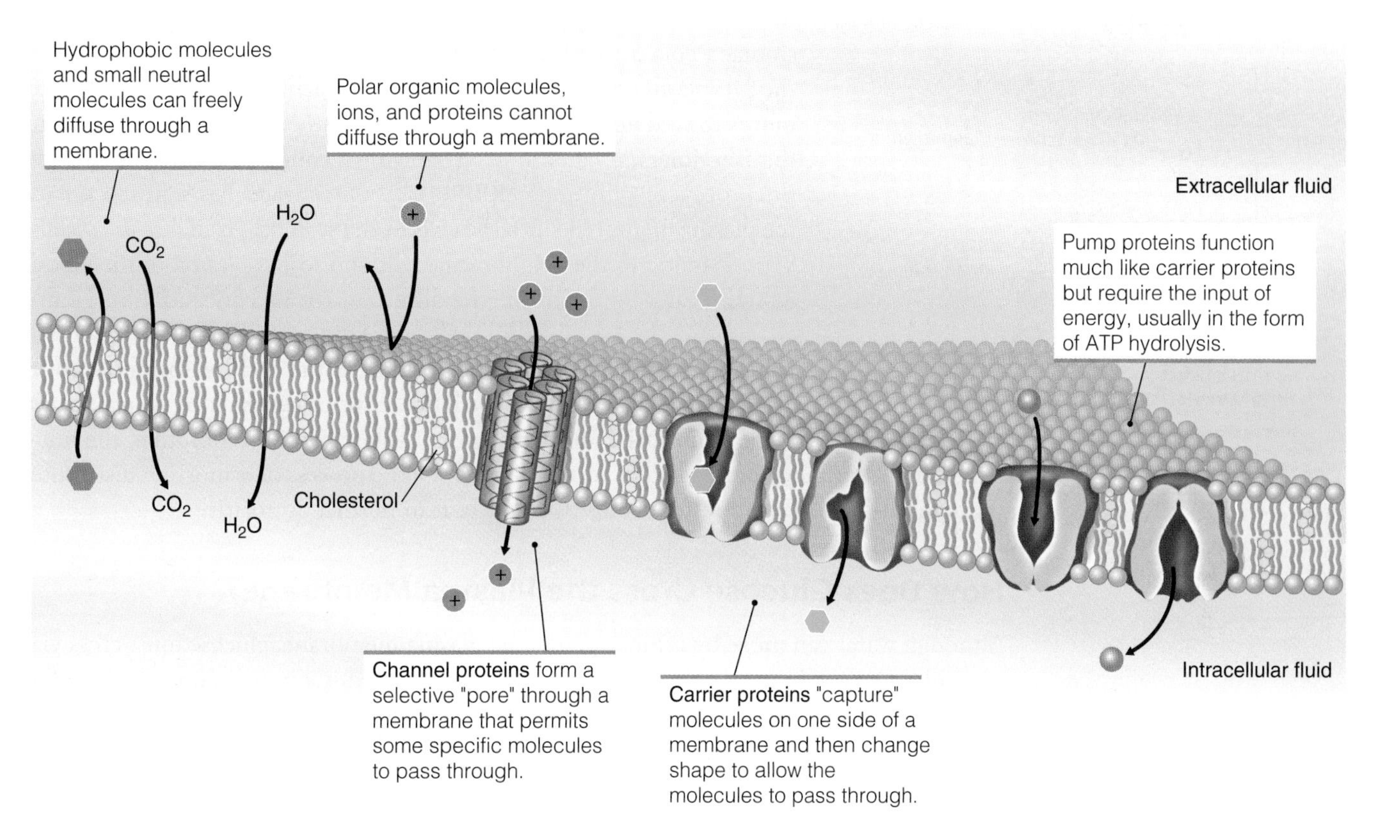

FIGURE 1.1 The plasma membrane and associated membrane proteins.

transporters. Even water, which can cross the lipid membrane slowly, crosses more efficiently through water channels. Thus, movement of nutrients and ions is closely regulated across the plasma membrane. Only gases such as O_2 or CO_2 can freely diffuse across membranes.

Compartmentalization: Extracellular Fluid Is Separated from Vascular ECF

Not only are cells separated from ECF, but there are separate compartments within the ECF. Cells and the ECF that surrounds them are organized into tissues and organs. These tissues are separated from the blood vessels that serve them by the cells that form the blood vessels. Thus, water, nutrients, and gases must move from blood plasma through the blood vessel endothelial cells, to the ECF of the tissue, and then into the tissue cell itself. Waste products and metabolites must make the reverse trip from the cell to the bloodstream. How is all this movement between compartments accomplished?

Water Moves Between Compartments by Osmosis

In our exercise scenario, you drank a large bottle of water after your workout. Where does this water go? How is it distributed between the body's compartments? Later, we will explore the mechanics of digestion, but for now, let's think about water moving from the

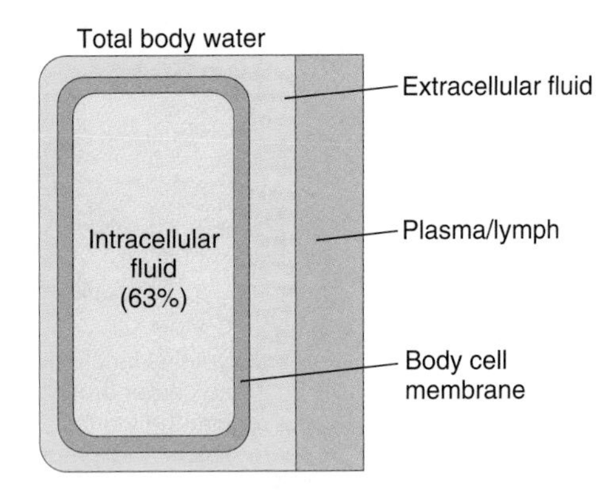

FIGURE 1.2 Body compartments and movement of water between them.

stomach to the blood vessels. Water in the stomach is hypo-osmotic to the cells within the stomach itself. That means that the water in the stomach contains fewer osmotically active particles than the water of the cytoplasm of the cells that line the stomach. During osmosis, water moves toward an area of higher solute. Another way of thinking about this is to imagine water moving "down" its concentration gradient, like the process of diffusion (**FIGURE 1.2**). Once water has diffused across the plasma membrane into the cells of the stomach, it makes these cells hypo-osmotic to the neighboring cells and to the plasma of the blood vessels. Water will continue to diffuse down its concentration gradient toward the blood, ultimately increasing your extracellular blood volume. Osmosis will continue until the entire body is isosmotic. Your exercise may have caused a slight dehydration, making your tissues hyperosmotic, i.e., having a greater solute concentration than the normal 300 mOsm. The bottle of water you drank will, by the process of osmosis, redistribute throughout the body and restore your ECF and ICF to normal osmolarity.

How Does Glucose Cross the Plasma Membrane?

Although water can move by osmosis across the plasma membrane, glucose must cross via a membrane protein, a transporter. The process is still based upon diffusion, moving of a substance from an area of higher concentration to lower concentration, but this time the movement must occur through an integral membrane protein. Because the number of proteins in the membrane will limit the amount of glucose transport possible, this process is called facilitated diffusion, because the diffusion must be facilitated by the membrane protein.

The glucose transporter exists in several isoforms, but the one we will discuss here is GLUT4, which facilitates movement of glucose across the plasma membrane of skeletal muscle, cardiac muscle, and adipose cells. The basic mechanism of this transporter is simple. Glucose binds to the extracellular side of the membrane protein, and its binding causes a conformational change in the transporter, exposing the bound glucose to the intracellular space (**FIGURE 1.3**). Glucose then diffuses down its concentration gradient into the ICF. The more transporters there are in the membrane, the faster the glucose can move into the intracellular space. The glucose in the candy bar you ate will be transported

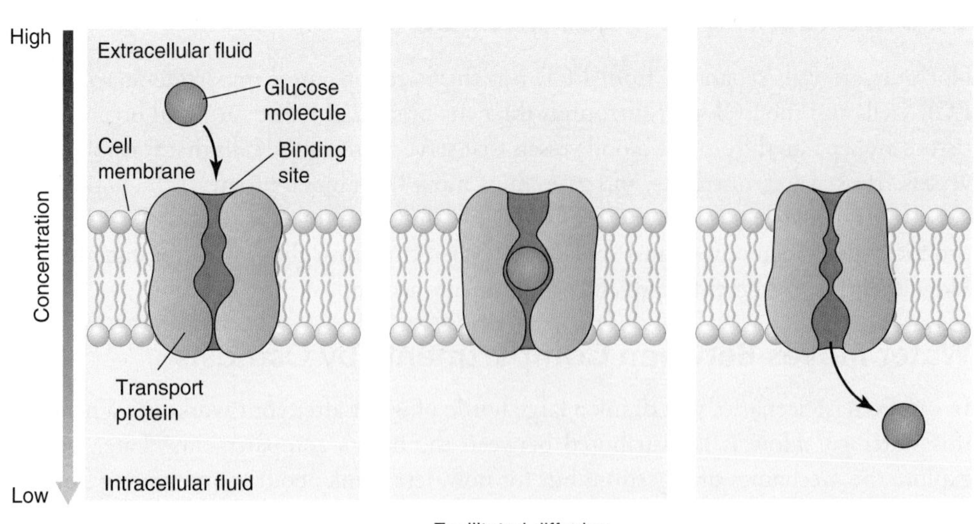

FIGURE 1.3 Glucose binds to the glucose transporter and is moved from the extracellular space to the intracellular space.

from the blood plasma, to ECF, to ICF via facilitated diffusion through membrane transporters. In the absence of transporters, glucose cannot enter the ICF.

Some Transport Requires Energy

Facilitated diffusion is a process that does not require energy. It is simple diffusion through a protein carrier. However, some membrane proteins engage in active transport, or transport that requires adenosine triphosphate (ATP) or physiological "work". The most ubiquitous of these is the Na^+/K^+ ATPase, also known as the Na^+K^+ pump. As you recall, the concentration of Na^+ in the ECF is much higher than in the ICF. At the same time, K^+ is in higher concentration on the inside of the cell and lower on the outside, in the ECF. The protein that helps to maintain this disequilibrium is the Na^+K^+ pump (**FIGURE 1.4**). An increase in intracellular Na^+ allows binding of Na^+ to the cytosolic side of the Na^+K^+ pump, a change in conformation, and a release of Na^+ to the extracellular space. K^+ binds to the extracellular face of the Na^+K^+ pump and is transported into the cell. However, both of these ions are moving against their concentration gradient, so diffusion is not possible.

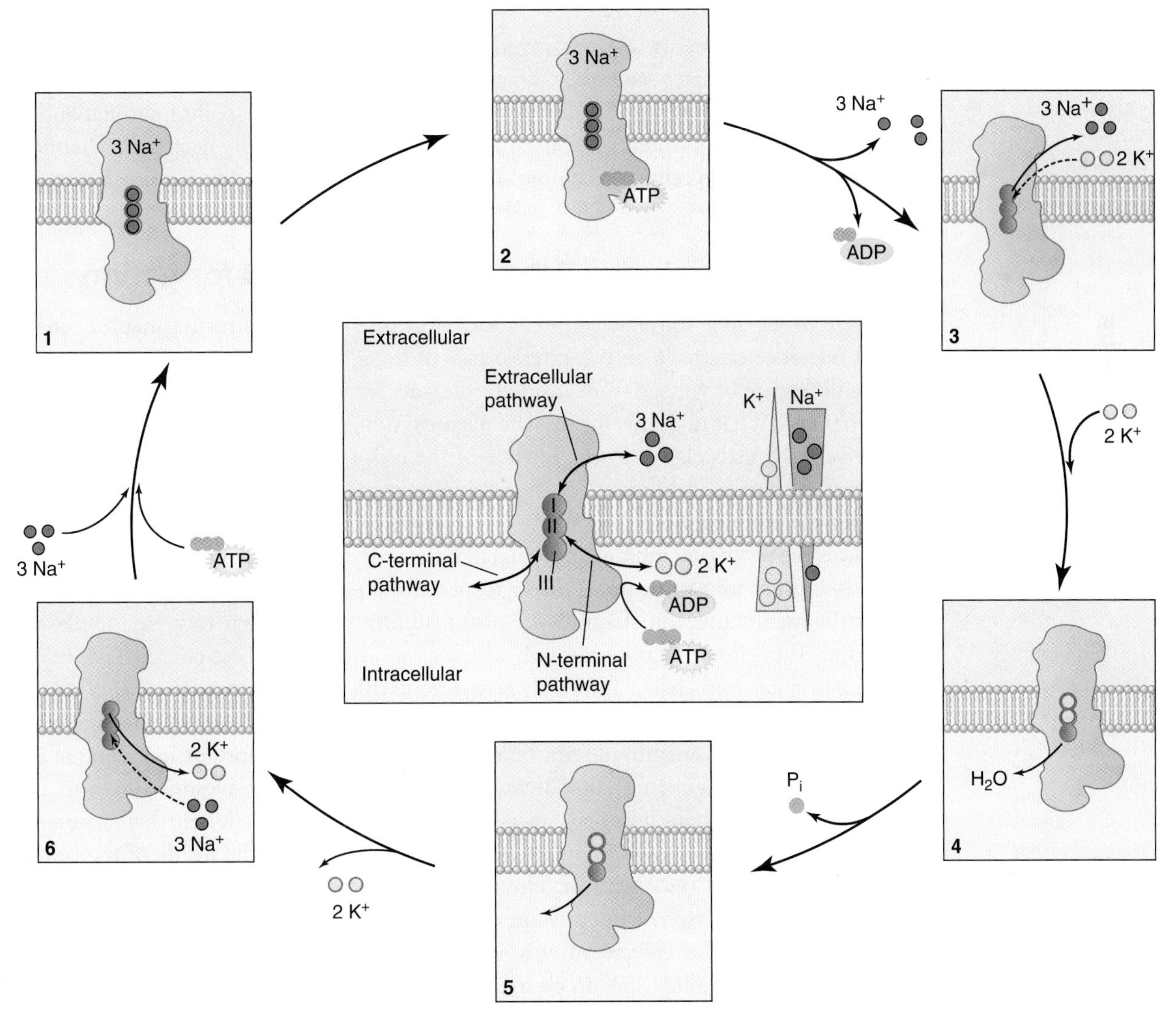

FIGURE 1.4 Na^+ and K^+ are moved across the membrane by a series of molecular conformational changes accompanied by ATP hydrolysis.

Movement of an ion against a concentration gradient, i.e., from an area of lower concentration to an area of higher concentration, requires energy in the form of ATP. For each ATP hydrolyzed, three Na^+ ions are pumped out of the cytosol and two K^+ ions are brought into the cytosol from the ECF. This imbalance of positive charges—fewer on the inside than on the outside—contributes to resting membrane potential, as we will see later. The Na^+K^+ pump is expressed in nearly every cell of the body and is so active that its operation accounts for 30% of our resting energy use.

While the Na^+K^+ pump is the most common active transporter, there are many others. In fact, any time an ion is moved against a concentration gradient, an ATPase or ion pump will be required. Ca^{2+} is moved against a concentration gradient by a Ca^{2+} ATPase within the endoplasmic reticulum (ER) membrane, or a different Ca^{2+}ATPase within the plasma membrane. H^+ ions are similarly moved by H^+ ATPases. The basic principle to keep in mind is that movement of an ion or molecule against a concentration gradient will always require energy.

Communication: How Do Cells Coordinate Activities or Change Function?

In order for organ systems to work together and tissues to perform the same function, cells must communicate—and they do so continuously. Communication can be fast and short-lived, or slower and sustained. Fast communication is usually neuronal and is accomplished by action potentials. Sustained communication usually occurs via chemical transmitters binding to cellular receptors. While very different, both forms of communication are essential.

The Resting Membrane Potential: Cells Poised for Action

If we were to set up a voltmeter, insert a fine electrode inside of a resting neuron, place another reference electrode on the exterior face of the plasma membrane, and then measure the difference in voltage from inside to outside, we would record a negative voltage, about −70 mV. What does this mean? The negative value means that the inside of the cell is negatively charged relative to the outside of the cell, i.e., it has fewer positive charges. This electrical potential difference is the resting membrane potential, which can provide energy for communication.

How is the resting membrane potential maintained? The plasma membrane is not permeable to ions, so ions must travel through specialized proteins—ion channels—in order to cross the membrane. Ion channels are multi-subunit proteins that traverse the plasma membrane. They provide a pore through which an ion can pass. These channels are selective for particular ions, being internally structured with a selectivity pore, which allows passage of only one ion. Ion channels are not always open, so, in this way, movement of ions across the plasma membrane can be regulated. The most important ion channel for maintaining the resting membrane potential is the K^+ leak channel. Despite its name, the K^+ leak channel is not always open but is open at about −70 mV. When this channel is open, K^+ ions can move down their concentration gradient from the inside of the cell to the outside of the cell (**FIGURE 1.5**). However, as more positive ions accumulate on the outside of the cell, an electrical repulsion occurs, slowing the movement of ions through the K^+ leak channel. This movement of K^+ ions down a concentration gradient, but against an electrical gradient, establishes an electrochemical equilibrium at about −70 mV.

K^+ ions are attracted to negatively charged proteins within the cell, which limits their movement through the K^+ leak channel. In addition, the Na^+/K^+ ATPase contributes to the magnitude of the resting membrane potential. Without the Na^+/K^+ ATPase, the resting membrane potential would be about 5 mV more positive. Certainly the distribution of other ions and the probability of other ion channel openings could affect resting membrane potential, and does so in disease or because of some drugs. However, in a normal, healthy person, the K^+ leak channel is the primary determinant of resting membrane potential. The resting membrane potential is a potential energy for opening of channels and generation of an action potential, the fastest form of intercellular communication.

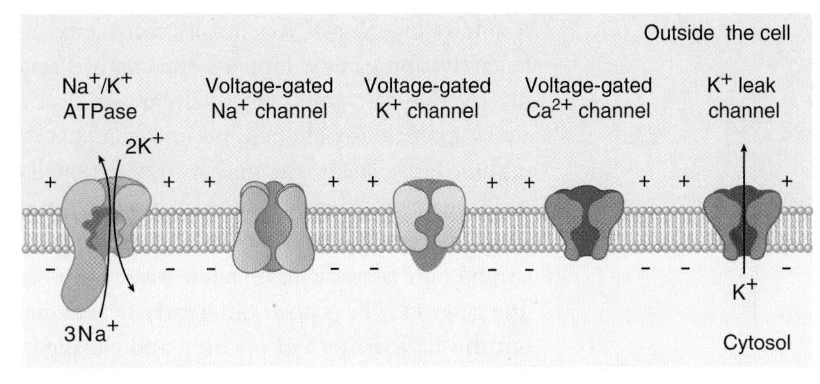

FIGURE 1.5 The resting membrane potential is maintained by the K^+ leak channel, which is voltage independent, open, and allows the passage of K^+ ions down a concentration gradient.

Remember that the ion distribution across a plasma membrane is asymmetrical: there is a high concentration of Na^+ on the outside of the cell, but low concentration inside and a high concentration of K^+ ions inside the cell, but a low concentration outside. There is also a high extracellular $[Ca^{2+}]$ relative to the intracellular space. Each of these ions passes through specific ion channels that are voltage-gated. The existence of voltage-gated ion channels is simple to understand as long as we recall protein structure. Amino acids are decorated with side chains, many of which are charged. The primary amino acid chain folds into a secondary α-helix or β-sheet, then forms into a tertiary structure, and finally assembles with other protein chains for the quaternary structure. These side chains attract or repel each other, forming stable ion channel conformations and a voltage sensor. The voltage sensor is simply an area of the protein, generally embedded in the intramembrane section of the channel, that responds to changes in local voltage. The protein's response is simply to flex toward or away from the nearby charge, thus changing the overall protein conformation. The opening of voltage-gated ion channels is regulated by these voltage sensors. Each ion channel type has a specific range of voltages that cause it to assume an open conformation.

The Na^+ Channel: A Typical Voltage-Gated Ion Channel

Let's use the voltage-gated Na^+ channel as an example. The voltage-gated Na^+ channel is a multi-subunit channel that allows Na^+ to move from outside the cell to the inside of the cell (**FIGURE 1.6**). The Na^+ channel has an activation gate that opens or closes the channel to the outside of the cell, and an inactivation gate that opens and closes the channel to the inside of the cell. Each of these gates is regulated independently. At resting membrane potential, the Na^+ channel will be closed, with the activation gate blocking the ion channel pore. The inactivation gate, however, will likely be open. In this state, the Na^+ channel is closed, but ready to open. If positive charges accumulate near the Na^+ channel on the interior face of the membrane, the voltage sensor will cause a change in protein conformation. This will increase the probability of the activation gate opening. As the membrane potential depolarizes (becomes more positive) from −70 mV toward −55 mV, the likelihood of Na^+ channels opening increases.

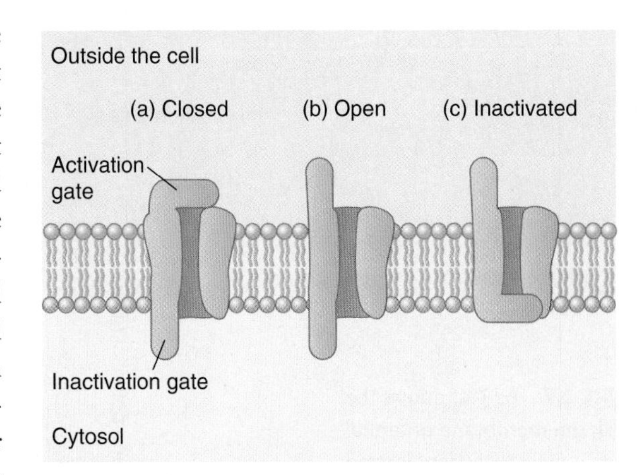

FIGURE 1.6 The Na^+ channel has an activation gate on the extracellular side of the protein and an inactivation gate on the cytoplasmic side of the protein. The activation gate opens in response to changes in voltage, while the inactivation gate closes over time.

A voltage of −55 mV is generally considered a threshold voltage at which most Na^+ channel activation gates will open. The channel stays open for about 2 milliseconds before the inactivation gate on the intracellular side closes, inactivating the channel. Even though the activation gate is still open, no ions can pass through the protein pore. Over time, several milliseconds usually, the inactivation gate will return to its open or resting state, the activation gate will close, and the channel will return to its original state—closed, but ready to open. The time required for recovery from inactivation is important in action potential conduction, as we will see later. Most ion channels behave similarly to the Na^+ channel. The gates may be shaped differently or may have different voltage sensitivities or kinetics, but this basic pattern of opening and closing is common to most ion channels.

Generation of an Action Potential

Electrical signaling between cells, with the electrical current carried by ions instead of electrons, is the fastest and most common form of intercellular communication. An action potential can carry a signal a long distance in a very short time. When you wiggle your toes, that action potential began in your brain. Yet, the time between thinking about wiggling your toes and doing it is extremely short! How is an action potential generated and continued over this long distance?

Remember that the resting membrane has potential energy, as indicated by the −70 mV resting membrane potential. Any depolarization of the resting membrane will move the voltage toward −55 mV, the threshold potential for Na^+ channels. Once the first Na^+ channel opens, allowing Na^+ ions into the cell, the membrane potential will become even more depolarized (**FIGURE 1.7**). Thus, opening even a few Na^+ channels can begin a feed-forward event that depolarizes the membrane and causes many Na^+ channels to open.

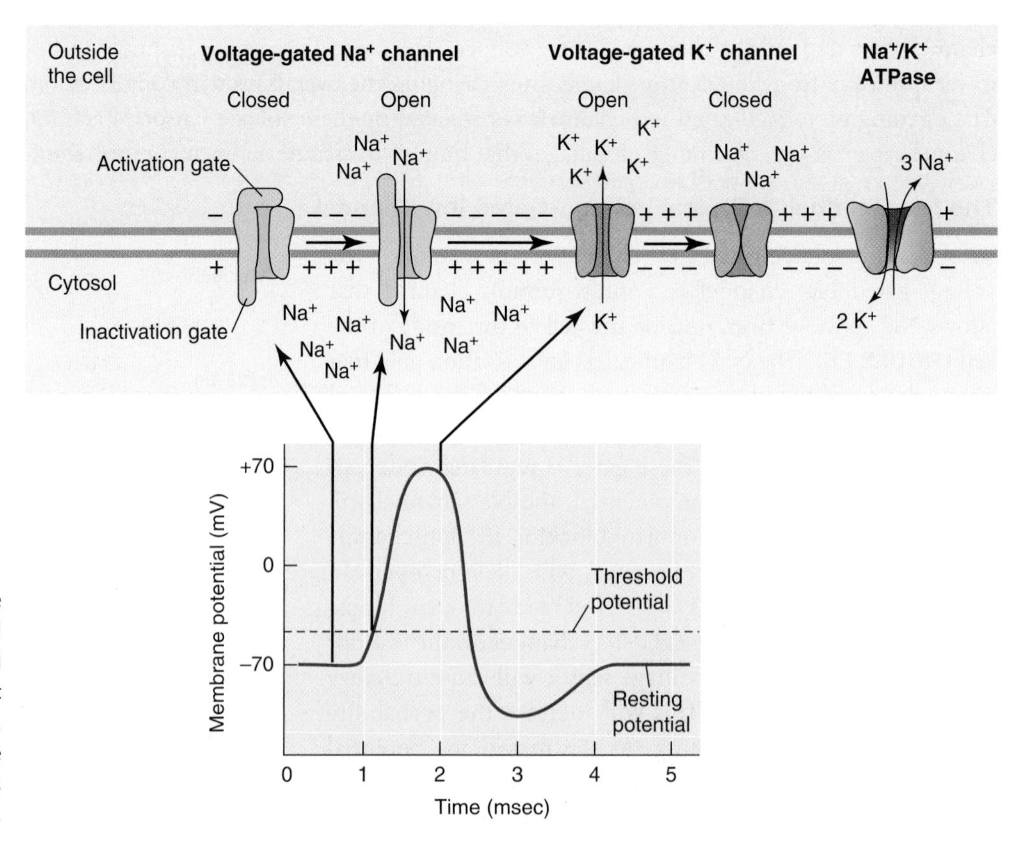

FIGURE 1.7 As Na^+ enters the cell, the membrane potential becomes increasingly depolarized until threshold is reached. At positive voltages, K^+ channels open, allowing positive K^+ ions to leave the cell, repolarizing it. Na^+ channels inactivate during repolarization.

This is the beginning of an action potential. The membrane actually depolarizes to +30 mV in response to this increase in intracellular Na^+ ions. Once opened, the Na^+ channels will inactivate and become refractory, meaning they will fail to open again until they are reactivated. This is a short time, but a finite time. New Na^+ channels farther along the neuron can open, but the ones already opened will become unavailable. This phenomenon "moves" the action potential along to new portions of the neuron and gives the action potential a direction. The initial section of the neuron, where the action potential began, has reached 0 mV to +30 mV, a voltage range at which voltage-gated K^+ channels can open. The electrochemical gradient is favorable for K^+ flow out of the cell into the extracellular space. Once again, the opening of K^+ channels proceeds down the neuron as the membrane potential enters the voltage range of these channels, 0 mV or more positive. The movement of positive charges out of the cell restores the resting membrane potential of −70 mV before the K^+ channels close. Only a small number of Na^+ and K^+ must cross the membrane to create the necessary change in voltage, and these are returned to respective spaces by the Na^+K^+ pump.

How Does Cellular Communication Result from an Action Potential?

An action potential is a simple change in membrane voltage that propagates along a neuron, muscle, or any excitable cell. How does that qualify as communication? By itself, the action potential serves to move a potential signal from one place to another but doesn't usually convey information on its own. Let's use an action potential in an α-motor neuron, connecting to a skeletal muscle cell, as an example. The α-motor neuron starts in the spinal cord and connects to skeletal muscle cells. We can think specifically about muscle cells in the legs, because you have gone out for a run this afternoon.

An action potential began in this neuron at a portion of the neuron near the neuronal cell body where there is a dense concentration of Na^+ channels. The action potential propagated from the spinal cord to the muscles of the leg. The neuromuscular junction is the point where the neuron meets the skeletal muscle cell; it is a specialized area of communication between the nerve and the muscle (**FIGURE 1.8**). As the action potential reaches the end of the neuron, the synaptic bulb, it depolarizes the membrane as usual—but in the synaptic bulb, there is another type of ion channel, a Ca^{2+} channel, which opens in response to the depolarization, allowing Ca^{2+} into the cell. Calcium ions inside the synaptic bulb bind to synaptogamin, a protein with a Ca^{2+} binding region, which initiates a series of protein interactions that will move vesicles from the intracellular space of the neuron to the plasma membrane, where they will fuse and exocytose their contents into the synaptic space. These vesicles contain a neurotransmitter, acetylcholine. Thus, the action potential—a simple change is membrane voltage—has opened an ion channel and caused the release of a powerful chemical (acetylcholine) into the space between the neuron and the muscle. The acetylcholine then binds to acetylcholine receptors on the skeletal muscle surface. These receptors are actually ligand-gated ion channels, which open when two molecules of acetylcholine bind. When they open, they allow Na^+ to flow into the skeletal muscle. This begins the depolarization of the membrane in the immediate vicinity of the neuromuscular junction. Voltage-gated Na^+ channels located on the skeletal muscle membrane will now begin to reach threshold and propagate a new action potential along the skeletal muscle membrane, so that the communication of signal continues, ultimately resulting in muscle contraction. This example, where electrical signaling via an action potential causes a chemical signal, is a common theme in intercellular communication.

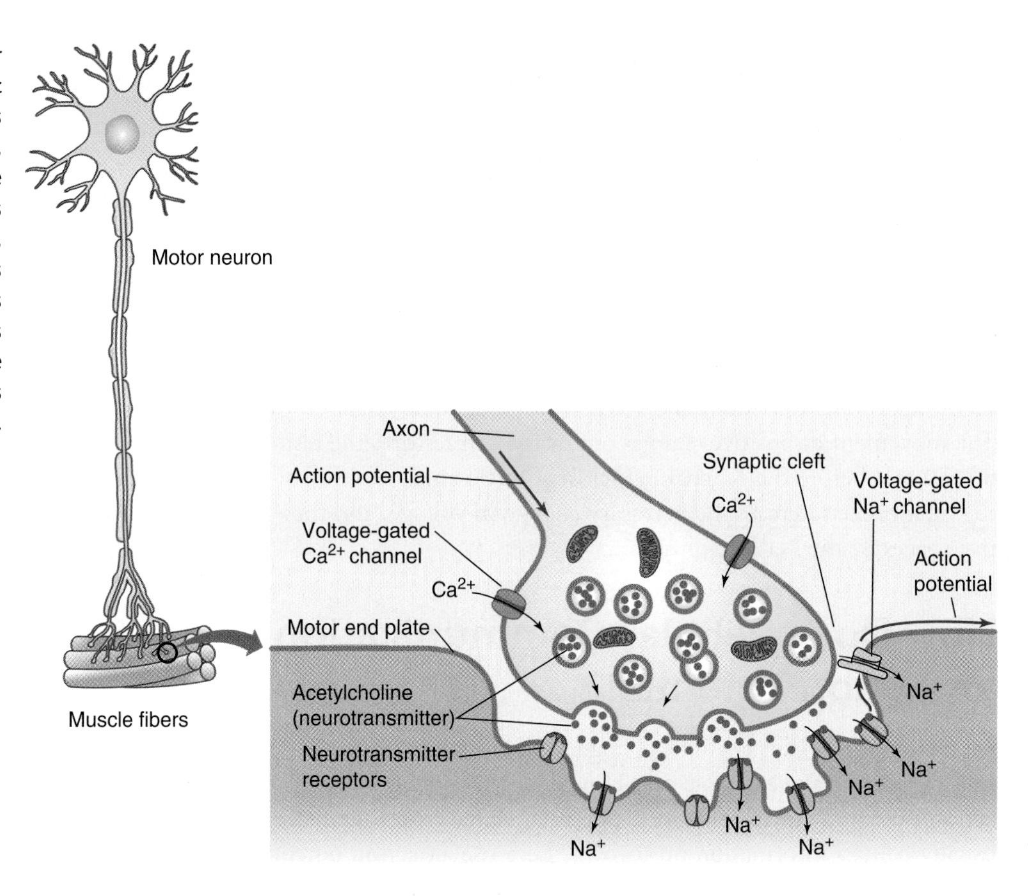

FIGURE 1.8 The neuromuscular junction. As Ca^{2+} enters the synaptic bulb, it causes fusion of vesicles with the plasma membrane, releasing acetycholine into the synaptic space. Acetylcholine binds to the ligand-gated ion channel, allowing Na^+ ions to move across the membrane. Na^+ entry changes the membrane potential and opens voltage-gated Na^+ channels in the skeletal muscle membrane, thus initiating an action potential.

Cellular Receptors Transduce a Signal Across the Cell Membrane Without Permitting Molecules to Cross the Membrane

So far in our discussion, water, glucose, and ions have all crossed the plasma membrane, providing intracellular water, glucose for ATP production, or ions for action potentials. Sometimes, however, intracellular actions occur without any molecule crossing the membrane. That is, a "signal" may pass from one cell to another without any movement across the membrane. Hormones binding to receptors are an example of this type of cellular communication. Receptors are membrane proteins that possess an extracellular binding site for a hormone or neurotransmitter and are bound on the intracellular side to integral proteins or enzymes. When a molecule generally known as a ligand binds to the extracellular side, the integral membrane protein changes conformation. You can think of protein conformations as being analogous to your body positions. Standing and sitting are both human positional conformations. Each position facilitates some functions. Proteins also have conformations, each of which exposes different binding sites or redistributes charge along the protein structure. When a ligand binds to the extracellular side of the protein, it changes the binding affinity of the intracellular side of the protein for the attached signaling molecules or enzymes.

Hormonal Signals Are Slower and Sustained

Not all intercellular signals are as fast and discrete as neuronal action potentials. Hormones, chemicals released by cells into the blood supply, can also bind to receptors and cause changes in cellular function in tissues far distant to the cells that released them. This type of chemical signaling takes longer to have its effect, but the effects are generally longer-lived, lasting from minutes to days instead of milliseconds. The most important thing to remember about hormonal signaling is that the hormone will bind to a receptor, and it is the receptor that determines the intracellular response to the binding. Let's use epinephrine, also known as adrenaline, as an example.

Epinephrine is released from cells of the adrenal glands, which are located just above the kidneys. Once in the blood supply, epinephrine can bind to β-adrenergic receptors, located on skeletal muscle (**FIGURE 1.9**). β-adrenergic receptors are part of a large family of membrane receptors called guanine nucleotide-binding protein (G-protein) coupled receptors, with seven transmembrane loops that link to intracellular heterotrimeric G-proteins α, β, and γ. These proteins are called G-proteins because they are inhibited by a bound guanosine diphosphate (GDP) molecule.

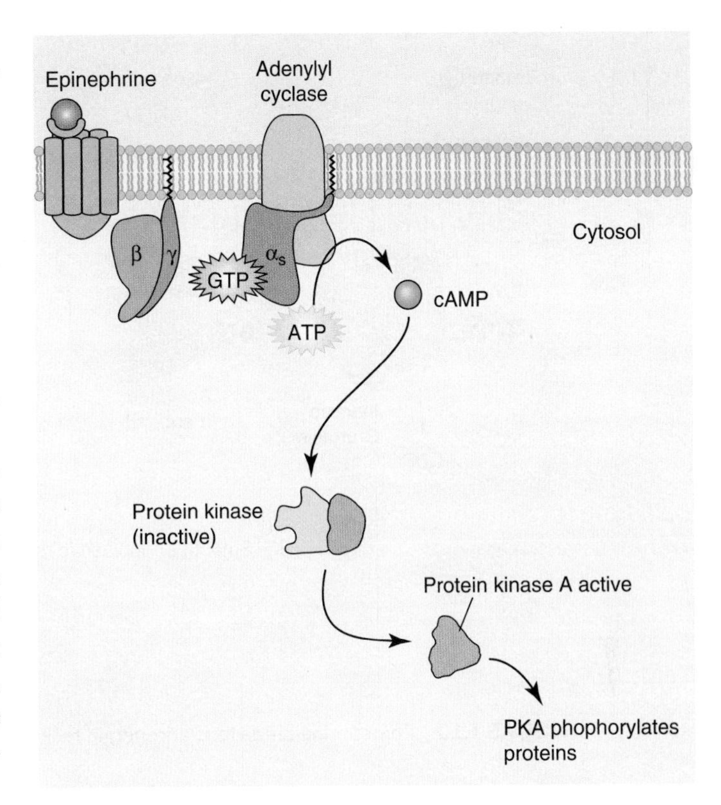

FIGURE 1.9 Epinephrine binds to β-adrenergic receptors and activates protein kinase A.

When epinephrine binds to this receptor, it changes the conformation of the transmembrane protein. The conformational change reduces the binding affinity of GDP, which releases and is replaced by guanosine triphosphate (GTP), thereby activating and releasing the α-subunit. The α-subunit detaches and binds to an enzyme, adenylate cyclase, attached to the inner leaflet of the plasma membrane. Adenylate cyclase converts ATP to cyclic adenosine monophosphate (cAMP), a signaling molecule. cAMP binds to protein kinase A, which can phosphorylate (add a phosphate group to) proteins. Phosphorylation is a common "switch" for regulating cellular proteins. Protein kinase A can phosphorylate many proteins, but the one we are interested in is glycogen phosphorylase, the enzyme that initiates the hydrolysis of intramuscular glycogen into glucose. Thus, epinephrine binding to a β-adrenergic receptor on skeletal muscle will increase the amount of free glucose available for energy, assisting you in your run.

Epinephrine can also bind to a different receptor, an α-adrenergic receptor. It binds with less affinity than at the β-receptor, but it can activate α-receptors (**FIGURE 1.10**). The α-receptors are also G-protein coupled receptors, but the G proteins are different and the intracellular action is different. When epinephrine binds to an α-receptor, the conformation change and substitution of GTP for GDP occurs as with the β-receptor. The activated subunit is α_q, and it binds to a different membrane-bound enzyme—phospholipase C. Phospholipase C cleaves the phospholipids of the membrane itself to create two signaling molecules: IP_3 and diacylglycerol. IP_3 binds to receptors on the endoplasmic or sarcoplasmic reticulum (SR) and causes the release of Ca^{2+}, itself a signaling molecule. Diacylglycerol activates a kinase, protein kinase C (PKC), which will phosphorylate proteins. While PKC can phosphorylate many proteins and initiate changes in gene expression, the protein target we will look at is the L-type Ca^{2+} channel, the same voltage-gated Ca^{2+} channel we saw in the neuron. When phosphorylated by PKC, this channel increases its conductance, thus allowing more calcium to enter the cell. Activated α-receptors, well distributed in smooth muscle, will increase smooth muscle contraction by two methods:

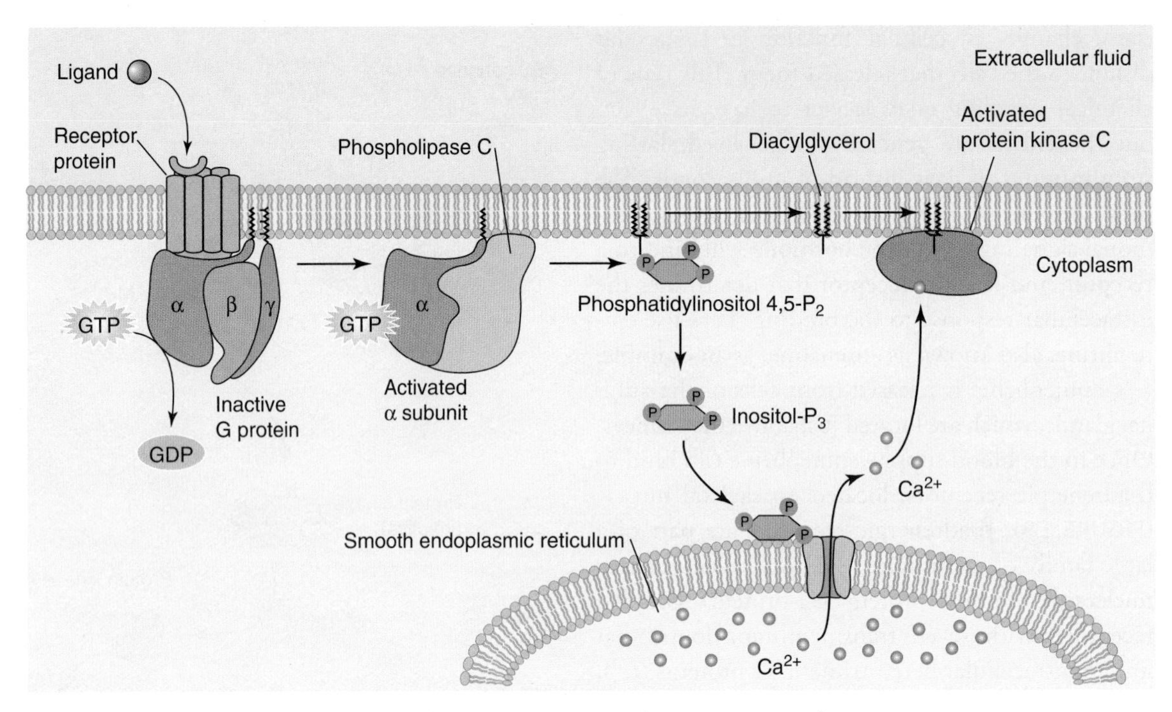

FIGURE 1.10 Epinephrine binds to α-adrenergic receptors and activates protein kinase C.

an increase in intracellular Ca^{2+} from the action of IP_3, and a similar increase resulting from greater flow of Ca^{2+} ions through the L-type Ca^{2+} channel. During your run, this will mean smooth muscle contraction of the venous vasculature, which increases blood flow into your heart. One hormone, therefore, can have multiple actions, depending on the receptor to which it binds.

Another class of membrane receptor proteins, from a very different family, are the receptor tyrosine kinases. These receptors are simple dimers of α helices within the plasma membrane and as such cannot create a conformational change in their basic structure. When a hormone binds to one of the receptor tyrosine kinases, the two subunits of the receptor phosphorylate each other on tyrosine residues of the helices. This phenomenon gave this class of receptors their name. The dimerized receptor then begins a complex series of signal transduction beginning with membrane associated proteins and ending at the nucleus. An example of this type of receptor is the insulin receptor (**FIGURE 1.11**). When you finished your run and ate a candy bar, it was the binding of insulin to tyrosine kinase receptors that allowed glucose transporters to be moved to the plasma membrane. In this way, glucose could get from the blood supply into the cell for metabolism and energy production.

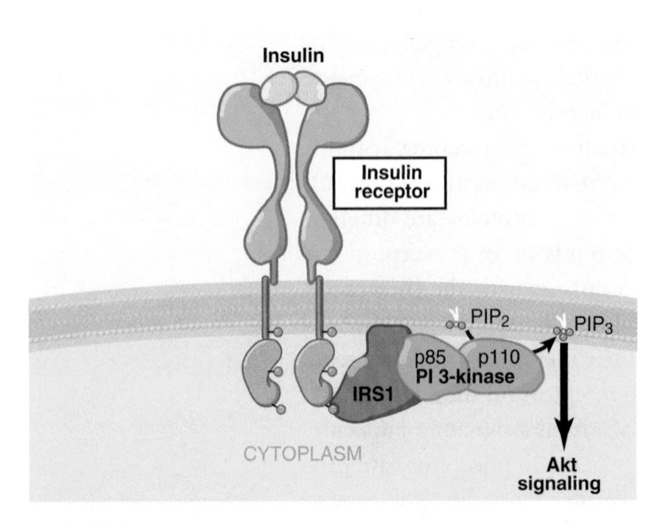

FIGURE 1.11 Insulin binds to an insulin receptor, which is a dimer of two α-helices in the plasma membrane that will phosphorylate each other once insulin binds. Binding and phosphorylation initiate a cascade of signaling events that ultimately end at the nucleus.

Some Hormones or Neurotransmitters Can Cross the Plasma Membrane Freely!

The most recently discovered class of signaling molecules are gases, which can freely cross a plasma membrane.

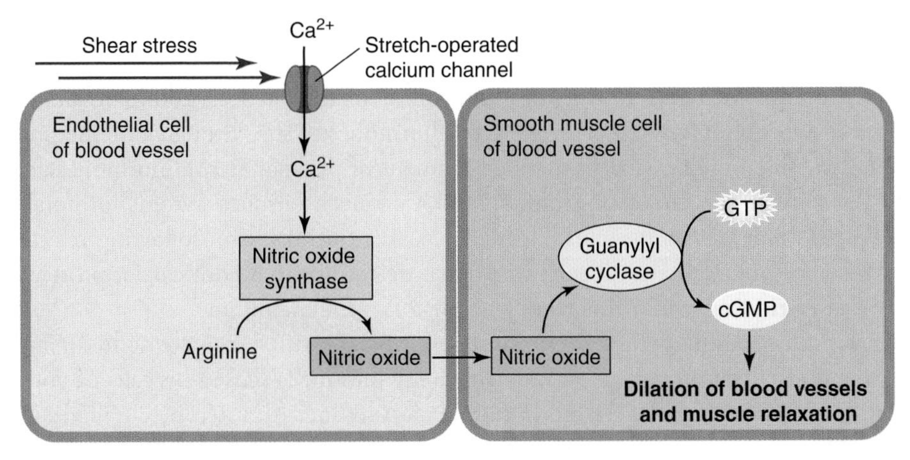

FIGURE 1.12 Nitric oxide, a gas, is created from the amino acid, arginine within the endothelial cell. NO diffuses into smooth muscle, where it activates guanylyl cyclase, converting GTP into cGMP. cGMP stimulates protein kinase G, ultimately relaxing smooth muscle.

Several gaseous molecules have been attributed with signaling properties, including CO_2 and H_2S, but the one we know the most about is nitric oxide (NO). Nitric oxide is enzymatically formed from the amino acid, arginine, and is a lipophilic gas (**FIGURE 1.12**). Being lipophilic, it can diffuse freely through plasma membranes, but its half-life is so short (seconds) that its effects are limited to tissues in the immediate area of its synthesis. For example, endothelial cells that line the interior of blood vessels synthesize NO, which diffuses into the blood and from there into the smooth muscle surrounding the blood vessel. Inside the smooth muscle cell, NO converts GTP into cGMP, activating protein kinase G (PKG). This protein kinase ultimately inhibits myosin and prevents smooth muscle contraction, thus allowing vasodilation.

Steroid Hormones Bind to Receptors Inside the Cell

Hormones derived from amino acids, like epinephrine, cannot cross a plasma membrane. Steroid hormones, however, have cholesterol as their parent molecule, and cholesterol is itself an important component of the plasma membrane. Therefore, steroid hormones can freely pass into a cell and bind to receptors either in the cytoplasm or in the nucleus (**FIGURE 1.13**). The receptor–hormone complex functions as a transcription factor, stimulating gene transcription. Steroid hormones, which include testosterone, estrogen, progesterone, cortisol, and aldosterone, cause changes in gene expression and therefore the production of new proteins. This process takes longer than signal transduction through a G-protein coupled receptor—hours rather than minutes—but the creation of new proteins that occurs as a result will be a much

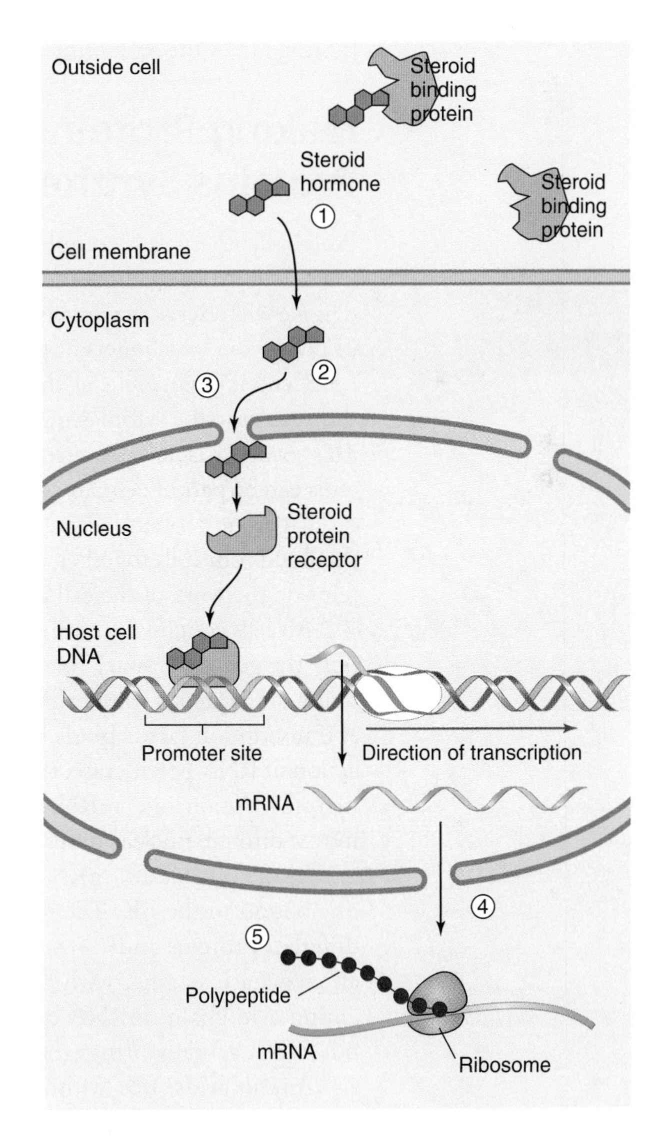

FIGURE 1.13 Steroid hormones dissociate from their protein carriers, cross the plasma membrane, and initiate protein synthesis.

more sustained response. Thus, steroid hormones can regulate cellular function for hours, days, and weeks.

If steroid hormones can pass freely into a cell, why aren't their actions dominant all the time? The same hydrophobicity that allows steroid hormones to slip through a membrane makes them insoluble in water, the primary component of plasma. Steroid hormones are carried by proteins in the circulating blood. At low concentrations, steroid hormones remain bound to their carrier proteins. At higher concentrations, e.g., following release from their tissue of origin, they will exist in a dynamic equilibrium with carrier proteins and will unbind periodically, allowing their passage across the membrane.

In our workout example, you became thirsty during the run. An increase in apparent K^+ plasma concentration, which occurs with dehydration, is sensed by cells of your adrenal glands, which release the steroid hormone aldosterone. Aldosterone circulates in the blood and has as its target the tubules of the kidney. Once inside the cells of the kidney tubule, aldosterone causes an increase in Na^+ ion channel expression, which results in concentration of urine, providing a means for conservation of water. The end result is that you produce less urine and maintain fluid balance. We will explore this mechanism more thoroughly in the endocrine chapter.

Making Proteins to Do the Work: How Are Proteins Synthesized?

Protein synthesis is a complex event, with regulatory steps at each junction. While it is important for us to know the fundamentals of this process in our study of human physiology, we will reserve the details of protein synthesis for cell and molecular biology texts. Let us review the basic tenets of protein synthesis.

Central dogma tells us that DNA is transcribed to mRNA, which is translocated from the nucleus to the cytoplasm, where it is translated by the ribosome and tRNA into proteins. This sequence is never reversed but always flows from DNA to proteins (**FIGURE 1.14**). Proteins can be part of cellular structure, like the cytoskeleton, or enzymes, kinases, receptors, or chaperone proteins. Proteins form the framework of our bones, the connective tissue that binds our cells together, and the intracellular regulatory and signaling molecules. Proteins do the work of the cell but are dependent upon DNA expression for their creation.

All cells contain the entire genome of DNA within their nucleus, but each cell expresses only the genes necessary for the function of that particular cell. Something must initiate that gene expression, and this is a transcription factor, as we saw in the previous section. A transcription factor binds to a particular region of DNA and stimulates or inhibits the action of RNA polymerase, the enzyme that unwinds the double helix of DNA and effects its transcription into mRNA (**FIGURE 1.15**). DNA remains in the nucleus, but mRNA moves through nuclear pores into the cytoplasm.

In the cytoplasm, mRNA binds to a ribosome, either one free in the cytoplasm, or one bound to the ER. The ribosome is a large structure, with two subunits made of 50 different proteins and ribosomal rRNA. These proteins and rRNA act cooperatively as an enzymatic engine, with binding sites for mRNA and tRNA, as well as the growing amino acid chain. mRNA carries the codons that determine the amino acid sequence, but tRNA actually brings the amino acids to the mRNA-ribosome complex.

Amino acids, free within the cytoplasm, must be bound to their matching tRNA. This is accomplished by aminoacyl-tRNA synthetases, enzymes that can match an amino acid, like alanine, to its specific tRNA (**FIGURE 1.16**). The binding process requires ATP, with ATP being converted to AMP, thus using two high-energy phosphate bonds for the

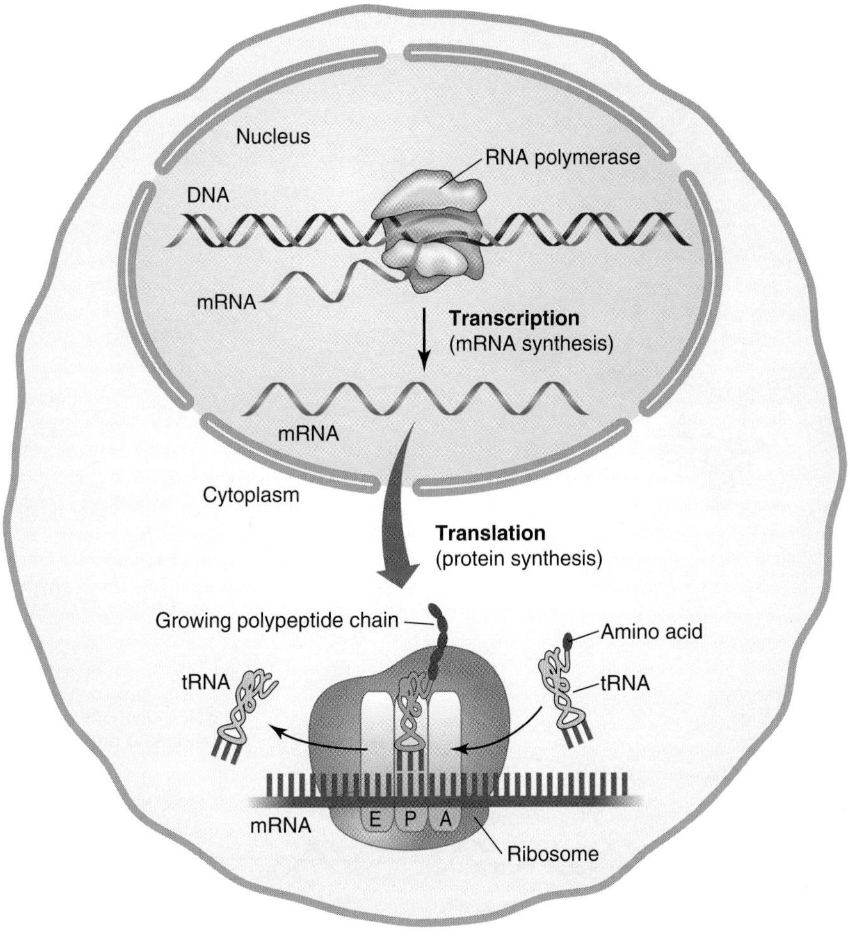

FIGURE 1.14 DNA transcription and translation create proteins that do the work of the cell.

binding of one amino acid. The amino acid plus tRNA complex, known as aminoacyl tRNA, must be bound to the ribosome, which is done by an elongation factor, using one molecule of GTP in the process. The newly arrived aminoacyl tRNA sits in one of the three binding sites for tRNA on the ribosome, the A site. The amino acid from the P site is added to the amino acid from the A site by peptide bond formation. The energy to make the peptide bond comes from the high-energy bond created by the aminoacyl-tRNA synthetase in the previous step, so no additional ATP is required. However, moving the growing chain from the A site to the P site requires another elongation factor and another molecule of GTP. This process repeats with every amino acid added to the primary structure of the protein. The process is quick—40 amino acids can be added to the chain each second!—but energetically expensive. Creation of the primary structure, the simple string of amino acid, requires four ATP equivalents per amino acid.

Once the primary structure is complete, the protein is released from the ribosome-mRNA complex by a releasing factor, using another GTP. To become a functional protein, the simple amino acid chain must be folded. This occurs in part because of the intermolecular charge interactions of the amino chain, but in the cytoplasm, this process is facilitated by chaperone proteins, which use ATP to bind or unbind from the newly made protein as it assists that protein in folding (**FIGURE 1.17**). Not all folding occurs correctly the first time, and chaperone proteins will fold and unfold the newly made amino acid chain until it is its correct formation. Again, chaperone activity uses ATP, but in a variable amount, depending on binding frequency.

(a)

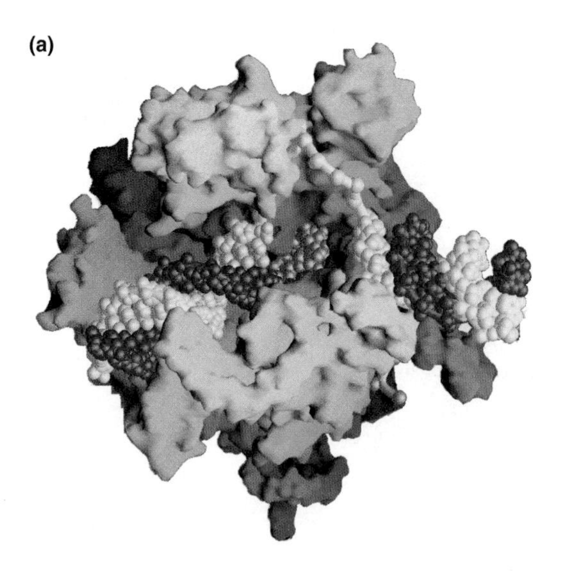

(b)

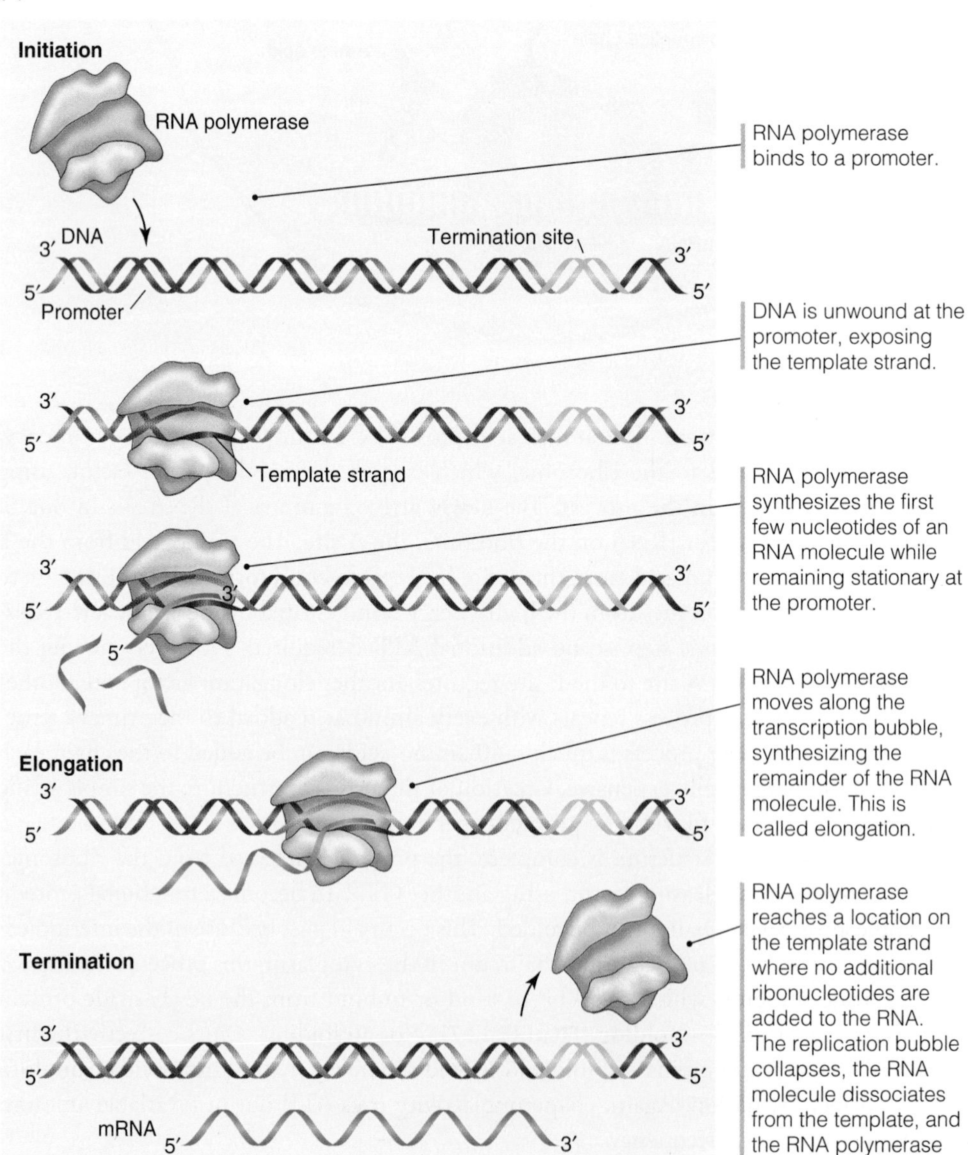

FIGURE 1.15 RNA polymerase (a, a three-dimensional structural model) initiates transcription and starts the process of protein synthesis (b). (Photo courtesy of Seth Darst.)

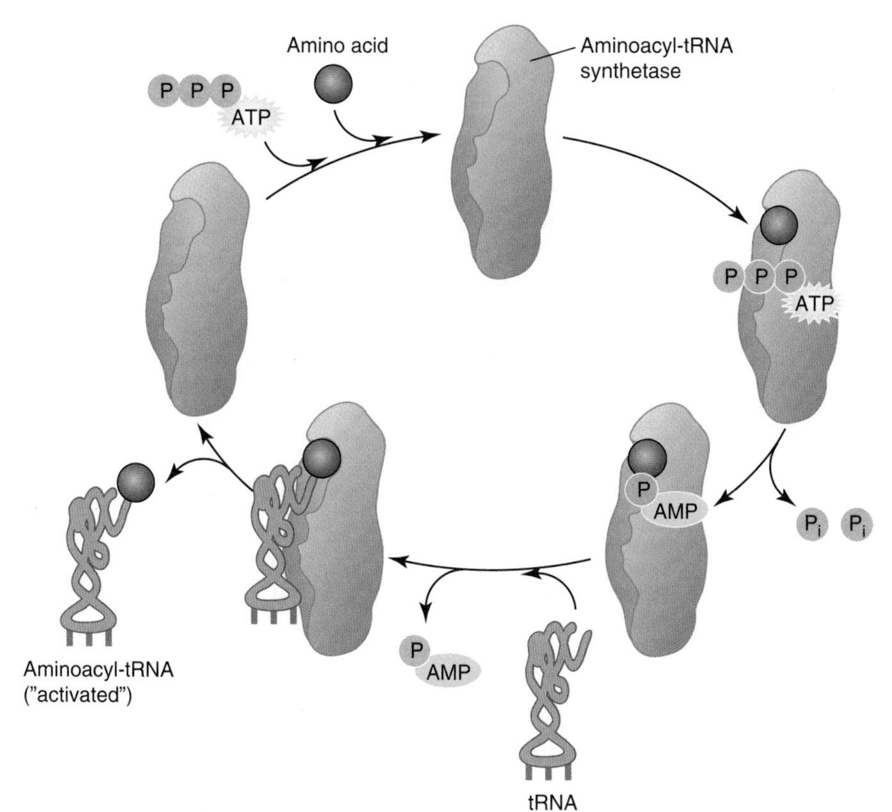

FIGURE 1.16 The enzyme aminoacyl-tRNA synthetase uses ATP to bind an amino acid to its appropriate tRNA.

Proteins that will remain free in the cytoplasm are made in the fashion described above and can be made by free ribosomes. Proteins that will be inserted into the plasma membrane, like ion channels or G-protein coupled receptors, contain hydrophobic regions that span the plasma membrane. These regions, being hydrophobic, cannot be exposed to the aqueous environment of the cytoplasm. These proteins are made on ribosomes attached to the membrane of the ER. As the chain elongates off the ribosome, it enters the lumen of the ER through a protein pore. Chaperone proteins like binding immunoglobulin protein bind to the hydrophobic regions of the forming protein within the ER, protecting it from the aqueous environment of the ER lumen (**FIGURE 1.18**). Once completed, these proteins will be glycosylated within the ER and then transported by vesicle to the Golgi apparatus for further processing. Vesicles bud off the Golgi apparatus and fuse with the plasma membrane, inserting the new membrane proteins in their place. Glycosylation is so common in membrane proteins that under high magnification, as seen in electron microscopy, the entire exterior of a plasma membrane appears to have a sugar "halo." Glycosylation is important in cellular self-recognition and charge distribution across the membrane.

Hormones and signaling mechanisms within the cell regularly stimulate or inhibit protein synthesis as a method of regulating cell function or allowing cellular adaptation to changing environmental conditions. Protein synthesis takes some time, as we saw from the steroid hormone series of events, but it is a longer-term form of response to changing conditions. Protein formation is also energetically expensive, requiring four ATPs per amino acid, in addition to the ATPs required for initiation, termination, and folding. If we consider that an average-sized protein is 450 amino acids long, and the largest protein known, titin, is 27,000 amino acids long, the energy required for protein synthesis is significant. Where does this energy come from?

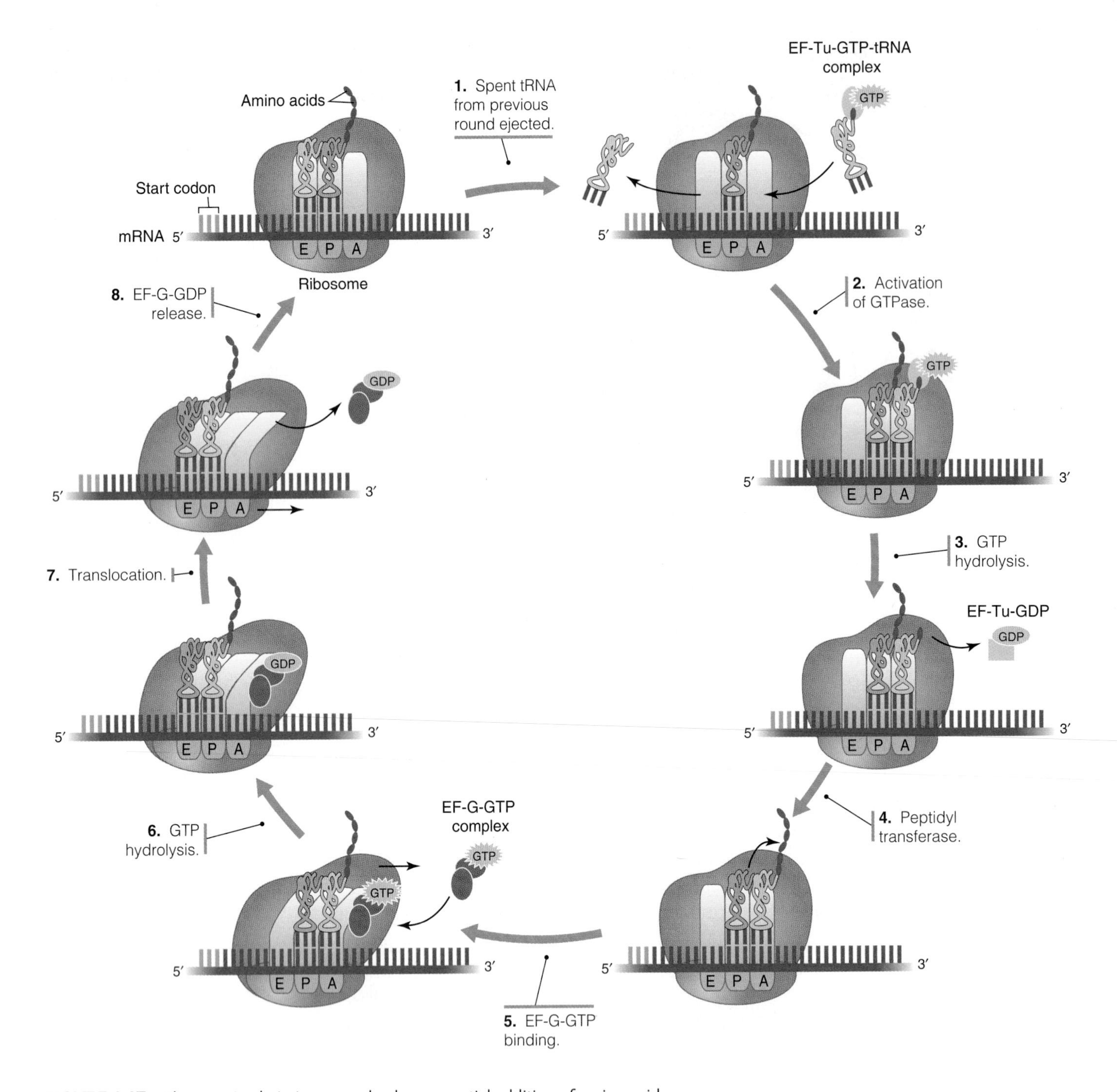

FIGURE 1.17 The protein chain increases by the sequential addition of amino acids.

Energy Production: Fueling Cellular Work

ATP production is a constant process, driven by chemical reactions between glucose and glycolytic enzymes and continuing through the citric acid cycle and the electron transport chain. Each reaction is driven by a change in energy level, or ΔG. Because all of our activity—from membrane transport to protein synthesis to muscle contraction—requires ATP, we will look briefly at how this vital molecule is generated. Let's begin with glucose, a six-carbon sugar, and the process of glycolysis.

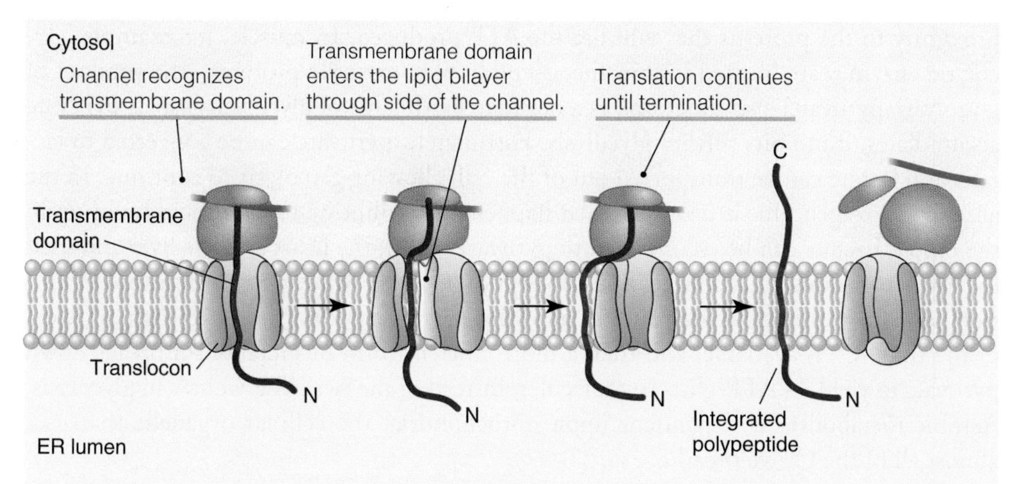

FIGURE 1.18 Membrane proteins with hydrophobic regions are made within the endoplasmic reticular membrane.

Glycolysis

Glycolysis is the enzymatic transformation of glucose to pyruvate through the sequential activity of 10 cytosolic enzymes. These enzymes are not floating free in the cytosol but are scaffolded together in large, ordered complexes so the reactions can occur quickly. The starting materials for glycolysis are glucose, ATP, and nicotinamide adenine dinucleotide (NAD). ATP is used in the initial step to phosphorylate glucose and prime it for rearrangement and breakdown (**FIGURE 1.19**). The phosphorylation step also traps glucose within the cell, so it cannot travel down a concentration gradient out of the cell. Once phosphorylated, the glucose molecule is serially rearranged; the energy from these rearrangements is collected as ATP (a gain of two molecules of ATP per glucose molecule) and as NADH, the electron carrier. The final substrate product is two molecules of pyruvate, a three-carbon molecule. Note that in this entire process, no carbons are lost, and no oxygen is used. By tradition, we show the metabolism of glucose. However, fructose, lactose, galactose, and mannose also can enter this pathway with a few additional steps. Therefore, this is the initial metabolic pathway for all dietary sugars.

As in many systems, there are regulatory feedback loops in which the product of one reaction inhibits the enzyme that creates that product. An example is the enzyme phosphofructokinase (PFK), which converts fructose-6-phosphate to fructose 1,6,-bisphosphate, early in the glycolytic cycle. PFK is inhibited by ATP, an end product of cellular metabolism, and is activated by AMP, a low-energy phosphate derived from ADP. When we are active and using ATP in quantity, such as during exercise, there is little build-up of ATP, and the synthesis of ATP continues at maximal rates. When we are less active, ATP inhibits PFK, and cell metabolism slows. Thus, ATP cellular concentrations are carefully regulated to remain constant and to change immediately with need. The result is a close coupling between ATP production and use.

By itself, glycolysis is a very fast source of ATP. Glycolytic enzymes are located in the cytoplasm, often in close

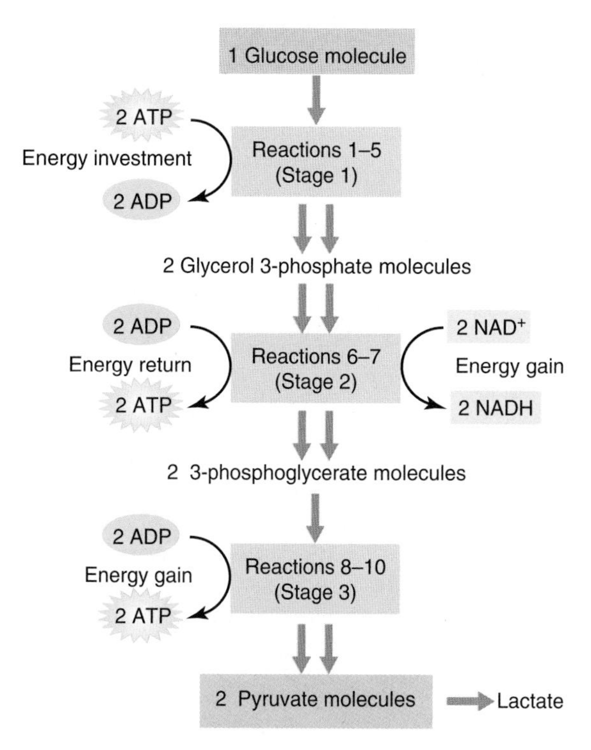

FIGURE 1.19 Glycolysis begins with glucose and ends with pyruvate or lactic acid.

proximity to the proteins that will use the ATP produced. In muscle, for example, glycolytic enzymes are located within the assembly of contractile proteins. The product of glycolysis, pyruvate, also functions as a negative modulator of glycolysis—so as pyruvate accumulates, it inhibits further glycolysis. Fortunately, pyruvate can be converted to lactate, and lactate can be transported out of the cell, allowing glycolysis to continue. In the absence of oxygen, this is precisely what happens, contributing to lactic acid build-up in the blood. Lactate can be reconverted to pyruvate within the heart and the liver and used for further metabolism, so lactic acid production is a way of preventing pyruvate inhibition of glycolysis within the cell and allowing energy recycling by other tissues. Most of the time, O_2 is available, and then a more efficient form of metabolism breaks down pyruvate to yield 34 ATP/glucose molecules, instead of the two ATP we saw in glycolysis. Aerobic metabolism is dependent upon mitochondria, the cellular organelle that uses almost all of the O_2 we breathe.

Mitochondria: Organelles That Produce Most of Our ATP

What do we know about mitochondria? Mitochondria are double-membraned organelles that exist in all human cells except for mature red blood cells. These organelles are thought to have originated a billion years ago as free-living organisms, which later became incorporated into a host organism, producing the eukaryotic cell we now recognize. The number of mitochondria within a cell is uncertain. It not only depends on cell type, but also on mitochondrial morphology, which is still unclear. The classic oval-shaped mitochondrion portrayed in scientific illustrations may be an artifact of histological section (**FIGURE 1.20**). There is evidence that mitochondria are tubular, dynamic organelles that change shape, twist, and undergo fission and fusion. Mitochondria may proliferate in active, aerobic tissues. Mitochondria are not evenly distributed within a cell but are concentrated at sites of ATP use. In muscle, they are found near contractile fibers and in nerves at the sites of protein synthesis in the cell body and in the synaptic bulb, where neurotransmitters are made and stored.

Mitochondria contain their own circular DNA, known as mtDNA, which we inherit from our mothers only. mtDNA does not encode all the proteins required for mitochondrial function. In fact, it codes for only 13 genes, all related to the electron transport chain proteins. There are many genes within the cell's nuclear DNA that encode structural and regulatory proteins necessary for mitochondrial function. Cooperative expression between two genomes is necessary for proper mitochondrial operation. Mutations within either genome can affect organelle function and therefore ATP production.

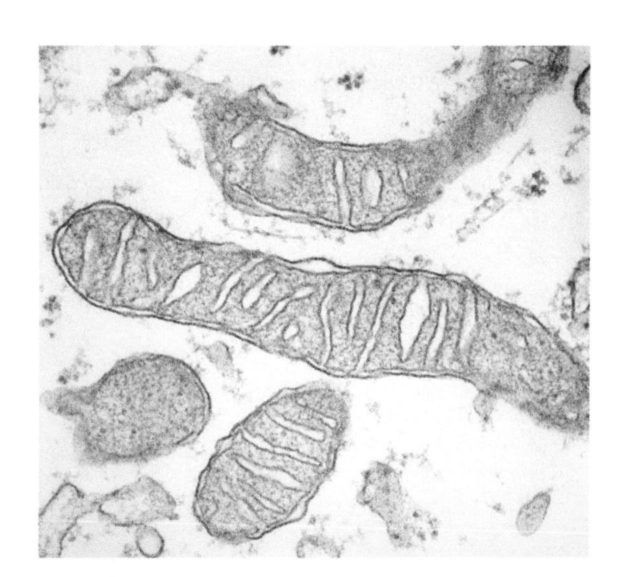

FIGURE 1.20 Mitochondria may resemble tubular arrays within the cytoplasm that appear oval when cells are sectioned. (© Dr. Gopal Murti/Science Source.)

While mitochondria lack the sophisticated DNA repair mechanisms of the nucleus, they are capable of fission (division) and fusion (joining of two or more mitochondria). Fission and fusion allow genetic mixing between mitochondria and isolation of damaged mtDNA that can be eliminated from the organelle. Fission and fusion also allow more homogeneity of mtDNA within a large cell, such as a neuron. Mitochondria also move, which is linked to the processes of fission and fusion. In neurons, for example, mitochondria must be located both within the cell body of the neuron and at the distant synaptic bulb. Without movement, mitochondria fail to distribute to the synapse, causing neuronal malfunction. Genes that regulate fission and fusion are within the cell's nuclear DNA, highlighting the interdependence of mitochondria and the cell they inhabit.

Energy Production Within the Mitochondria

Mitochondria are famous for their role in ATP synthesis. Indeed, life as we know it is not possible without the ATP provided by mitochondria. Glycolysis alone cannot supply the ATP demands required for human life. Once glucose or other sugars are converted to pyruvate, the rest of the metabolic process must proceed within a mitochondrion. Pyruvate, produced in the cytoplasm, is transported into the mitochondria through pyruvate transporters. Once inside the mitochondrial matrix, pyruvate is decarboxylated and linked to coenzyme A to become acetyl coenzyme A. This reaction requires several cofactors known as vitamins, including pantothenic acid (vitamin B5), which is a component of CoA, and thiamine (vitamin B1). The conversion produces NADH as a product. NAD^+ is also formed from a vitamin, niacin (vitamin B3). Acetyl-CoA can also be formed from a two-carbon product of β-oxidation, which is how fat metabolism feeds into this metabolic framework.

Fatty Acids Are a Source of Acetyl-CoA

Most of the fat we ingest is stored in the form of triglycerides composed of a glycerol backbone linked to long-chain fatty acids. Triglycerides are broken down by removal of glycerol, which is metabolized along the same pathway as glucose. The remaining fatty acids are broken down within the mitochondria, in a cycle known as β-oxidation. The final product of β-oxidation is acetyl-CoA, which feeds into the citric acid cycle just as acetyl-CoA from pyruvate metabolism does (**FIGURE 1.21**). The energy yield from fatty acids is

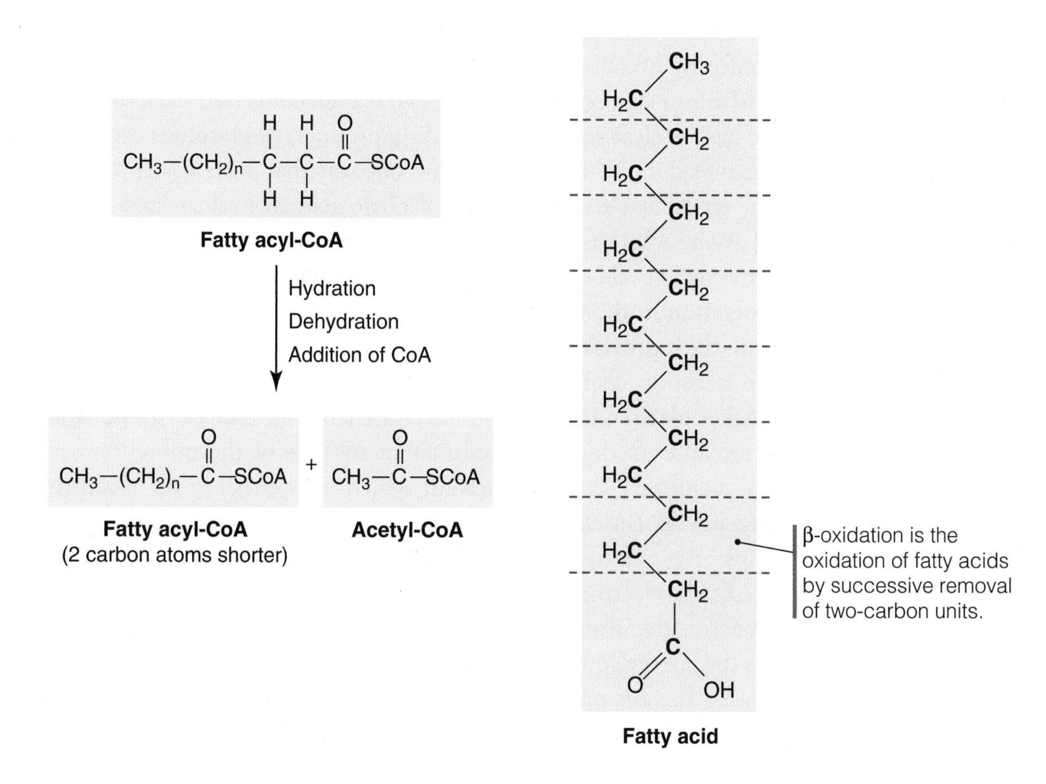

FIGURE 1.21 β-oxidation involves the sequential oxidation of fatty acid chains two carbons at a time, creating acetyl-CoA.

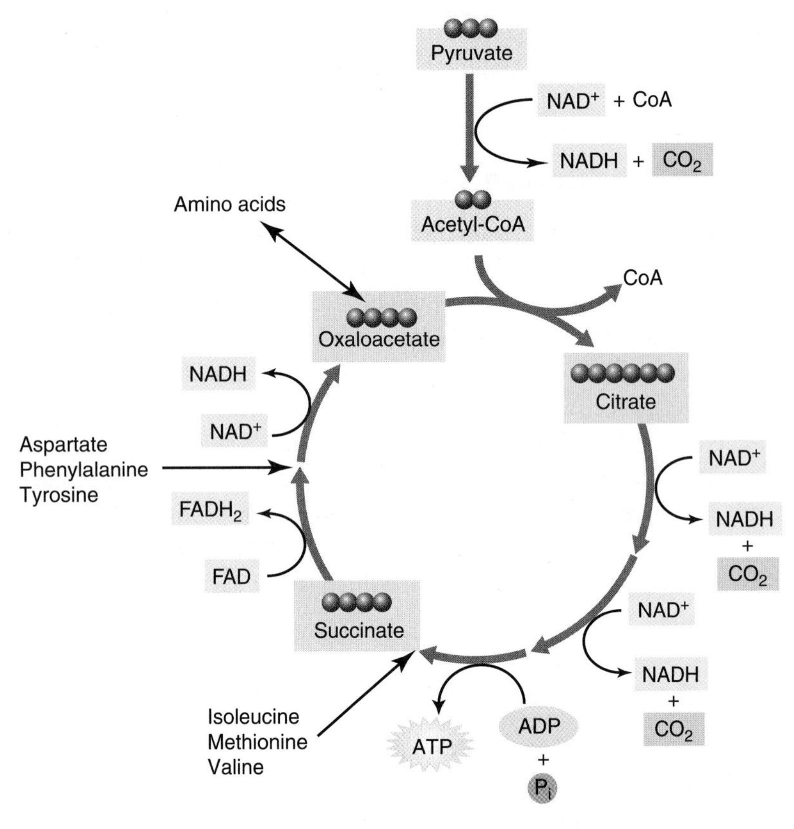

FIGURE 1.22 Amino acids can be deaminated and enter the citric acid cycle as intermediates. In this way, amino acids can be utilized to make ATP.

very high, making fat our most efficient energy source. However, fatty acids require O_2 for metabolism.

Whatever its source, the two-carbon acetyl-CoA joins with oxaloacetate to form citric acid. This is the starting point of the citric acid cycle, also known as the Krebs cycle. The citric acid cycle can be confusing because it begins with citric acid, a six-carbon molecule, and ends with oxaloacetate, a four-carbon molecule, which is then converted back to citric acid. This nonlinear cycle has no clear starting and ending products as glycolysis did, and deaminated amino acids can feed into this cycle, which is one way that proteins are metabolized (**FIGURE 1.22**). The molecular rearrangement of citric acid to oxaloacetate yields energy, which is captured as one ATP, three NADH, and one $FADH_2$. The important part to remember about the citric acid cycle is that NADH and $FADH_2$ will be valuable starting materials for ATP generation in the next step of metabolism. So, while the citric acid cycle generates little ATP itself, it provides the electron carriers needed for the final step in metabolism—the electron transport chain.

The citric acid cycle takes place in the mitochondrial matrix, but the enzymes of the electron transport chain are all embedded on the inner membrane of the mitochondria (**FIGURE 1.23**). The inner membrane is tightly sealed, which is essential if the electron transport chain is to produce ATP. The electron transport chain is a chain of proteins, containing metal-sulfur groups or cytochromes, that can bind electrons. NADH or $FADH_2$ are oxidized to NAD or FAD, the electrons are carried on the cytochromes, and the H^+ ions are pumped, via proteins, across the inner membrane into the intermembrane space using the electron movement as the energy source. The result is a high concentration of H^+ ions in the intermembrane space. There is only one path for movement of H^+ ions back into the mitochondrial matrix, and that is through the ATP synthase. Movement of three H^+

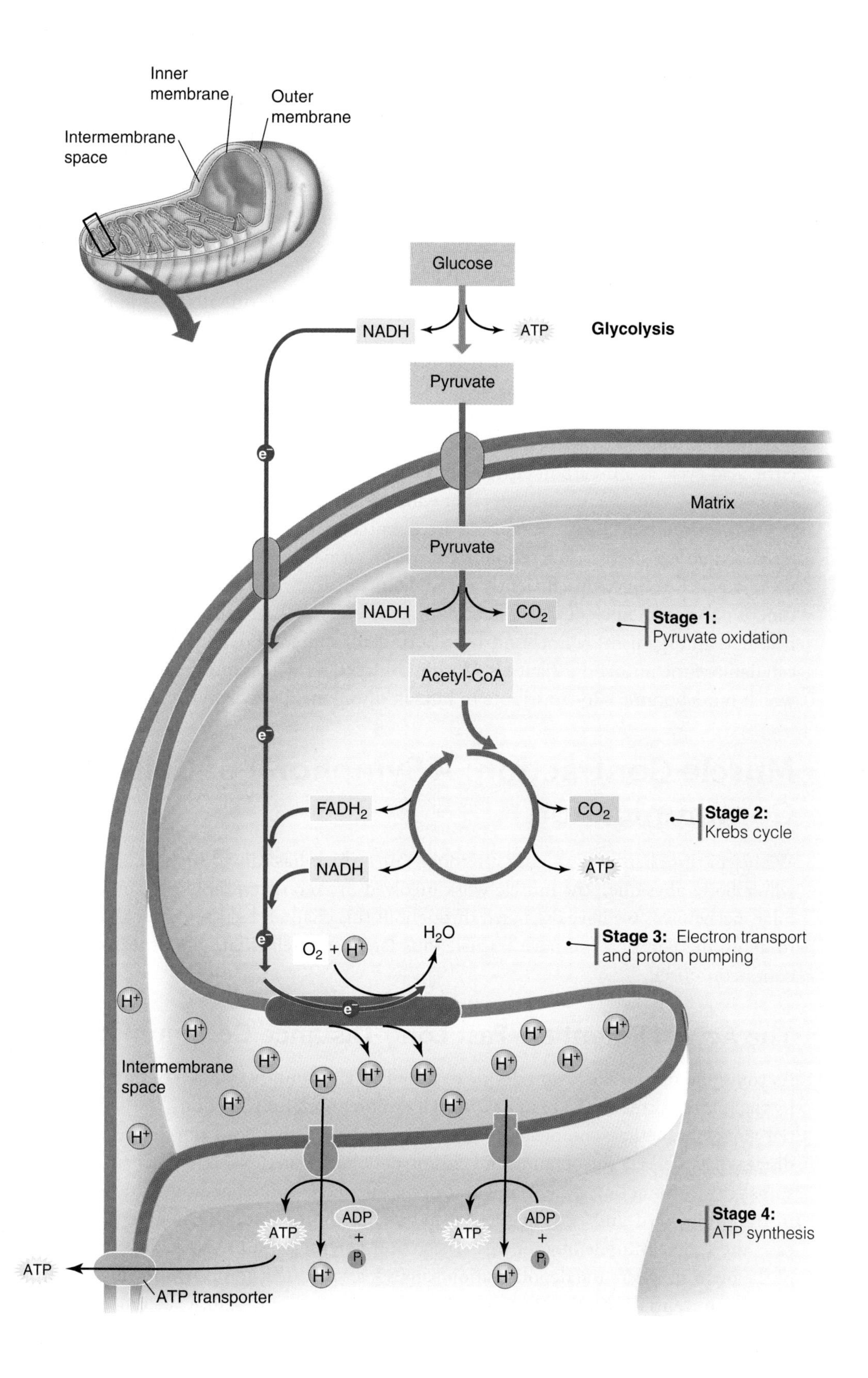

FIGURE 1.23 ATP production from glucose begins in the cytoplasm, continues in the mitochondrial matrix, and finishes with ATP being transported back into the cytoplasm.

ions down their concentration gradient through this ion pore provides enough energy to generate one ATP from one ADP. Thus, the H^+ ions carried by NAD and FAD are the primary producers of ATP within the mitochondria. ATP produced within the mitochondria is transported through adenine nucleotide transporters into the cytosol for use.

At the end of the electron transport chain, electrons are transferred to O_2. This is the ultimate job of an oxygen molecule—to accept the electrons from oxidized NADH (Figure 1.23). Without O_2, the movement of electrons ceases and ATP production stops. This is why we are obligate aerobic animals. Our ATP production depends on O_2 as the final electron acceptor. All of your respiratory efforts as you run are for one purpose: to provide enough O_2 to fuel this process.

However, acquiring O_2 is only half of the breathing process. As you run, you inhale O_2, but you also exhale CO_2. This CO_2 is generated as a by-product of metabolism and is a significant waste product that must be eliminated. The first CO_2 comes from pyruvate metabolism, when it loses a carbon to become acetyl-CoA. The next two come from the citric acid cycle as a 6-carbon citric acid becomes a 4-carbon oxaloacetate. Each pyruvate thus generates three CO_2 molecules that will need to be eliminated.

As you can see, the mitochondria are essential to our lives. The metabolic processes described above are how mitochondria function in health. We would like to think that these organelles function efficiently throughout our lives. However, mitochondrial malfunction, or an inefficiency in function, often occurs, with serious consequences for our health as an organism. Mitochondria are a primary source of oxygen free radicals, which can damage the mitochondria itself. Within this text, we will examine some of the ways in which mitochondria can contribute to human disorders, disease, and aging.

Muscle Contraction: A Symphony of Cellular Communication

We began this chapter with your afternoon run, which has caused such a change in your whole body physique. The muscle work involved in exercise exemplifies many of the cellular mechanisms we have discussed throughout this chapter. Let's use muscle contraction to apply the concepts we have learned thus far, and to elucidate the process of muscle contraction itself.

The Action Potential—Fast Long-Distance Communication

As you stand on the track, you decide to run. This is voluntary muscle contraction, driven by the motor cortex of the brain, as we will explore more fully later on. The action potential along the nerve running from your brain to the spinal cord travels precisely as we discussed earlier, via the opening and closing of voltage-gated Na^+ and K^+ channels. At the spinal cord, this action potential is transferred to the α-motor neuron, which originates in the spinal cord and ends at skeletal muscle, at the neuromuscular junction—in this case, the skeletal muscle of your legs. The action potential terminates at the synaptic bulb of the neuron, where the depolarization causes Ca^{2+} channels to open, allowing the intracellular concentration of calcium to rise within the synaptic bulb. Calcium triggers the exocytosis mechanism, and vesicles of acetylcholine fuse with the plasma membrane of the neuron and release acetylcholine into the synaptic space. Acetylcholine binds to nicotinic acetylcholine receptors on the plasma membrane of the muscle cell. Nicotinic acetylcholine receptors, also known simply as acetylcholine receptors, are ligand-gated ion channels. Once acetylcholine binds, the ligand-gated ion channel undergoes a conformational change and opens, allowing Na^+ ions to flow into the muscle cell. These Na^+ ions cause a local depolarization of the plasma membrane, opening voltage-gated Na^+ channels in the membrane and initiating an action potential in the skeletal muscle membrane. Thus, the action potential, an electrical signal, which began in your brain, continues chemically to the muscle where the electrical action potential propagates down the muscle fiber (**FIGURE 1.24**).

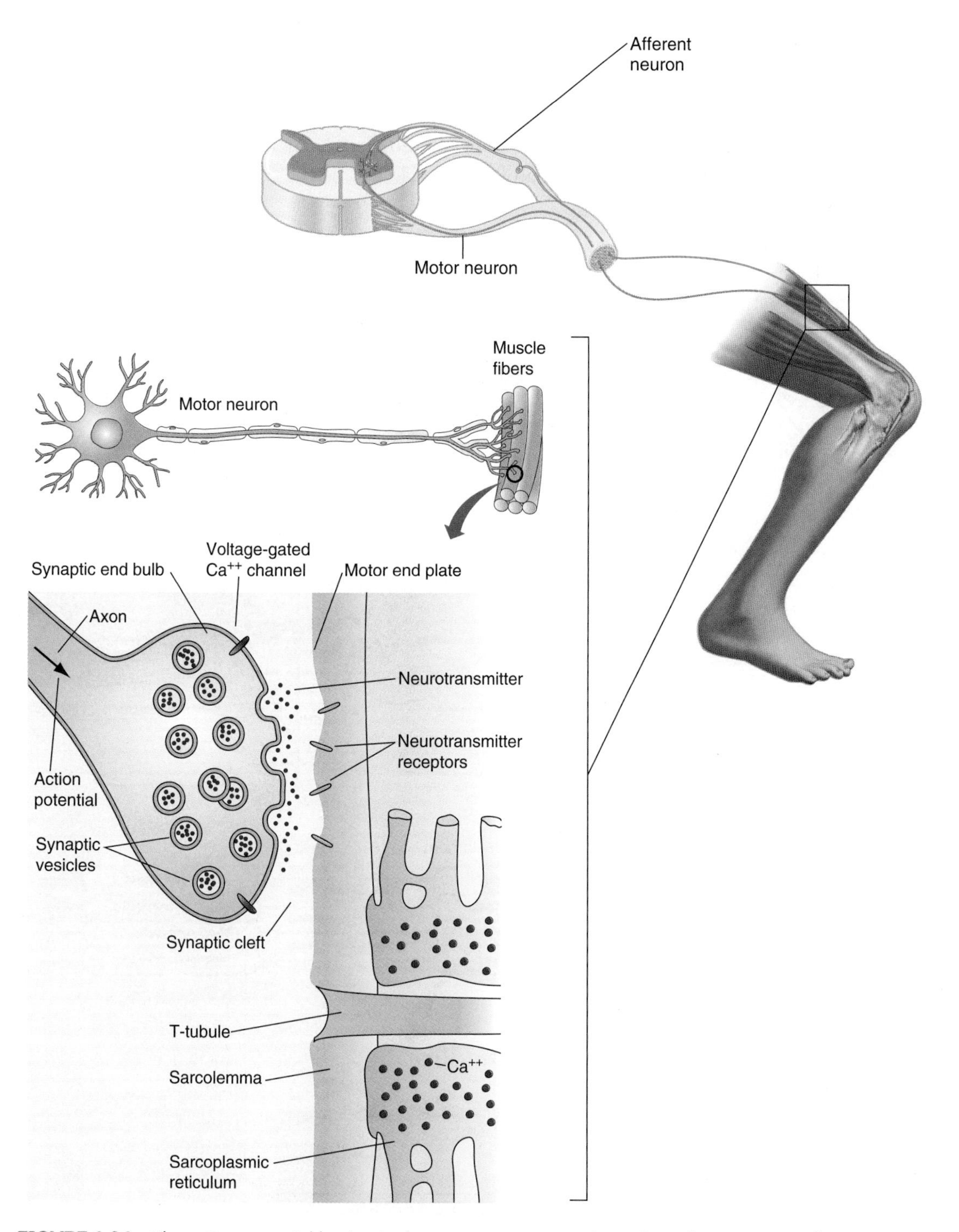

FIGURE 1.24 The action potential begins in the α-motor neuron and travels to the neuromuscular junction, where it initiates muscle contraction in the muscles of your leg.

Excitation-Contraction Coupling—Linking an Electrical Signal to Mechanical Work

For many years after the cellular mechanism of muscle contraction was understood, it was unclear how the depolarization of the skeletal muscle fiber caused myosin and actin to interact. It was clear the depolarization always caused muscle contraction, but why? The connection between membrane depolarization and muscle contraction is formally known as excitation-contraction coupling, or E-C coupling. Let us walk through these events. The plasma membrane of skeletal muscle cells contain deep invaginations known as T-tubules. T-tubules are continuous with the plasma membrane and contain ion channels like the rest of the membrane. During an action potential, the T-tubular membranes will depolarize. Here is a case where understanding anatomy is essential. Lying very close to the T-tubule membrane, inside of the muscle cell, is the SR, a specialized ER of the

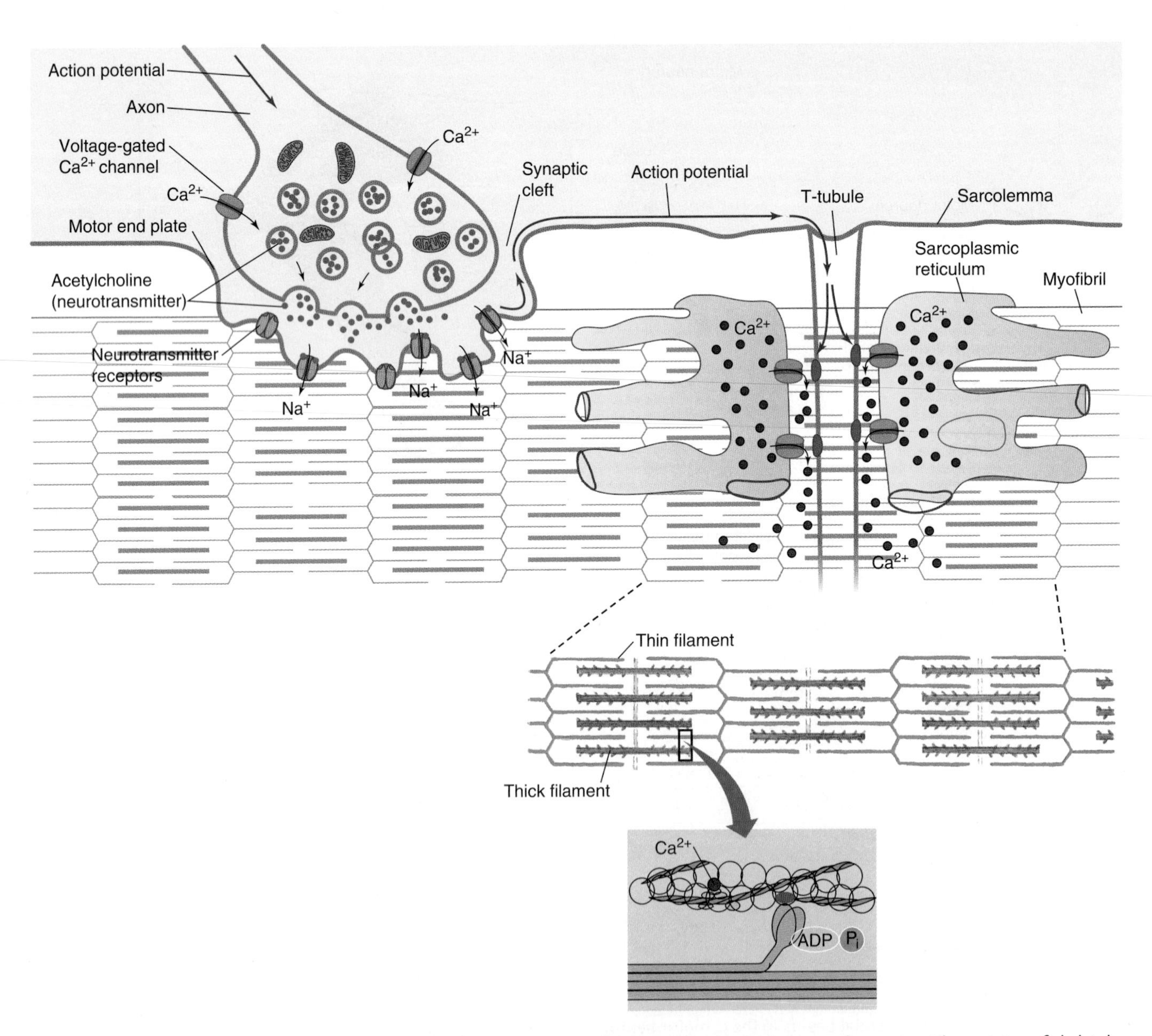

FIGURE 1.25 Excitation-contraction coupling links the action potential at the plasma membrane to the contractile proteins of skeletal muscle. The molecule that links the two processes is the Ca^{2+} ion.

muscle cell. The SR lies so close to the T-tubule that proteins within the SR membrane will experience a voltage change during the action potential. There is no action potential along the SR membrane, but there is a change in SR membrane voltage sufficient to open Ca^{2+} release channels in the SR. SR is a Ca^{2+} storage vesicle within skeletal muscle, and when the Ca^{2+} release channel opens, the intracellular $[Ca^{2+}]$ of the skeletal muscle cell increases (**FIGURE 1.25**). The increase in skeletal muscle calcium concentration is the signal transduction mechanism that links an action potential to mechanical work. Calcium ions were the mystery molecule of E-C coupling!

How Does Skeletal Muscle Contract?

Skeletal muscle is called striated muscle because of its orderly "striped" appearance. Within a skeletal muscle cell, myosin and actin proteins are aligned next to each other, both connected to Z-disks that link to the plasma membrane. The space between Z-disks, a sarcomere, is the repeating protein structure of the fiber. Each sarcomere is arranged as myosin proteins (thick filaments) lying between actin proteins (thin filaments).

Myosin proteins have several sections: the tail, the hinge, and the head. The head contains the actin binding site, while the hinge can assume several stable conformations, each of which is important to muscle contraction. Actin filaments are composed of globular actin polymerized into chains. Actin filaments also possess a binding site that can be occupied by myosin. If actin and myosin were left in this simple state, our skeletal muscle cells would be contracted all of the time. However, actin filaments are encircled by tropomyosin, which covers the binding site on actin, making it inaccessible to the myosin head. Attached to tropomyosin are the troponin proteins, the most important of which is troponin C. Actin, with its associated tropomyosin and troponin, is the regulator of skeletal muscle contraction.

How is exposure of the actin-myosin binding site regulated? Remember that Ca^{2+} is the link between the action potential and muscle contraction. Troponin C has a binding site for the Ca^{2+} ion. As $[Ca^{2+}]$ rises intracellularly, troponin C binds the calcium ion (**FIGURE 1.26**). This causes a conformational change in this protein, moving tropomyosin away from the actin-myosin binding site. Once exposed, myosin can bind to this site, initiating the molecular events of muscle contraction.

Myosin Binding, Cross Bridge Cycling, and ATP Hydrolysis

As you have probably noticed, to this point, none of the physiological events of muscle contraction have required ATP. The action potential, neurotransmitter release, opening of ligand-gated ion channels, the initiation of an action potential in the muscle cell, and E-C coupling have all been accomplished with virtually no ATP hydrolysis. Yet, we know that exercise and muscle contraction take work, which means ATP consumption. Cross-bridge cycling of myosin is where ATP is used. Now, let's look in detail at how it is used.

Let's begin where we left off, with an elevated intracellular $[Ca^{2+}]$ and an exposed binding site on actin. Myosin will quickly bind under these conditions. The myosin head is now connected to actin, with an associated ADP and Pi. Myosin is not only a structural protein and an important part of your skeletal muscle, but it is also an enzyme, an ATPase capable of cleaving ATP into ADP and Pi. Instead of diffusing away from the myosin head, these hydrolysis products remain attached for a time. Pi leaves first, and when it does, it causes a conformational change in the position of the hinge, inducing a 45° bend. This is the power stroke of skeletal muscle contraction; it is how myosin pulls along actin and shortens the sarcomere, causing what we see, grossly, as muscle contraction. ADP diffuses away next, and the myosin head remains attached to actin. The now "naked" myosin head binds a new molecule of ATP, which allows detachment from actin.

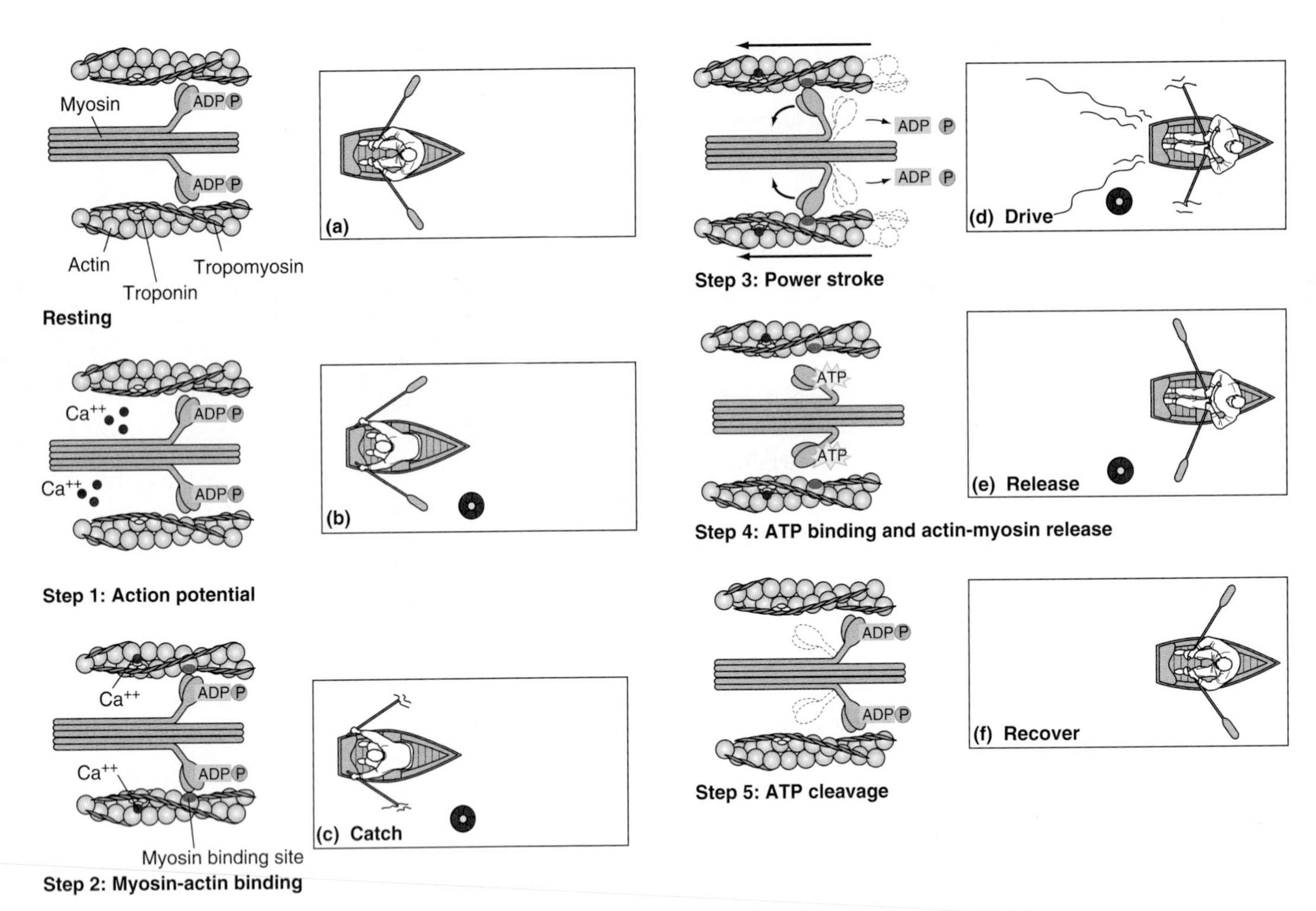

FIGURE 1.26 The events of muscle contraction.

ATP is immediately hydrolyzed to ADP and Pi, and the cycle begins again. Notice that ATP is required for relaxation, and the power stroke is a result of Pi detachment from the myosin head (Figure 1.26). The easiest way to remember this unexpected mechanism for muscle contraction is to consider the events of rigor mortis. Rigor mortis sets in when muscle cells have exhausted their supplies of ATP. When no ATP is available for muscle relaxation, skeletal muscle contracts, making the body rigid.

We have illustrated muscle contraction with a single myosin head for clarity. But in life, myosin heads are arranged in circular arrays, and cross-bridge cycling occurs thousands of times during a simple muscle contraction. This is where the expenditure of energy occurs.

Summary

Your afternoon run and simple snack actually require complex physiological mechanisms to accomplish! We will build on these mechanisms through the text as we learn about each of the organ systems.

Key Concepts

Osmosis
Diffusion
Facilitated diffusion

Na^+K^+ ATPase
Voltage-gated Na^+ channel
K^+ leak channel
Resting membrane potential
Action potential
Ligand-gated ion channels
Receptor signaling
Kinases
Receptor tyrosine kinases
Nitric oxide
Steroid hormones
Protein synthesis
Transcription
Translation
Glycolysis
Citric acid cycle
Electron transport chain
Excitation-contraction coupling
Muscle contraction

Key Terms

Acetylcholine
β-adrenergic receptor
α-adrenergic receptor
Transcription factor
Chaperone protein

Application: Pharmacology

1. Muscle strains are sometimes treated with dantrolene sodium, a drug that prevents calcium release from the SR. How would this reduce muscle work?
2. Botulinum toxin, known commonly as Botox, prevents the release of acetylcholine into the synaptic space. Why might this be used to reduce wrinkles?

Clinical Case Study

Type 2 Diabetes Mellitus, Part I

BACKGROUND

Type 2 diabetes mellitus (T2DM), also sometimes known as metabolic syndrome or insulin resistance syndrome, is a set of clinical conditions that were once thought to be associated and are now known to be interrelated. The key elements of this syndrome are defined as:

- Insulin resistance/diabetes
- Hyperinsulinemia
- Hyperlipidemia

- Obesity
- Hypertension
- Atherosclerosis
- Endothelial dysfunction

Insulin is a hormone released from the β-cells of the pancreas into the blood, where it then binds to receptors on insulin-dependent tissues, primarily skeletal muscle, adipose tissue, and liver. These receptors are not G-protein coupled but are a pair of simple α-helices in the plasma membrane that are capable of mutual phosphorylation and coupling. Once the hormone insulin binds to the receptor and the phosphorylation reaction begins, a tyrosine kinase phosphorylates and activates intracellular proteins, including insulin-receptor substrate 1, which is essential for muscle glycogen synthesis. Thus, insulin binding immediately begins the process of glucose storage within muscle and, by a related kinase reaction, in the liver as well. Insulin receptors also stimulate another kinase, phosphotidyl-inositol-3-kinase (PI-3 kinase), which stimulates glucose transport by moving intracellular vesicles of glucose transporters to the plasma membrane, thus increasing the number of glucose transporters (GLUT4) in the membrane and the number of glucose molecules that can transit them. Finally, a different signaling pathway is stimulated by the insulin receptor, which will promote GLUT4 transcription and translation, thus increasing the total number of glucose transporters.

In T2DM, the insulin receptor becomes resistant to signaling by the hormone, especially at the PI-3 kinase step. Thus, insulin binds the receptor, but the intracellular signaling pathway fails to increase the number of glucose transporters in the membrane. Exercise is generally prescribed for people suffering from T2DM, in part because exercise increases the concentration of glucose transporters in muscle membrane in an insulin-independent manner.

Normally, insulin release from the pancreatic β-cells is stimulated by glucose in a closely coupled cellular mechanism. As glucose levels fall, insulin release ceases. However, high circulating glucose levels cause continuous release of insulin, resulting in hyperinsulinemia and insulin resistance. If blood glucose levels remain high long enough, it will cause overt T2DM. While the high circulating insulin levels fail to increase glucose uptake by cells, they have effects on other tissues that are uninhibited by T2DM. We will see some of these effects in other sections of this clinical case study.

THE CASE

Your aunt has been diagnosed with T2DM because of her elevated fasting glucose levels (140 mg/dL). She has been placed on a low-sugar, low-carbohydrate diet and a regular regime of exercise. After two weeks of carefully following her diet and exercising, she has noted that her fasting glucose level has dropped to 110 mg/dL. Normal is 90–100 mg/dL, so she still has some work to do but, at this point, she is encouraged.

THE QUESTIONS

1. Why is glucose remaining in her blood so long after a meal?
2. How does sugar normally get into a cell?
3. Describe the action of insulin and how it facilitates sugar transport. Be specific about the signaling pathway.
4. How does daily exercise lower blood sugar?

© gorillaimages/ShutterStock, Inc.

2

Autonomic Nervous System

Case 1

One Saturday, you are relaxing on the couch, eating chips and drinking soda while reading an engaging novel. Suddenly you hear the squeal of brakes and, looking quickly around, realize that your cat is not in the house. Fearing she has been hit by a car, you rush outside. What is your autonomic state (1) while lying on the couch and (2) after you heard the squeal of brakes? Where do the autonomic signals originate? What happens at each organ under the predominant influence of each system? At each organ, which receptors mediate this action?

Introduction

We like to think of ourselves as intelligent creatures, but most of our physiological functions are managed without any conscious input. In this chapter's case, for example, you are eating and drinking, requiring digestion and metabolism, all while your conscious mind is occupied with a novel. Your heart beats, you breathe, your body temperature is maintained, and your blood pressure is managed when you jump off the couch. All of these functions, and many more, are managed by the autonomic nervous system (ANS). The ANS regulates our physiology using portions of the brain over which we have no conscious control, so the ANS does not control skeletal muscle, only cardiac muscle, smooth muscle, and some glands. We cannot decide to increase our heart rate, just as we cannot decide to increase the speed at which we digest food. All these "vegetative" functions are managed unconsciously by the ANS. However, emotions can have a profound effect on ANS function, as you realize when you feel your heart pounding because of fear for your cat. Let's learn about the functions of the ANS, along with some of the molecular mechanisms through which it works.

There are two branches of the ANS, with slightly different anatomies, neurotransmitters, and functions. The parasympathetic nervous division regulates most of our digestive functions and maintains our resting heart rate and breathing rate. It is sometimes called the "rest and digest" system. The sympathetic nervous division regulates our "fight or flight" responses, causing an increase in heart rate, breathing rate, muscle metabolism, alertness, and vascular control. The sympathetic nervous system is dominant not only during anger and fear responses, but also during normal exercise. Both of these systems are active at all times, but during your rest on the couch, the parasympathetic nervous system (PNS) will be dominant. As you run down the stairs searching for your cat, the sympathetic nervous system will become dominant.

We will begin with some background on the nervous system; then we'll examine the anatomy and actions of each of the two branches of the ANS.

Background on the Nervous System

We saw the cellular actions of the nervous system in the last chapter, when we looked at the action potential, ending at a neuromuscular junction. We will revisit the somatic nervous system and how it allows us to move in a future chapter. Now we need to familiarize ourselves with the basic structure of a nervous system.

Our nervous system is divided anatomically into the central nervous system and the peripheral nervous system. The central nervous system includes the brain and spinal cord. The peripheral nervous system is all the rest, including all the nerves that go to the target organs.

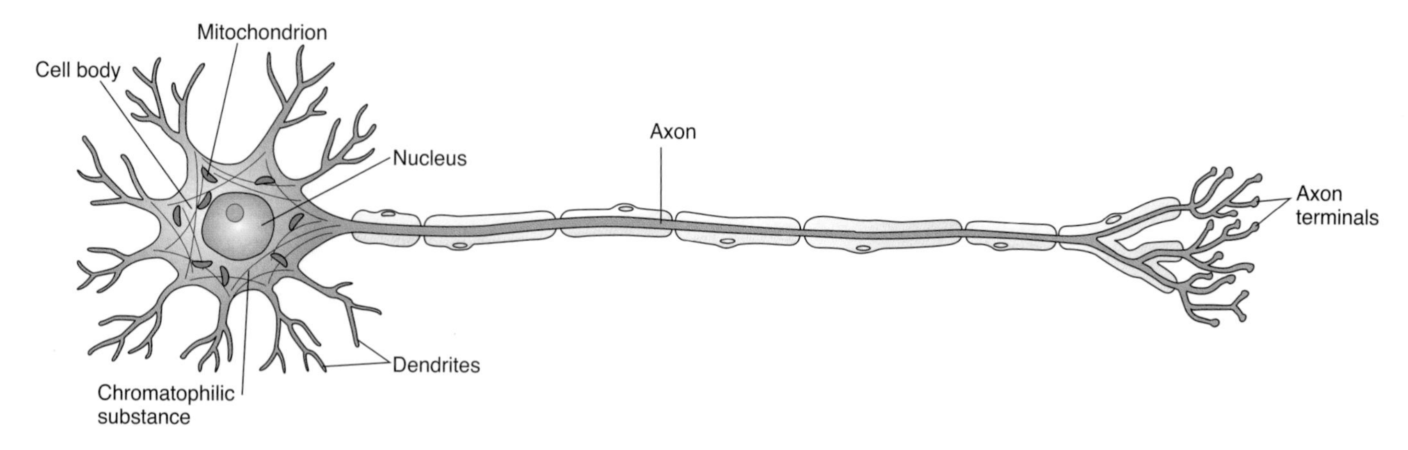

FIGURE 2.1 The basic organization of a multipolar neuron.

The primary cell type of any nervous system is the neuron. Neurons are unusually shaped cells, having a cell body decorated with dendrites on one end and a long axon on the other (**FIGURE 2.1**). Of course, there are variations in this neuronal shape, but the basic structural plan of dendrites, which synapse with other neurons, and an axon, which carries an action potential over distances, stays the same. The axon can be nonmyelinated, having little more than a plasma membrane to cover it. However, most neurons are myelinated, covered by a fatty layer of myelin, produced by either a Schwann cell or an oligodendrocyte. The myelin sheath allows faster conduction of the action potential. Neuronal cell bodies frequently cluster in an area, which is known as a nucleus when it is within the central nervous system and a ganglion when it is in the peripheral nervous system.

Rest and Digest—The Parasympathetic Nervous System

Organizationally, the parasympathetic division of the ANS is the simpler of the two divisions. The parasympathetic nervous system (PNS) is sometimes called the craniosacral system, because of its physical locations. Nuclei in the brainstem give rise to the preganglionic neurons in the cranial portions of the PNS (**FIGURE 2.2a**). Nuclei within the sacral portion of the spinal cord are the origin of the preganglionic neurons of the lower spinal cord. Preganglionic neurons of the PNS synapse with postganglionic fibers, at one of several ganglia. At the synapse of the pre- and postganglionic neurons, neurotransmitters are released, and receptors receive them as in the neuromuscular junction. In the PNS, the preganglionic neuron releases acetylcholine from its nerve terminal into the synaptic space. This occurs by an identical mechanism to that of the acetylcholine release from the α-motor neuron. Once in the cleft between the pre- and postganglionic neurons, acetylcholine binds to nicotinic receptors on the surface of the postganglionic neuron. This is the same ligand-gated ion channel present in the neuromuscular junction. So, the same communication mechanism we observe in the neuromuscular junction is repeated in the first action potential transfer between preganglionic neurons and postganglionic neurons of the parasympathetic system. The action potential, which is generated in the postganglionic neuron, terminates on tissues, the ultimate target of the parasympathetic system. At the synapse with the target tissue, acetylcholine is once again released, but this time it binds to a different type of acetylcholine receptor, a muscarinic acetylcholine receptor.

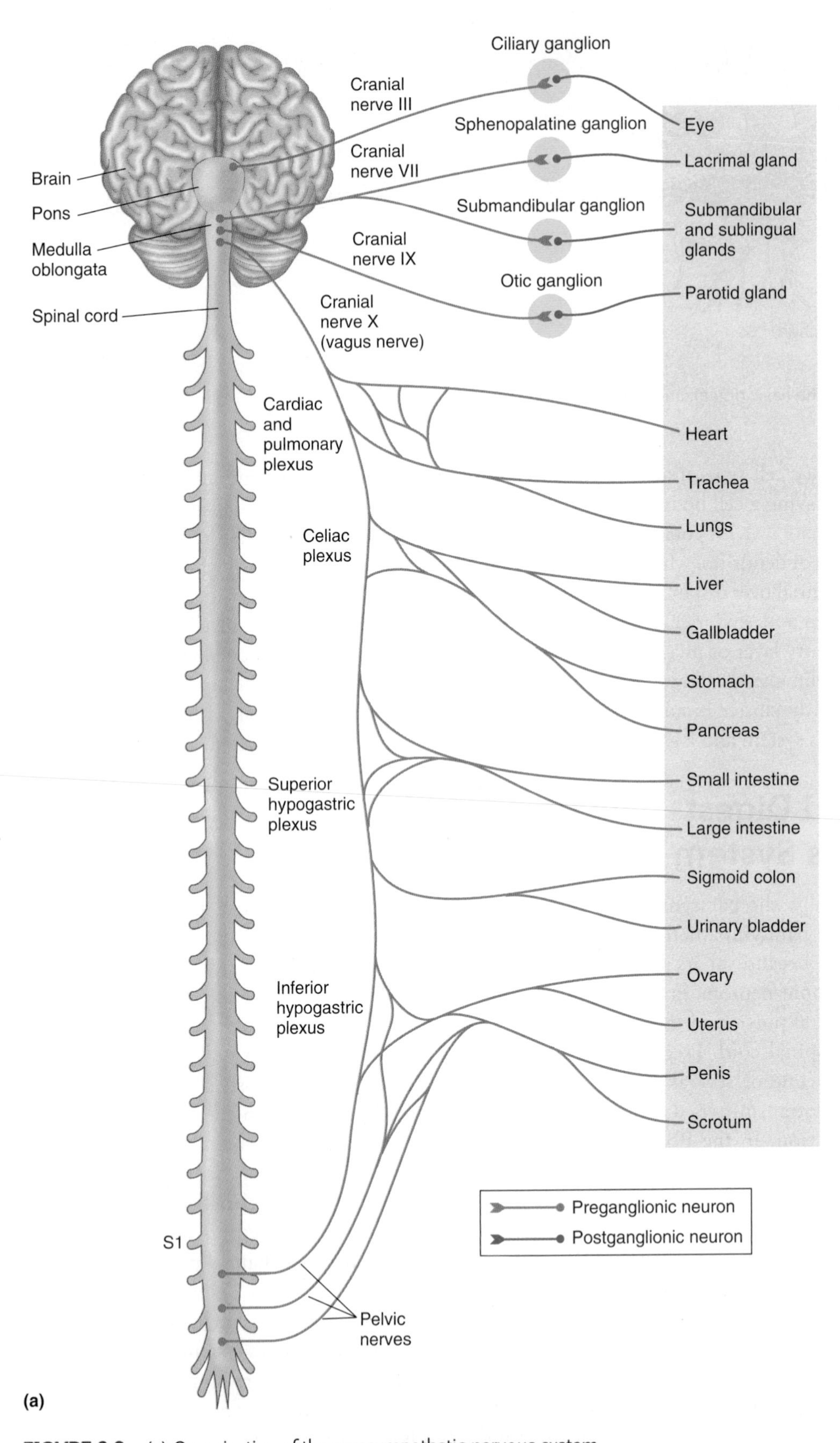

FIGURE 2.2 (a) Organization of the parasympathetic nervous system.

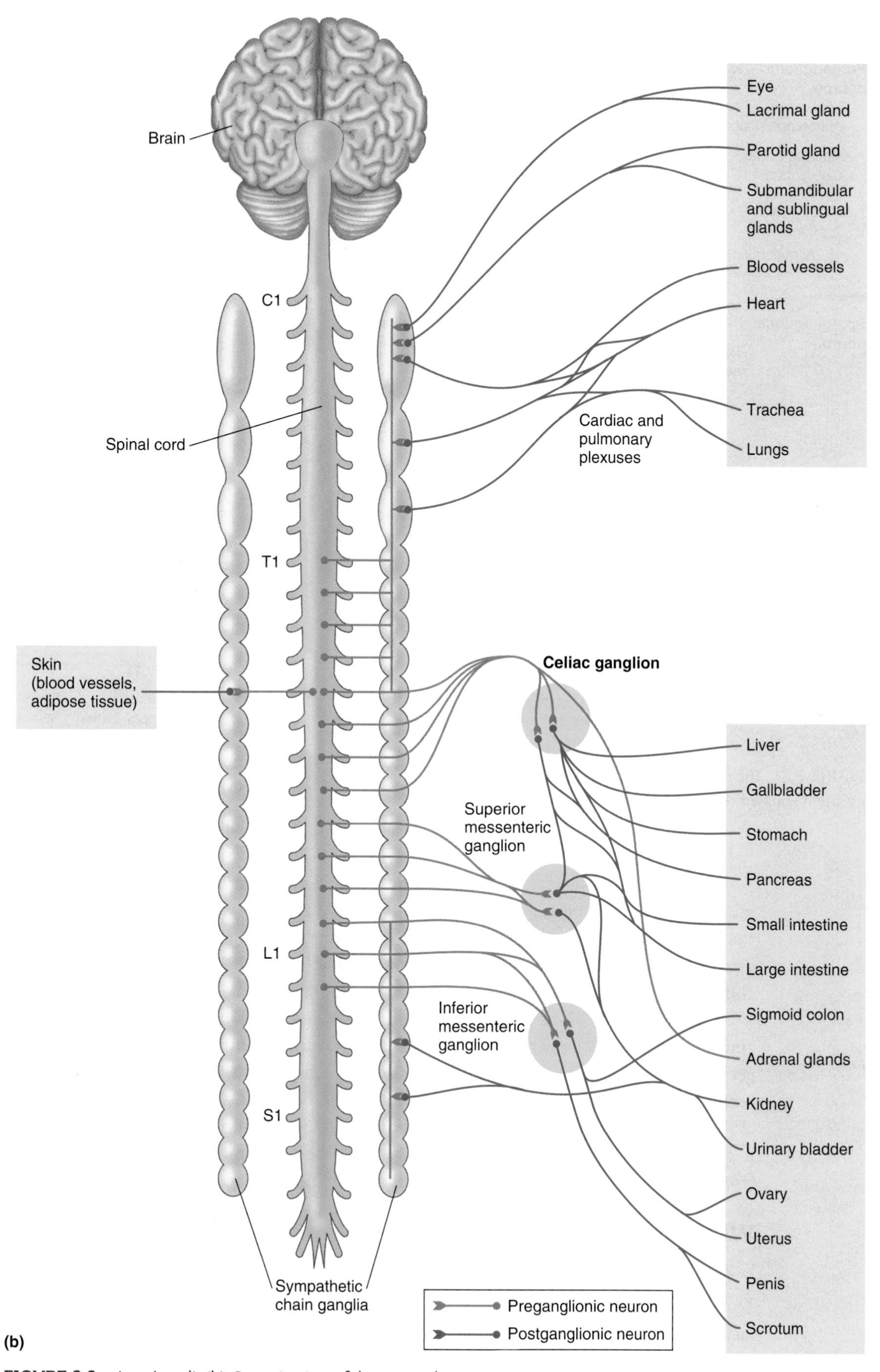

FIGURE 2.2 *(continued)* (b) Organization of the sympathetic nervous system.

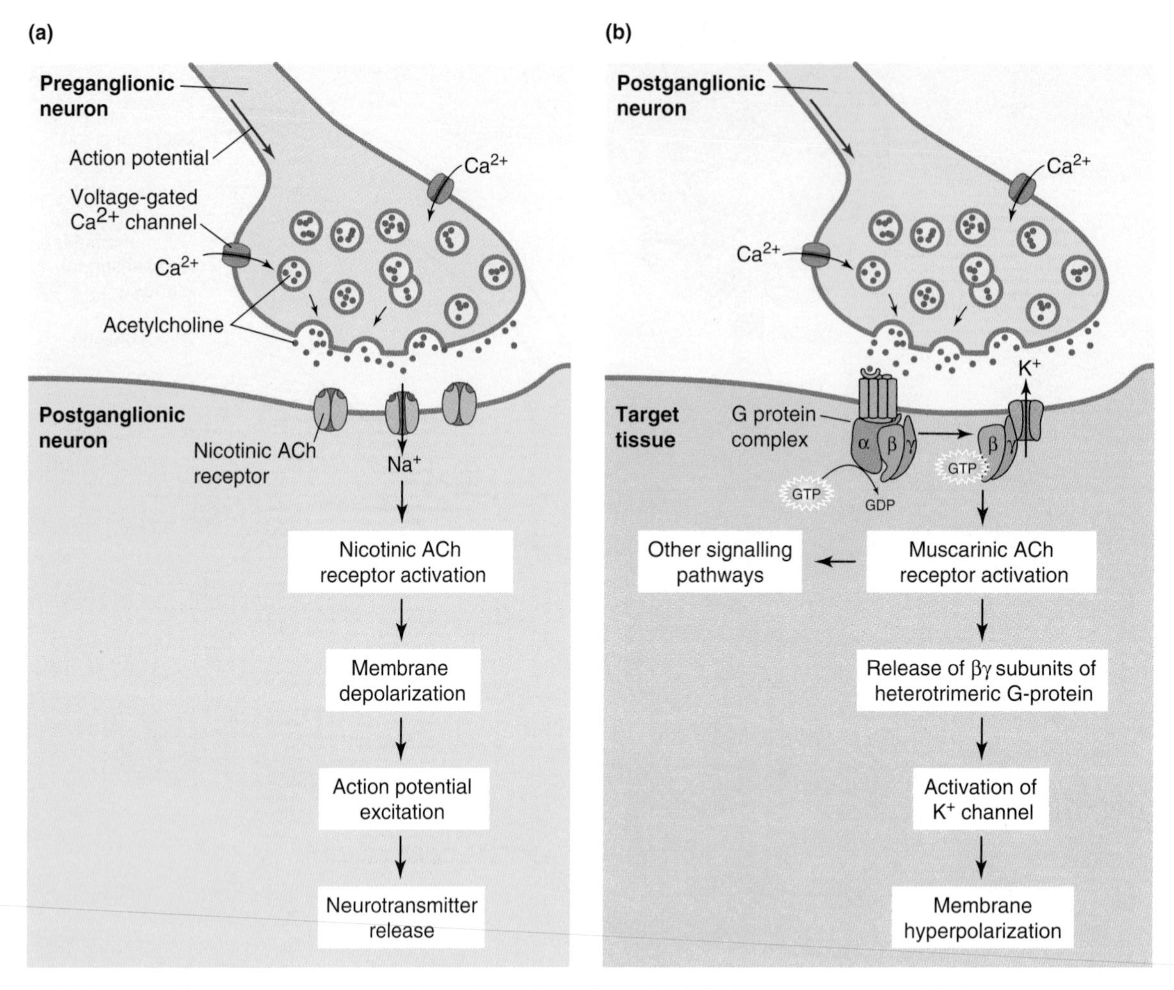

FIGURE 2.3 The nicotinic receptor is a ligand-gated ion channel, while the muscarinic acetylcholine receptor is a G-protein coupled receptor.

The muscarinic acetylcholine receptor is so named because it is activated by the mushroom-derived poison, muscarine. This receptor is a G-protein coupled receptor in the same family as the adrenergic receptors, so it is very different structurally from the nicotinic receptor at the preganglionic neuronal synapse (**FIGURE 2.3**). Like many G-protein coupled receptors, it is bound to a heterotrimeric (α, β, and γ) complex of G proteins. When acetylcholine binds to this acetylcholine receptor, the β- and γ-subunits are released and can bind to K^+ channels or begin a signaling cascade within the cell. Thus, at the target tissues, the postganglionic neuron releases acetylcholine, which binds to a muscarinic acetylcholine receptor and causes changes in cellular behavior within the tissue.

There is one other important point at this junction. The preganglionic neuron, when it synapses at the ganglion, stimulates only a few postganglionic neurons. Thus, signals from the parasympathetic system are well localized and discrete.

What Are the Effects of the Parasympathetic Nervous System?

The effects of the PNS, mediated by the cranial portion of the system, include pupillary constriction, stimulation of the lacrimal glands that constantly produce watery tears to cleanse the eye, stimulation of nasal mucus production to cleanse the nasal passages, production of saliva to keep the mouth clean, and additional salivary secretions when we are engaged in eating. Cranial nerves, particularly the vagus nerve, also serve as the preganglionic neurons for the thoracic cavity. The vagus nerve slows the heart rate and reduces the contractility of the atria, causes bronchoconstriction within the lungs, stimulates bile release from the gallbladder, and increases stomach and intestinal motility to maintain movement of food through the digestive system.

The sacral portion of the parasympathetic system, mediated by the pelvic splanchnic nerves, contracts the muscles of the bladder and relaxes the internal sphincter of the bladder, allowing urination. These nerves also innervate the penis and clitoris of the reproductive organs to allow engorgement and erection.

Thus, when the parasympathetic system is dominant, our heart rate is slow, our breathing is slower and more shallow, we produce copious secretions to keep our eyes, nose, and mouth clean and to aid in digestion, and we have increased gut motility to assist in digestion of food. This state of rest and repose is also conducive to reproduction. Clearly, the parasympathetic system is essential to our daily lives, yet we rarely recognize its valuable contributions.

Fight or Flight—The Sympathetic Nervous System

Anatomically, the preganglionic sympathetic nerves arise from the thoraco-lumbar sections of the spinal cord (Figure 2.2b). From the spinal cord, the preganglionic neurons can travel to the chain ganglia, which lie near the spinal cord itself and synapse there. Alternatively, the preganglionic neurons can travel through the white ramus and join a spinal nerve. Or, the preganglionic neurons can move through the chain ganglia to synapse at other levels of the spinal cord. Some sympathetic preganglionic neurons pass through the chain ganglia without synapsing and synapse instead within collateral ganglia. Finally, preganglionic neurons also travel directly to the adrenal medulla and synapse within that organ. Clearly, the pathways by which the sympathetic preganglionic action potentials travel are very diverse. Wherever they synapse, the sympathetic preganglionic neurons release acetylcholine as a neurotransmitter, which binds to a nicotinic receptor on the surface of the postganglionic neuron. Notice that this is exactly the same communication pattern as exists between the pre- and postganglionic neurons in the parasympathetic nervous system!

The preganglionic sympathetic neurons synapse with many postganglionic neurons. This makes the tissue response to sympathetic stimulus diverse and widespread throughout the body. While parasympathetic responses are closely controlled and localized, the sympathetic response is far-reaching and diverse. Postganglionic neurons of the sympathetic system release norepinephrine (NE) from their nerve terminals, which binds to G-protein coupled adrenergic receptors on the target tissue.

Unlike the nerve terminals of the neuromuscular junction, which is a discrete junction at the end of the neuron, NE from sympathetic neurons and acetylcholine from parasympathetic neurons are released to target tissue from enlargements along the postganglionic neurons (**FIGURE 2.4**). These enlargements are arranged serially, making the postganglionic neurons resemble a string of pearls, where each of the pearls can release neurotransmitter. This structure allows neurotransmitter release along tubes of smooth muscle, as we find surrounding the bronchioles or the blood vessels.

The sympathetic preganglionic neuron that synapses within the adrenal medulla is the longest of the preganglionic neurons in the sympathetic system, while the postganglionic

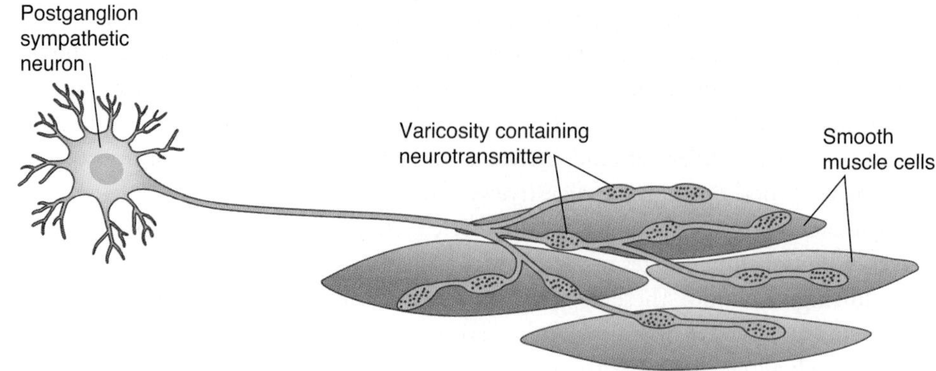

FIGURE 2.4 The nerve endings in the autonomic nervous system are serial boutons that can release neurotransmitter along the nerve's length.

neuron with which it synapses is the shortest. Once stimulated, the postganglionic neuron within the adrenal medulla stimulates the enzymatic conversion of medullary NE to epinephrine, which is released into the blood supply as a hormone. This is the "adrenaline rush" we experience during times of fear or anger. So, the sympathetic nervous system is a neuronal communication system that also causes the release of a hormone into circulation. Of course, this hormone will bind to receptors on many organs and will also contribute to the diverse effects of the fight or flight response.

What Are the Effects of the Sympathetic Nervous System?

We have all experienced the effects of the sympathetic nervous system. Perhaps it happened when you gave an oral presentation in front of a class, or when you were frightened by a particularly large spider, or when someone made you extremely angry—and you have certainly experienced it during aerobic exercise. All of these situations are stimuli for activation of the sympathetic nervous system.

The most obvious sign of sympathetic stimulation is an increase in heart rate and contractility. Your heart beats faster and more strongly, seeming to pound within your chest. Both of these tissue responses are a function of adrenergic receptors. Simultaneously, your breathing rate and depth increases, moving more oxygen-rich air into the lungs through bronchioles that are dilated by sympathetic stimulation. So just as you are about to "fight or take flight," you increase moment-by-moment your O_2 supply and distribution via the circulation. Sympathetic stimulation also causes constriction at the afferent arteriole of the kidney, reducing the amount of blood going through the nephron to be filtered as urine. Therefore, your vascular blood volume is preserved. Muscles lining the bladder relax, allowing the bladder to fill maximally. Blood vessels to the gut are constricted, while blood vessels serving skeletal muscle dilate, providing increased blood flow to voluntary muscle. Metabolism is also affected by the sympathetic nervous system. Sympathetic stimulation increases glycogen hydrolysis through G-protein coupled receptors. It also increases fatty acid release from adipose cells and prevents circulating nutrients from being stored. You have just been given the O_2, circulatory power, and metabolic fuel to allow you to run fast and escape attack by a Bengal tiger!

There are other, less obvious effects as well. For example, your eyes dilate, increasing your distance vision; your awareness and alertness are heightened; and your salivary secretions are inhibited, causing your mouth to feel dry. All of these functions have been mediated by adrenergic G-protein coupled receptors. There is one final important effect of the sympathetic nervous system—it increases sweat production. Sweating is our most important mechanism for maintaining our body temperature of 37°C when we are exercising. Without the evaporative cooling provided by sweat, our body temperatures would soar as adenosine triphosphate (ATP) hydrolysis and production create heat through wasted energy. Indeed, sweating is one of the great physiological talents of our species, allowing us to seek food during the heat of the day when other animals must rest in order to remain cool. Sweat production is stimulated through an acetylcholine release from postganglionic neurons that release acetylcholine to muscarinic receptors on the surface of sweat glands. Thus, while NE is the primary neurotransmitter of the sympathetic nervous system, it is not the only one.

How Can Norepinephrine Cause Vasoconstriction in Some Vessels and Vasodilation in Others?

Perhaps you have already noticed that the blood vessels of the afferent arteriole of the kidney are constricted, as are blood vessels serving the gut, while blood vessels serving skeletal muscle are dilated by sympathetic stimulation. How can this occur when NE is the only neurotransmitter released? As you may know, there are two types of adrenergic G-protein coupled receptors: α and β. Each of these receptors has a unique intracellular signaling cascade and influences different proteins. The action of NE will depend upon the receptor it binds to—and receptors are **not** homogeneously distributed throughout tissues. Quite the contrary, receptors are frequently tissue-specific or even specific to regions of a cell membrane. This alone is sufficient to provide for differential tissue responses to the same neurotransmitter.

Norepinephrine release from postganglionic neurons in the adrenal medulla causes the enzymatic conversion of NE to epinephrine and the release of these hormones from the adrenal gland. Epinephrine binds preferentially to β-adrenergic receptors, while NE binds more avidly to α-receptors, although each molecule can bind to both receptors. So, during exercise, or when you feel threatened, blood epinephrine concentrations will increase, allowing increased activation of β-receptors. Blood vessels that serve skeletal muscle have a higher concentration of β-receptors than most vessels, and β-receptor activation will dilate blood vessels. In contrast, the afferent arteriole of the kidney possesses α-receptors, which increase vasoconstriction (**FIGURE 2.5**). Thus, a multiplicity of actions at a variety of tissues can be expected during sympathetic stimulation.

Dual Innervation of Tissues by Parasympathetic and Sympathetic Branches

By now you have realized that some tissues, notably the heart, lung, bladder, and external genitalia, receive input from both the parasympathetic and sympathetic nervous systems. Dual innervation means that postganglionic neurons from both systems terminate on these organs. At rest, there is a balanced release of neurotransmitters by each system, giving us a resting autonomic tone. Both systems are active all the time, just at different levels of

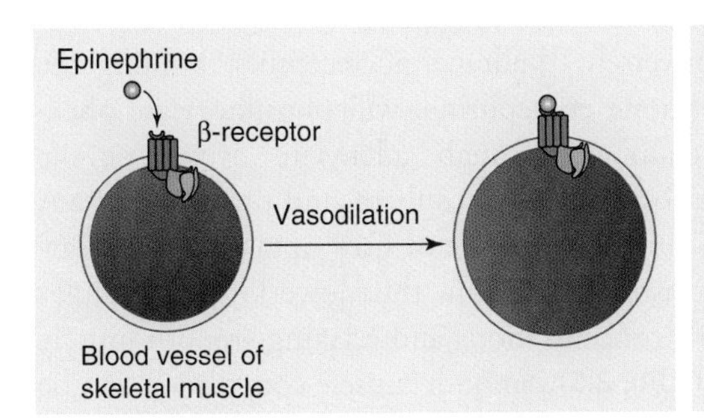

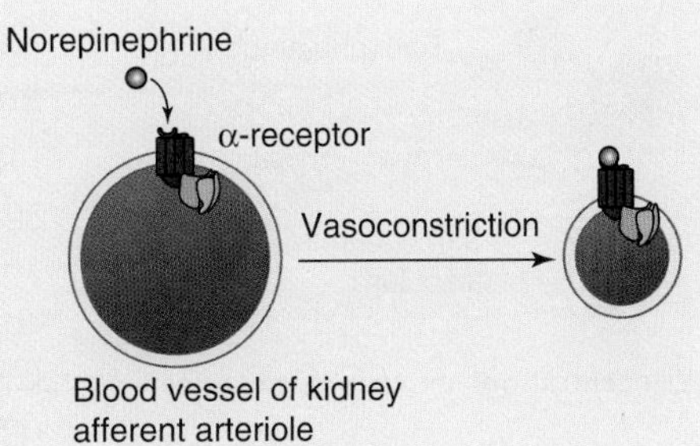

FIGURE 2.5 Blood vessels are dilated by the action of β-adrenergic receptors and constricted by the actions of α-adrenergic receptors.

activity. We can use a lighting dimmer switch as an analogy: the intensity of the light may vary, but there is no on-off switch. The properties of tonic signaling and dual innervation make tissues extremely responsive to changes in our situation, allowing rapid increases in heart rate as you run up the stairs, and rapid slowing of the heart when you come back downstairs, for example. In organs where there is dual innervation, the parasympathetic and sympathetic systems generally have opposing effects. While the parasympathetic system constricts the pupil within the eye, the sympathetic system dilates it. Parasympathetic stimulation constricts smooth muscle of the bladder, while sympathetic stimulation relaxes these same muscles. Dual innervation provides for the maximal range of control for the smooth muscle of bronchioles and bladder as well as cardiac muscle.

Smooth Muscle Contraction and Its Control

One of the primary targets of the ANS is smooth muscle. Smooth muscle surrounds our blood vessels, bronchioles, and the ducts of exocrine glands, and it also forms the smooth muscle sphincters of the eye. Smooth muscle is an essential part of the gastrointestinal system, being the muscle that contracts the esophagus, the stomach, and the intestines. In fact, smooth muscle surrounds virtually all of the tubes of our body. We shall examine the structure and contraction of this tissue so we can understand its regulation.

Smooth muscle is nonstriated muscle, lacking the regular linear structure we see in skeletal muscle. While smooth muscle contracts using actin and myosin, the actin is attached to the plasma membrane and to dense bodies within the cell, not to a Z-disc as occurs in skeletal muscle. Furthermore, it is not an actin-binding site that regulates contraction or relaxation, but phosphorylation of myosin. Thus, smooth muscle is generally said to be myosin-regulated. The excitation-contraction coupler is still calcium, so an increase in intracellular Ca^{2+}, either from extracellular stores or intracellular stores, can elicit smooth muscle contraction. Thus, depolarization of the smooth muscle membrane potential by neural transmission can cause contraction through the opening of voltage-gated Ca^{2+} channels. Similarly, smooth muscle contraction can be stimulated by α-adrenergic receptors, the signaling pathway of which releases Ca^{2+} from the smooth muscle sarcoplasmic reticulum through phospholipase C and inositol phosphate. This occurs without a change in membrane voltage. So NE, released from postganglionic neurons of the sympathetic nervous system, can stimulate smooth muscle cell contraction.

Smooth muscle cells are lavishly supplied with gap junctions, which allows the rise in intracellular Ca^{2+} to be shared with surrounding cells, ensuring a coordinated contractile response. Conversely, β-adrenergic receptors, binding the hormone epinephrine, will phosphorylate phospholamban through adenylate cyclase, cyclic adenosine monophosphate, and protein kinase A and increase the rate of Ca^{2+} uptake into the sarcoplasmic reticulum, thus lowering intracellular Ca^{2+} concentrations and relaxing smooth muscle (**FIGURE 2.6**). Smooth muscle contraction can be

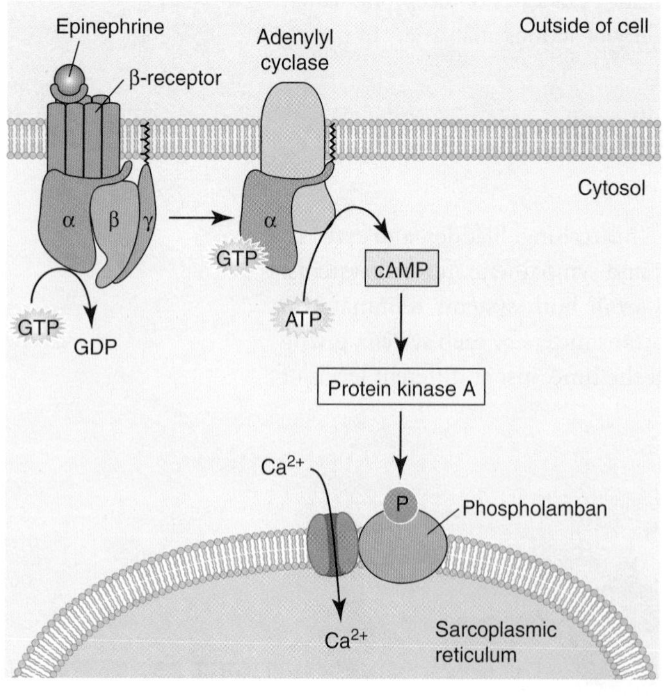

FIGURE 2.6 Intracellular Ca^{2+} concentrations determine vascular contraction and tone.

regulated either by changing membrane voltage or by ligand-stimulated changes in intracellular Ca^{2+}.

How Does Smooth Muscle Contract?

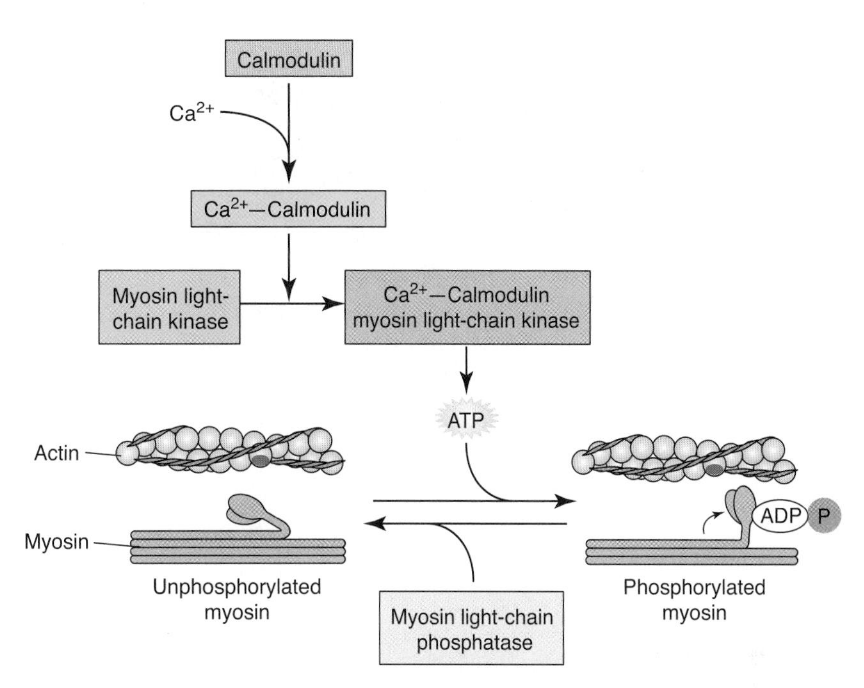

FIGURE 2.7 Smooth muscle contraction is myosin-regulated and depends upon myosin light chain phosphorylation for activation. Myosin phosphatase will dephosphorylate myosin.

Regardless of the mechanism by which intracellular Ca^{2+} rises, Ca^{2+} binds to the protein calmodulin, forming Ca^{2+}-calmodulin (**FIGURE 2.7**). Ca^{2+}-calmodulin is an activator of myosin light chain kinase. Myosin light chain kinase phosphorylates myosin light chain, which uncurls it, straightening the light chain and simultaneously causing a conformational change in the myosin head. This alteration increases its capacity for ATP hydrolysis and allows it to bind to actin. After binding to actin, the ADP and Pi on the myosin head detach while leaving the myosin head still bound to actin. At this point, smooth muscle can enter one of two states: if a new molecule of ATP binds to the myosin head, then it will detach from actin, allowing relaxation; alternatively, the myosin and actin may remain bound in a latch state, which allows smooth muscle to stay contracted for long periods with little ATP consumption. During relaxation, myosin phosphatase will dephosphorylate myosin light chain, ending contraction. These are the molecular events involved in smooth muscle contraction.

At the cellular level, when smooth muscle cells contract, they do not shorten linearly as skeletal muscle cells do. Instead, they contract in a more spiral fashion, because of the arrangement of actin and myosin within the cell. This is the perfect contractile pattern for narrowing the opening of a tube—which, as we have previously noted, is the shape of most structures composed of smooth muscle, whether they be the intestines, the blood vessels, or the bronchioles.

Single Innervation of Blood Vessels—How One System Does It All

While there are a few exceptions, in general, the circulatory system of blood vessels is innervated only by the sympathetic nervous system. I have included this discussion in a separate section, because it is so often misunderstood. After thinking about the opposing actions of the parasympathetic and sympathetic nervous systems, young physiologists often leap to the conclusion that dual innervation exists everywhere. This certainly is not true in the vasculature. The sympathetic nervous system is solely responsible for vascular tone, using the neurotransmitter NE and the hormone epinephrine, binding to α- or β-adrenergic receptors, respectively; α-receptors will cause vasoconstriction, and β-receptors will cause vasodilation. The distribution of the receptors and the concentration of the neurotransmitters and hormone will determine vascular diameter. Generally, there is a vascular sympathetic tone, generated by tonic binding to each of these receptors.

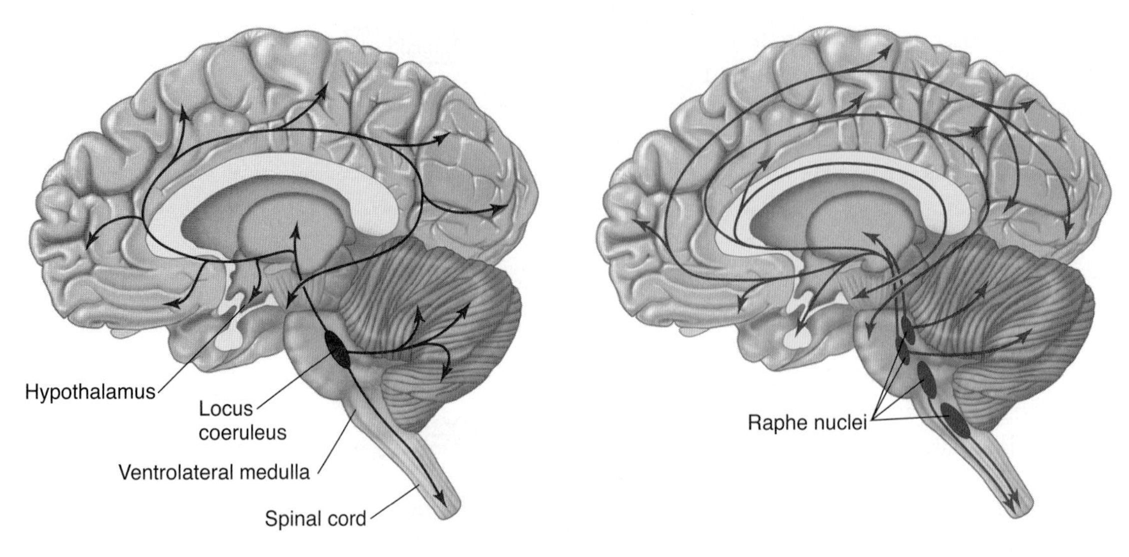

FIGURE 2.8 Areas of the brain that contribute to sympathetic nervous stimulation.

In the same way that dual innervation provided a large scope of tissue activity, sympathetic stimulation of vasculature is modulated by these two receptor types.

Does the Brain Play a Part in the Autonomic Nervous System?

While both the parasympathetic and the sympathetic nervous systems arise from the spinal cord and lower brain stem, there are certainly neural connections to the brain. There are several areas of the brain and brain stem that contribute to autonomic function, including the medulla, raphe, and the locus coeruleus (**FIGURE 2.8**). In the ANS, each of these areas contributes to overall autonomic tone. How do they do that?

Each of the areas projects neurons to the spinal cord, or parasympathetic brain stem nuclei, and can initiate specific responses. So, for example, the neurons from the ventrolateral medulla can increase sympathetic stimulation to the cardiovascular system, without causing a whole-body sympathetic response. In this way, the parasympathetic and sympathetic systems can have more local, discrete responses.

Each of these brain areas also projects neurons to the hypothalamus of the brain. The hypothalamus is responsible for a host of physiological functions, but one of its jobs is to coordinate autonomic activity. Unfortunately for students of physiology, the brain pathways are many and interrelated. However, we do not need to fully detail all of the pathways to understand that the ANS, which initiates at the spinal cord and brain stem, is in part directed by other brain areas, adding complexity and plasticity to this system. If other brain areas, including cognitive pathways, were not integrated with the autonomic system, how would we experience a sympathetic fear response to a garden spider or a zombie? Our emotions, memories, and experiences all contribute to what situations we find comforting or threatening.

Body Temperature Regulation

Temperature regulation involves the ANS, primarily because the sympathetic nervous system stimulates evaporative cooling through sweating. Even though the primary regulator of body temperature is the hypothalamus, we are considering body temperature regulation in this chapter because of the important role sweating plays in body temperature regulation.

The Sensors: How Do We Know If It Is Hot or Cold?

This is actually a complex question, because two important pieces of information are needed to make a physiological adjustment: what is the temperature outside in the environment, and what is the temperature inside of us? Our internal temperature is the most important driver of thermoregulation, but external sensors give the brain information about the outside environment.

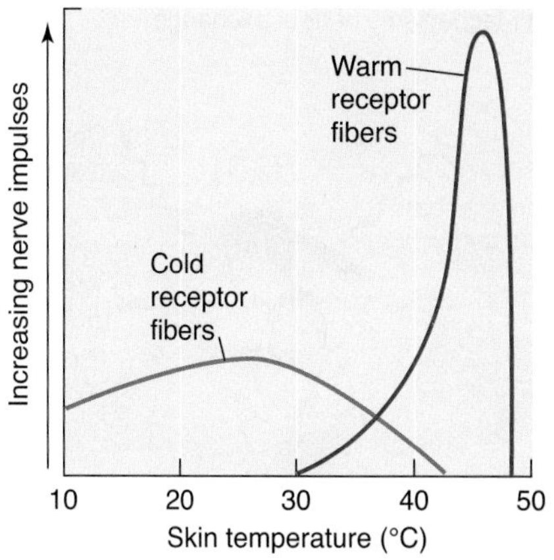

Within the skin, we have separate temperature receptors for hot and cold that monitor surface temperature. These receptors connect to neurons, which project to the conscious brain, allowing us to behaviorally respond to cold by putting on more clothing or to heat by removing some. These neurons also connect to the hypothalamus (**FIGURE 2.9**). So, by action potential signals, the hypothalamus—a portion of our unconscious brain—has an indication of external temperature. The temperature of the blood serving the hypothalamus serves as the carrier of information about internal temperature. If blood temperature increases even a tiny amount, thermosensitive neurons within the hypothalamus react very responsively. Thus, the hypothalamus can coordinate our thermoregulatory responses.

FIGURE 2.9 Warm sensors and cold sensors in the skin respond to different temperatures ranges.

What Are the Sources of Body Heat?

We already know that ATP production and hydrolysis produce heat as waste energy. No energy conversion process is 100% efficient, and the energy lost takes the form of heat. This same principle causes your car engine to become hot. When we are using ATP rapidly, as we do during a run, this generates body heat. If we are running outdoors, we may also experience the radiant heat from the sun's rays beating down on us. If we are running on a dark-colored running track, the sun's radiant heat may have warmed the running surface, so that we are getting heat conducted to our feet from the ground. Metabolic heat, radiant heat, and conductive heat are all potential sources of heat during exercise (**FIGURE 2.10**). The environmental heat will be sensed by warm receptors in the skin, but all of the heat will translate to higher blood temperatures reaching the hypothalamus.

How Do We Cool Our Blood Temperature?

By far the most robust method for cooling is evaporative cooling. The chemical transition of water from sweat evaporating off of our skin into gaseous form removes heat more effectively than any other method. The lungs also provide evaporative cooling, as we exhale moist air. Because we are running, we already have a sympathetic stimulation of sweat glands, but this can be augmented by increased signaling from the hypothalamus. Sweating causes us to lose total body water, but it allows us to maintain our homeostatic 37°C internal body temperature, to which our enzymes and body processes have adapted. If evaporation is made less efficient, as it is on a very humid day, it seriously impairs our ability to thermoregulate.

An increase in body temperature also causes vasodilation of surface blood vessels, increasing heat transfer from the body core to the surface and off to the environment. This is a much less effective method of heat transfer than is evaporative cooling, but it does offer another method of heat transfer. If body temperature rises to 39°C, we begin to experience

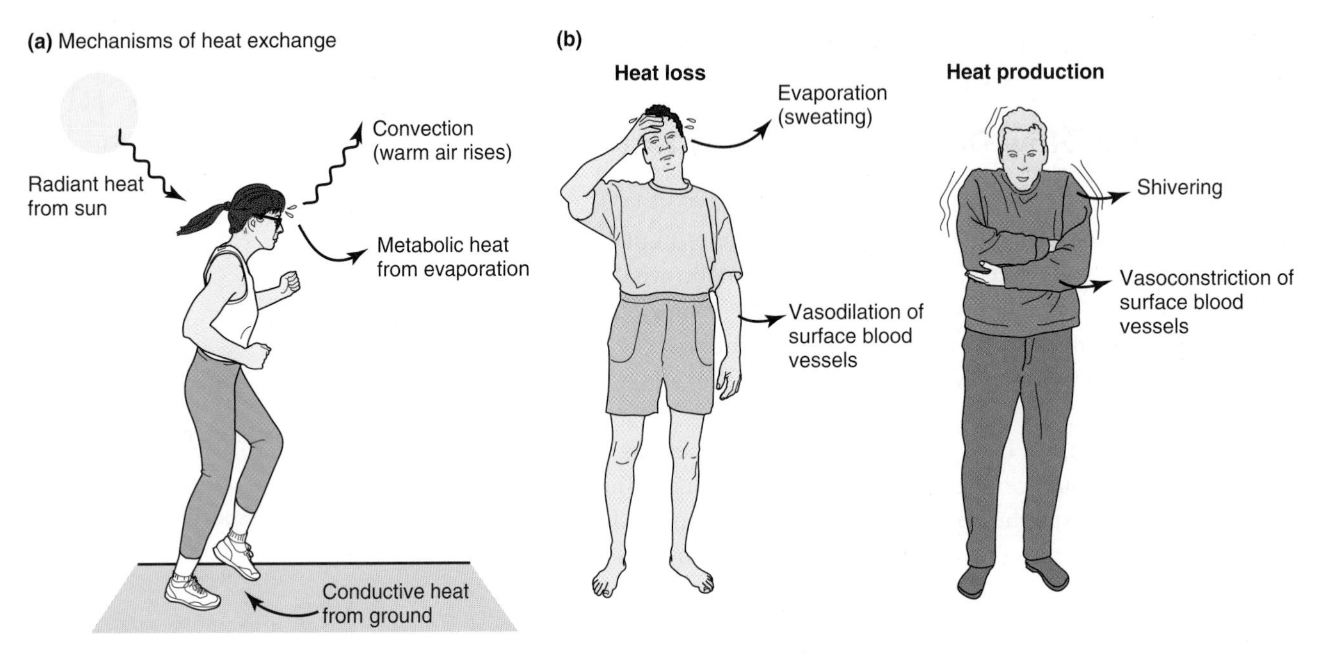

FIGURE 2.10 Sources of heat include conduction, radiation, and metabolic heat.

heat exhaustion, and at 41°C we experience heat stroke, a very serious dysregulation of body functions.

How Do We Stay Warm?

As a species, we are much better at cooling ourselves than we are at raising our body temperature. We are relatively hairless, which works to our advantage during evaporative cooling, but it means we lack an insulation layer in the cold. We can shiver, a nonvectorial muscle contraction, which consumes ATP and produces metabolic heat. Aside from those two mechanisms, we must use our conscious brain and alter our behavior. This generally means we must put on more clothing or move ourselves to a warmer environment. Hypothermia is as dangerous to us as hyperthermia.

Summary

The ANS and the brain areas that contribute to its function control much of our physiology. The difficult part of that for the physiology student is that many of these functions are nearly invisible, and we are forced to examine functions we rarely think about. The good part is that during our daily lives we do not have to think about our heart rate, our breathing rate, our temperature control, or how we digest food! We can save all that time for study and play.

Key Concepts

Parasympathetic nervous system
Sympathetic nervous system
Dual innervation
Smooth muscle contraction
Sources of heat and methods of cooling

Body temperature regulation
Mechanism of receptor action

Key Terms

Preganglionic
Postganglionic
Nicotinic receptors
Muscarinic receptors
α-adrenergic receptors
β-adrenergic receptors
Phospholamban
Myosin light chain kinase

Application: Pharmacology

1. Propranolol is a nonspecific β-blocker, meaning that it blocks the effects of epinephrine at β-receptors. This makes propranolol a β-adrenoreceptor antagonist. What physiological effects would you expect to occur if you were to take propranolol? What organs would be affected, and how?
2. One of the goals in drug development is to create drugs that have specific actions and do not affect multiple receptors. One such drug is the β-blocker atenolol, which is specific for β_1 receptors located in the atria of the heart. These receptors are responsible for an increase in heart rate. If you were to take atenolol instead of propranolol, how would the drugs' organ effects differ?
3. What effects might you see if you used the nonselective muscarinic antagonist, atropine? Which organs would be affected, and how?

Clinical Case Study

Heart Failure, Part 1

BACKGROUND

Heart failure has many causes, but all result in a decline of cardiac contractility. The most common cause of heart failure is myocardial infarction (loss of blood supply to a portion of the heart). Blood clots in coronary arteries that occlude blood flow can damage the heart muscle permanently, so it cannot participate in contraction. The heart contracts as a syncytium, i.e., as a single tissue instead of as individual cardiac muscle cells, so the loss of hundreds of cells in any region of the heart will reduce its contractile force. Because the heart is the pump that creates blood flow, a decline in cardiac pumping ability reduces blood pressure and signals an increase in sympathetic stimulus, i.e., norepinephrine from sympathetic nerve terminals and epinephrine from the adrenal medulla.

THE CASE

Your grandfather went to his primary care physician complaining of a lack of energy and exhaustion. After the exam, he was referred to a cardiologist for testing. Tests revealed that your grandfather had had several small myocardial infarctions that occurred after the one for which he had been hospitalized several years ago. The combined loss of cardiac tissue was enough to compromise the pumping ability of the heart. The blood pumped out of the heart at each beat, the stroke volume, had been reduced.

Of course, the entire family was concerned—heart failure is clearly serious. However, the cardiologist assured everyone that this condition could be managed. Your grandfather returned home with a list of prescriptions that would improve his condition.

You have just started your human physiology course and, while reading the chapter on the ANS, begin to see how this system is affecting your grandfather.

THE QUESTIONS

1. What are the receptor types for sympathetic stimulation?
2. Where are the different subtypes located?
3. What are the specific effects of sympathetic stimulation on each tissue of the body?
4. What are the effects of sympathetic stimulation on the heart?
5. What are the effects of sympathetic stimulation on the vasculature?
6. What is the effect of sympathetic stimulation on the kidney?

3

Endocrine Physiology

Case 1

After seasons of watching the *Survivor* series, a survival disaster has finally happened to you. During a vacation in northern Maine, your daily rental fishing boat sinks, and you are stranded on a tiny island with limited water and no food. To make matters worse, it is autumn, and temperatures are ranging from 60°F during the day to 30°F at night. Please describe the hormonal changes that occur as you begin your unplanned starvation in the cold. Never fear: the Coast Guard, notified of your absence by the boat rental shop, will rescue you two days later. Describe the hormonal changes that result from eating and drinking lavishly as you return to civilization.

Case 2

You got up late this morning and did not have time to eat breakfast before your morning run. It's 11:00 AM, and you are starving. At noon, you indulge in a 12-inch submarine sandwich and big piece of cake to make up for the missed breakfast. Which hormones maintain your energy state all morning, and what are their cellular actions? What happens to your energy state after lunch? Which hormones are involved, and what are their cellular actions?

Introduction

How do we maintain our energy supplies, fluid balance, mineral homeostasis, and cellular growth or proliferation? These long-term issues of homeostasis are managed by hormones. Hormones can be released regularly and rhythmically, or released in response to a unique stimulus. Unlike electrical stimulation, which is a focused communication between cells, hormones are carried in the blood and may target a number of tissues. Hormones are not tissue- or cell-specific, but receptor-specific. Just as with neural communication, we should think about regulation of hormonal control by asking such questions as: Where is the sensor for hormone release? What is sensed? From where is the hormone released? Where is its target? What does it do? These questions will help us understand how hormones regulate adenosine triphosphate (ATP) production, ion balance, or fluid balance, and how clinical conditions develop.

The case studies involve both starvation and feasting. Human beings are well adapted to periodic eating—that is, we do not have to eat continuously. As a species, we are able to gain weight and store energy as fat when food is plentiful, and to continue our normal organismal function for a long period of time (weeks) when food is not available. This feat is accomplished by regulating which energy substrates (carbohydrates, fats, or proteins) will be utilized for making ATP. This involves a variety of hormones and feedback mechanisms and will be a useful focus for studying the endocrine system.

How Do We Maintain ATP Production When We Eat Periodically?

As we know, ATP is not stored within cells but is made as needed. For example, when you are lying on the couch, your ATP usage and production are minimal. When you go for a run, ATP expenditure for muscle contraction, respiration, cardiovascular work, and so forth increases, and ATP production increases to match. ATP levels within the cell are maintained at constant levels, regardless of activity. If ATP is not stored, how do we store energy? Carbohydrates can be converted to glucose, which is stored as glycogen in the liver or in skeletal muscle. Protein can be stored in skeletal muscle as a component of the muscle itself. Finally, fat can be stored in adipose tissue, as fat. Glucose, protein, and fat can all be converted to fat and stored in adipose tissue. All of the foods we eat are broken down into one of these fuel types and either used immediately or stored for future use. Hormones regulate which fuels are used and how are they are released from storage. Hormones also regulate which fuels are transported across the cellular membrane for metabolism and ATP production within the cell.

Glucagon Regulates Release of Glucose from Glycogen Stores in the Liver

Each of our case studies begins with an episode of starvation. During starvation, whether it is short-term (like skipping breakfast) or longer term (like going for 2 days without food), the hormone glucagon helps us maintain our blood sugar at a constant level of approximately 5.5 mM. A stable blood sugar level is essential because the nervous system uses only glucose for ATP production. Without a continuous supply of glucose, our brain ceases to function, and we lose consciousness.

Glucagon is a peptide hormone produced by α cells in the pancreas and released into the blood supply, where it targets cells within the liver. Its action at the liver hepatocytes is to stimulate hydrolysis of glycogen into glucose. This glucose is then released into circulation, increasing blood glucose levels and thus re-establishing homeostasis. Glucagon release is triggered by (1) low blood sugar levels, (2) low blood insulin levels, (3) cholecystokinin (CCK), a hormone released from the digestive tract when fat and protein are digested, (4) increased amino acids, and (5) sympathetic nerve stimulation via β-agonists. Let's examine each of these conditions of glucagon release.

Several hours following a meal, blood glucose levels decline and stay low until the next meal. If glucose levels fall below the normal 5.5 mM, then glucagon release is stimulated. It's widely known that insulin facilitates glucose entry into insulin-sensitive tissues, such as liver, adipose tissue, and skeletal muscle. When blood glucose concentrations fall, insulin concentrations also decline. This acts as a secondary trigger for glucagon release, as do CCK and circulating amino acids. Amino acids can be converted into glucose by an enzymatic deamination reaction, but amino acids by themselves do nothing to maintain blood glucose concentrations, so glucagon is released. Finally, glucagon is released from α cells of the pancreas in response to sympathetic stimulation, via β-receptors. Exercise or stress stimulates glucagon release, causing an increase in blood glucose levels, which is essential for providing energy during your run or while you are searching your environment for food on the deserted island (**FIGURE 3.1**).

The α cells in the pancreas receive secretory signals from the blood perfusing them and—in response to low glucose, high amino acid concentration, or β-receptor agonists—release their stored hormone, glucagon, into general circulation. Glucagon binds to receptors and results in intracellular action in the form of hydrolysis of glycogen and release of glucose from the liver hepatocytes. While this is a sensitive response, it is slower than nervous transmission, taking minutes rather than milliseconds. Hormonal control, therefore, provides for longer-term communication between cells and tissues.

Insulin Regulates Glucose Uptake and Storage by Skeletal Muscle, Adipose Tissue, and Liver

If glucagon raises blood sugar concentrations, some mechanism must lower it. This hormone is insulin, which lowers blood glucose levels by signaling the insertion of glucose

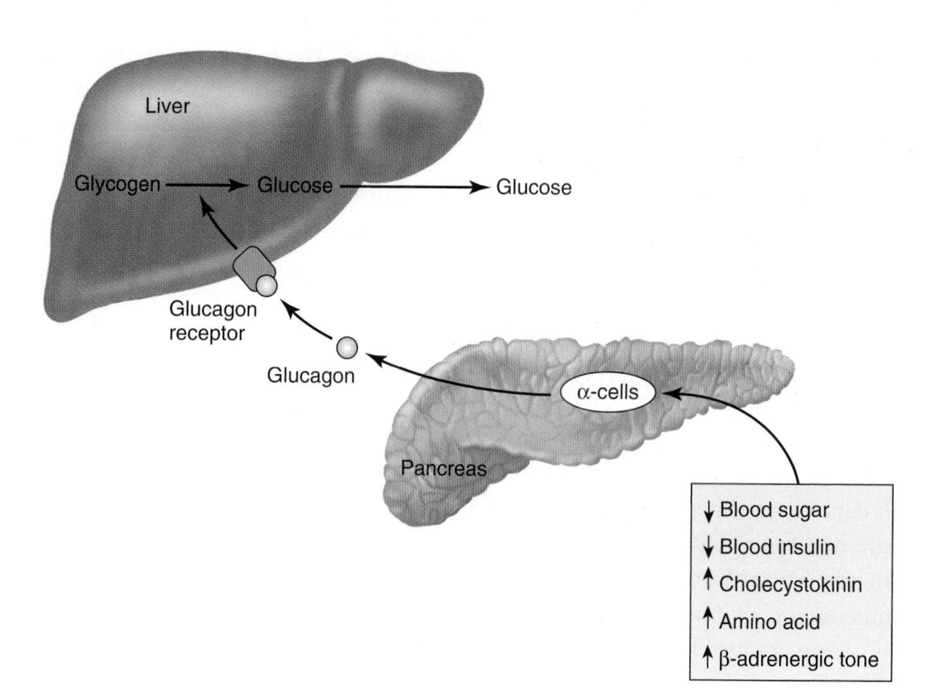

FIGURE 3.1 Glucagon is released from α cells of the pancreas in response to five different stimuli. The target cells for glucagon are the hepatocytes of the liver, where they will stimulate glycogen release.

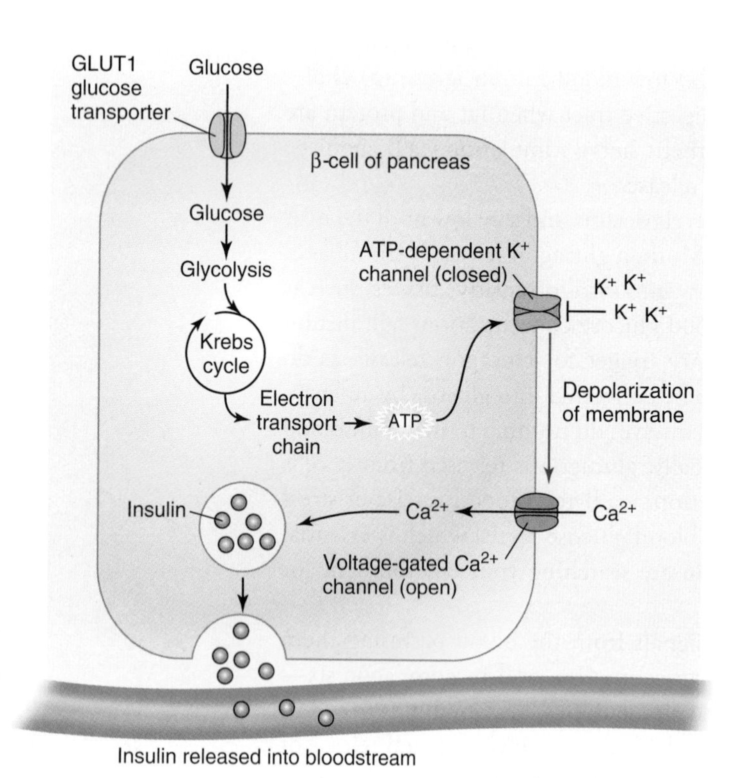

FIGURE 3.2 A rise in extracellular glucose will start the cascade of events that results in insulin release from the β cells of the pancreas. Glucose enters the pancreatic β cells through a noninsulin dependent transporter. In the cytoplasm, glucose is metabolized to ATP. ATP binds to an ATP-dependent K^+ channel, which closes, preventing K^+ from leaving the cell. This causes membrane depolarization as the positive charges accumulate inside the cell. Depolarization opens voltage-gated Ca^{2+} channels, allowing Ca^{2+} ions into the cell, where they initiate the process of exocytosis of vesicles containing insulin.

transporters into the plasma membranes of insulin-sensitive tissues. Remember that glucose transporters are not always present in the same number in a cellular membrane, and without them, glucose cannot cross the lipid bilayer. Insulin is released from β cells of the pancreas, also located within the pancreatic islets, in response to an increase in blood glucose. So, for example, when you eat heartily at lunch after your morning fast, carbohydrates are hydrolyzed into simple sugars and glucose enters the general circulation, increasing blood sugar levels. The β cells are exposed to this blood supply, as is every other tissue. Glucose is taken up by β cells and ATP is produced, which closes ATP-dependent K^+ channels, thus depolarizing the β-cell membrane. This opens Ca^{2+} channels, stimulating insulin-containing vesicle release from these cells (**FIGURE 3.2**). Thus, blood glucose concentration is directly linked to insulin release from β cells.

Insulin does not target all tissues; insulin receptors are primarily located in liver, skeletal muscle, and adipose tissue. In skeletal muscle and adipose tissue, insulin regulates the number of glucose transporters (GLUT4 type) within their plasma membrane. Skeletal muscle, a tissue that comprises the majority of our body mass, differs from liver in that the glucose crossing its plasma membrane cannot be released to circulation but is retained within skeletal muscle for ATP production. Adipose tissue will take up glucose for conversion into fatty acids and storage as fat. Glucose uptake by tissues serves to lower blood glucose levels back to a homeostatic level. Note that insulin-sensitive tissue is unlike neural tissue, which does not require insulin for glucose uptake. In neural tissue, glucose transporters remain in the plasma membrane and are not transient as they are in insulin-sensitive tissue. Hepatocytes can take up glucose via GLUT2 type glucose transporters even without insulin. However, insulin still has a significant effect on glucose metabolism within hepatocytes by activating enzymes involved in glycogen formation and inhibiting glycogenolytic enzymes. Thus, glucose in the liver is stored as glycogen for future release in the blood supply.

Although increasing glucose transport across cellular membranes is its key function, insulin plays other roles. Insulin also inhibits lipolysis at the adipose tissue by inhibiting the hormone-sensitive triglyceride lipase enzyme, while simultaneously increasing fatty acid transport into adipose tissue. Fatty acid transport into the adipocytes is facilitated by an increased expression of lipoprotein lipase, an enzyme that cleaves fatty acids from their protein carriers. Thus, insulin reduces fatty acid levels in circulation, as well as glucose. Insulin also increases amino acid uptake by muscle, stimulates protein synthesis, and inhibits protein degradation. Insulin stimulates the neutral amino acid transporter system A, thus increasing amino acid uptake in muscle and liver. However, the cellular mechanisms involved in insulin-enhanced protein synthesis and simultaneous reduction in proteolysis are not well understood. Within the brain, insulin acts at the satiety center in the hypothalamus to promote a sense of fullness, i.e. satiety (**FIGURE 3.3**). So, during your lavish lunch, insulin, released from the β cells, circulates to insulin-sensitive tissues, promoting glucose, amino acid, and fatty acid uptake, as well as providing an organismal signal for you to stop eating. Excess nutrients are saved for a time of fasting, and ATP production can remain at a normal level.

How is the action of insulin terminated? Why don't blood glucose levels decline precipitously? Like many receptors, the binding of insulin to the dimerized receptor complex

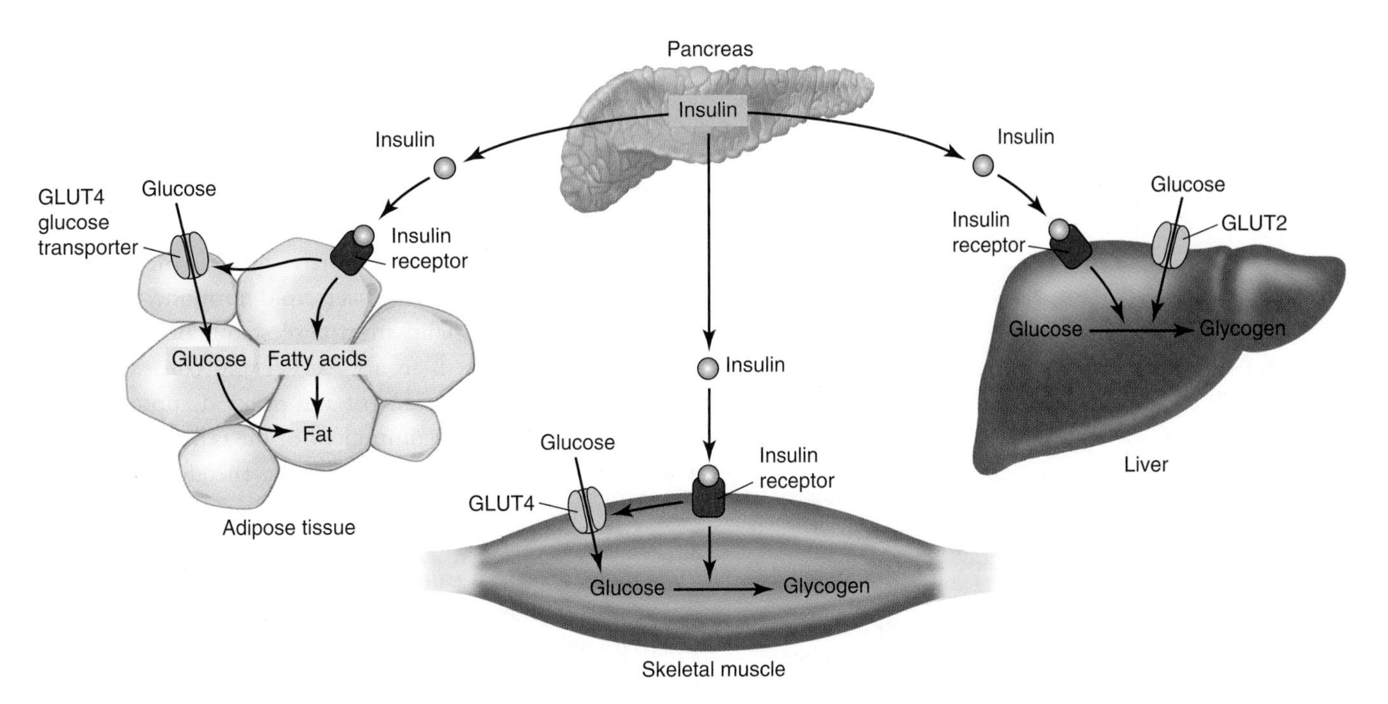

FIGURE 3.3 Insulin actions at its target tissues.

causes the receptor to be internalized via receptor-mediated endocytosis. Thus, insulin functions as its own negative feedback mechanism: the more insulin that circulates, the more insulin receptors are downregulated (internalized), and the fewer the receptors that are available to bind hormone. In addition, as glucose leaves circulation, less insulin is released from β cells. So insulin's action becomes self-limiting.

The pancreatic hormones, glucagon and insulin, are excellent examples of hormones with simple feedback loops to control their own release. We began with these two hormones because they have the simplest control mechanisms, and because glucagon and insulin are essential controllers of our nutrient balance. However, they are not the only controllers of glucose, fatty acid, and amino acid availability. Not all hormone release is signaled at the releasing tissue. Some hormone release is signaled by the hypothalamus.

Cortisol Release from the Adrenal Cortex Is Signaled from the Hypothalamus and the Anterior Pituitary

While insulin and glucagon respond rapidly to immediate changes in blood glucose concentration, other hormones act more slowly and tonically to regulate blood glucose, amino acids, and fatty acids. A few hormones are released at regular intervals, in association with our diurnal rhythm, rather than in response to a specific signal. The master rhythm controller of the brain is the suprachiasmatic nucleus (SCN) located within the hypothalamus. Neurons from the SCN have their own endogenous pacemakers. These pacemaker cells generate action potentials to stimulate specialized neurons within the paraventricular nucleus of the hypothalamus, which release corticotropin-releasing hormone (CRH). CRH, a peptide hormone, is released in small quantities directly into the hypophyseal portal circulation, which drains the hypothalamus and perfuses corticotrophic cells of the anterior pituitary. These corticotrophs release adrenocorticotrophic hormone (ACTH). Thus, tiny quantities of CRH in a small circulatory bed can target receptors on ACTH releasing cells within the anterior pituitary (**FIGURE 3.4**).

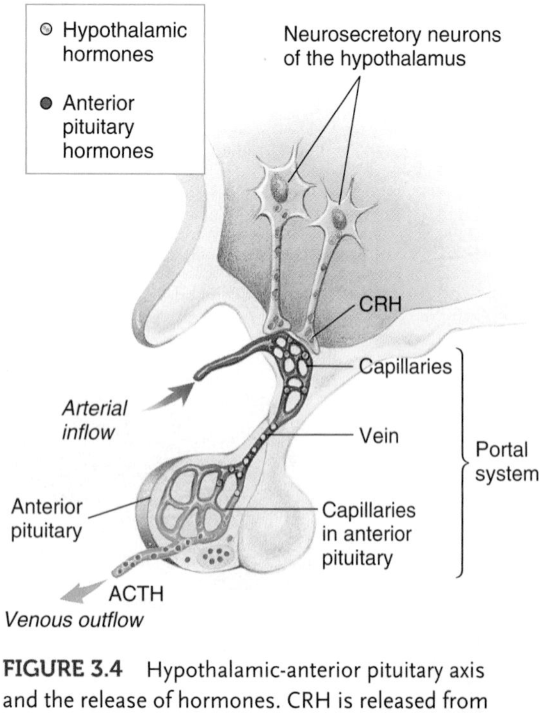

FIGURE 3.4 Hypothalamic-anterior pituitary axis and the release of hormones. CRH is released from the hypothalamus and circulates through the portal circulation into the anterior pituitary, where it targets specific secretory cells. These cells release ACTH into the systemic circulation. The target of ACTH is cells of the adrenal cortex.

ACTH is then released in larger quantities into the general circulation, where its target is cells within the adrenal cortex. ACTH, a peptide hormone, binds to receptors on cells of the adrenal cortex and signals synthesis and release of the hormone cortisol. Cortisol, a steroid hormone, is synthesized in the adrenal gland from cholesterol. Because CRH is rhythmically released, so are ACTH and cortisol. Cortisol release begins after midnight and peaks at approximately 8:00 AM. Cholesterol, a component of the cell membrane, is a neutrally charged, lipophilic molecule. Therefore, cortisol is membrane permeant and lipid soluble, not water soluble, and is carried in the blood supply bound to proteins (**FIGURE 3.5**).

Once dissociated from its carrier protein, cortisol crosses the cell membrane of target cells, particularly hepatocytes, where it stimulates enzymes of the Cori cycle, which deaminate amino acids. Once deaminated, the amino acids alanine, cysteine, glycine, serine, and threonine can be enzymatically

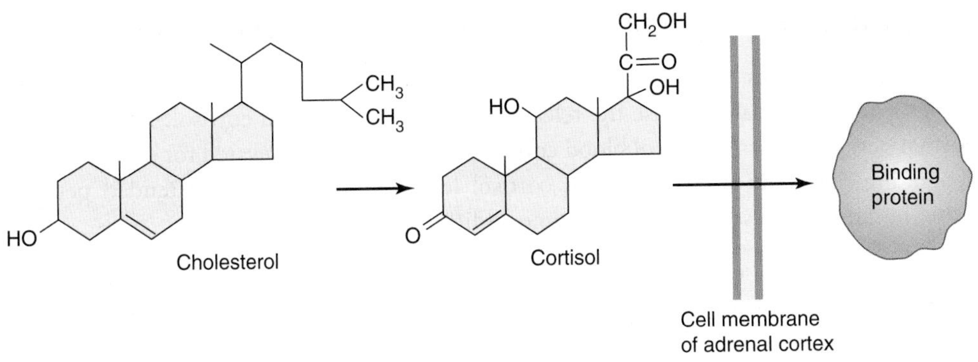

FIGURE 3.5 Cortisol is released from cells of the adrenal cortex and binds to carrier proteins in plasma.

converted into pyruvate. Cortisol also increases gene expression of gluconeogenic enzymes. Gluconeogenic enzymes can convert pyruvate back into glucose, thus creating new glucose from amino acids (**FIGURE 3.6**). Because this occurs in the liver, this glucose can be released into circulation. This is especially important during times of fasting or starvation, such as early in the morning before we have eaten.

Cortisol's effect on skeletal muscle and adipose tissue augments the effects at the liver. Cortisol stimulates skeletal muscle proteolysis and amino acid release, providing raw material for the gluconeogenic cycle in the liver. It also increases lipolysis in adipose tissue, producing an increase in circulating fatty acids, which can be used as fuel by the heart and muscle. Lipolysis also cleaves the glycerol backbone off fatty acids, so that it can be metabolized to glucose. Finally, cortisol reduces glucose uptake via the GLUT4 transporter. Thus, cortisol works in several ways to maintain circulating blood glucose and free fatty acids while we are in a fasting state, allowing for continual ATP production by tissues (**FIGURE 3.7**). This mechanism is so important that a decrease in cortisol secretion results in hypoglycemia. As with all hormones, there are feedback mechanisms that limit the quantity of circulating cortisol. The end product of the pathway, cortisol, inhibits release of ACTH from the anterior pituitary and CRH from the hypothalamus.

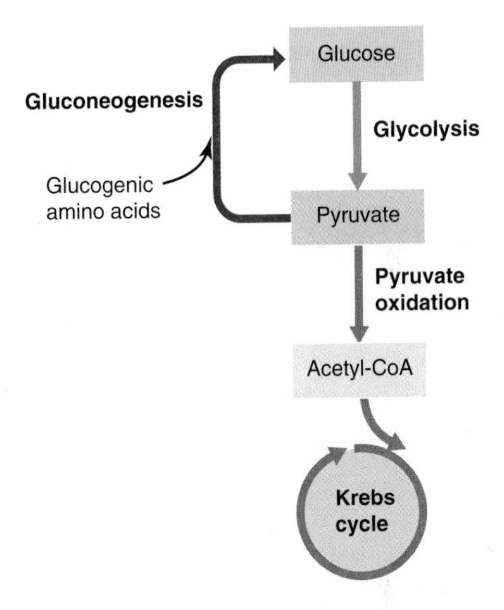

FIGURE 3.6 Gluconeogenesis forms glucose from pyruvate or amino acids or glycerol by reversing the process of glycolysis.

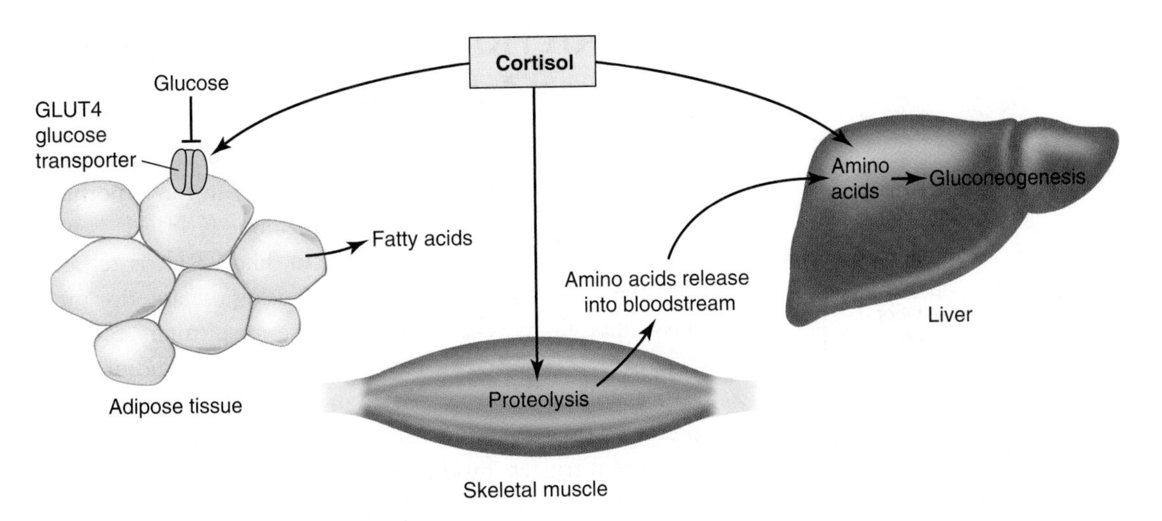

FIGURE 3.7 The target tissues of cortisol include skeletal muscle, adipose tissue, and liver.

How Does Stress Affect Cortisol Release and Function?

Physiological stresses, such as cold, severe exercise, and injury, or psychological stresses like fear or anxiety can increase the release of CRH, ACTH, and cortisol. Instead of the normal periodic release, cortisol blood concentrations may stay elevated for days to weeks, depending upon the stressors. When cortisol levels remain high for extended periods, either because of stress or pathology, some additional effects can observed:

1. Cortisol acts as an antiinflammatory, reducing interactions of the immune system that result in inflammation. This effect of cortisol is commonly used medicinally to reduce swelling and inflammation. Because inflammation itself can be life-threatening, downregulation of an inflammatory response may be protective.
2. Cortisol acts synergistically with epinephrine to increase cardiac output. Increased concentrations of cortisol are therefore helpful in promoting circulation during high-intensity exercise.
3. Cortisol promotes bone resorption. Ca^{2+} blood concentrations must be closely controlled for our survival. Bones are the primary reservoir of calcium, so releasing calcium salts from bone (bone resorption) maintains blood Ca^{2+} levels when we are not consuming calcium-containing foods. Over a long period of time, resorption would weaken bone, but during a physiological stress event (being stranded on an island without food or water), maintenance of our blood Ca^{2+} levels is most important.
4. Cortisol promotes muscle proteolysis. The amino acids freed from skeletal muscle are then available for transformation into glucose by the liver. However, this is done at the expense of muscle tissue. In the short term (a few days) there would be no noticeable loss of muscle strength, but blood glucose levels would remain stable.
5. Cortisol increases appetite. Certainly appetite, a desire for food, is essential for survival. While all of these actions of cortisol are beneficial during physiological stress, you can see how long-term psychological stresses, such as anxiety, could have detrimental effects: a compromised immune system, excess cardiac work, reduced bone and muscle strength, and weight gain.

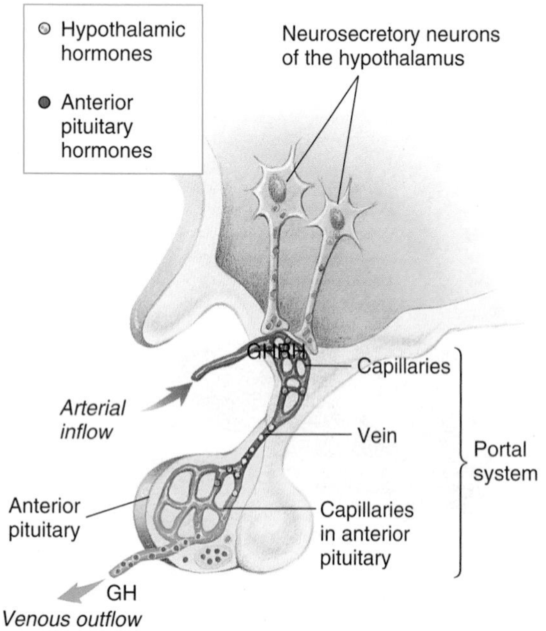

Growth Hormone Also Promotes Glucose Availability

Growth hormone (GH) is another of the hormones whose release is stimulated by releasing factors from the hypothalamus. Growth hormone–releasing hormone (GHRH) is produced within the hypothalamus and, like CRH, is released into the hypophyseal circulation perfusing the anterior pituitary. Specialized cells within the anterior pituitary produce growth hormone (GH), which has receptors on adipose tissue, muscle, and liver cells. Note that unlike ACTH, the product of the anterior pituitary is the active hormone, not a factor that triggers production of the hormone (**FIGURE 3.8**).

GH binding to these receptors has a suite of effects on metabolism and substrate availability. GH promotes fatty

FIGURE 3.8 GHRH is released by the hypothalamus to the anterior pituitary. The anterior pituitary releases GH into circulation, where it affects liver, skeletal muscle, and adipose tissue.

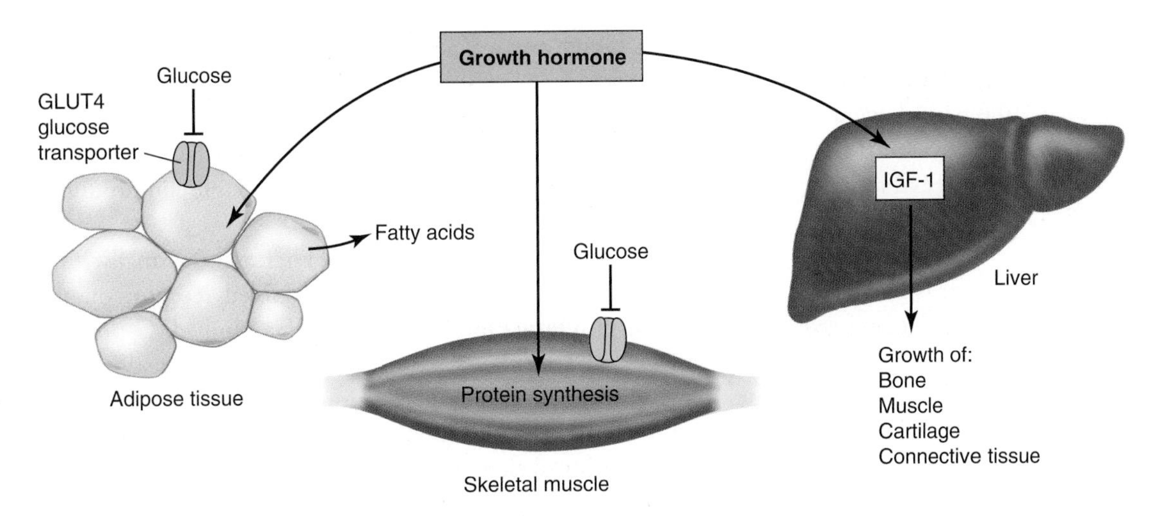

FIGURE 3.9 Growth hormone has its metabolic effects on many tissues. IGF-1 promotes tissue growth.

acid release from adipose tissue, increasing the available fatty acids, which can be used for ATP production. At the muscle, it increases protein synthesis while reducing glucose uptake. In insulin-sensitive tissues, such as muscle and adipose tissue, GH reduces the sensitivity of the insulin receptor for the hormone insulin, thus decreasing its effect. As a result, muscle takes up less glucose. Thus, GH preserves circulating glucose by preventing glucose uptake.

Release of GHRH is cyclical, as with CRH; however, it is linked to the wake-sleep cycle, not an endogenous diurnal rhythm. GHRH and GH peak early in the morning, just as cortisol release does, but changes in sleep cycles (such as shift work) have more influence on GH release. In addition to this rhythmic release, GH is released in response to physiological stress, exercise, starvation, and hypoglycemia. Episodes of starvation and exercise, which tend to reduce blood glucose levels, stimulate the release of GH.

Why is a hormone that regulates metabolic substrates called growth hormone? At the liver, GH signals the release of insulin-like growth factor (IGF-1). This peptide hormone is responsible for active growth within bone, cartilage, muscle, and connective tissue. Before IGF-1 was isolated, it was mistakenly believed that GH itself was causing proliferation within these tissues. Of course, an increase or decrease in GH will also affect the amount of IGF-1 that is released and therefore affect growth. So, a malfunction of GHRH release, for example, would reduce the amount of GH, which would affect both metabolism and growth, whereas lack of IGF-1 would affect growth, but not metabolism (**FIGURE 3.9**).

A unique aspect of GHRH release is that it is antagonized by another hypothalamic hormone, somatostatin. So, GH release can be inhibited at the point of origin, in the hypothalamus, by a tonic signal from somatostatin-releasing neurons. A similar pattern of end-product negative feedback regulation occurs with GH as we saw with cortisol. Growth hormone inhibits GHRH release from the hypothalamus, while IGF-1 inhibits GHRH release at the hypothalamus and GH release from the anterior pituitary. Hormonal release is thus self-limiting.

How Is Hormone Release Linked to Energy Balance?

How does the brain know you are starving? In response to a lack of food, the peptide hormone ghrelin—produced by cells of the stomach, pancreas, intestine, and neurons of the hypothalamus—is released. Ghrelin stimulates appetite and modulates release of many of the nutrient-regulating hormones we have discussed thus far. Specific nuclei in the brain have receptors for the hunger-stimulating hormone, ghrelin. Some areas of the hypothalamus are outside of the blood-brain barrier, allowing the hypothalamus to monitor the extracellular environment of the whole person via the blood. So, ghrelin produced peripherally can be sensed centrally.

Ghrelin release from hypothalamic neurons increases GH release. Normally, GH release is stimulated by GHRH via a Gs-coupled, G-protein-coupled receptor on the GH-producing cells of the anterior pituitary. Ghrelin binds to a Gaq-linked, G-protein-coupled receptor on these same cells (**FIGURE 3.10**). Thus, GH release can be stimulated via two distinct hormones, GHRH or ghrelin, each by a unique receptor. So, in times of hunger, ghrelin will promote all the effects of GH: lipolysis, insulin-receptor desensitization, and amino acid sequestration within skeletal muscle.

Ghrelin also affects the release of several other metabolic hormones. ACTH secretion is stimulated by ghrelin, so cortisol's actions of lipolysis at adipose tissue, gluconeogenesis at the liver, and proteolysis at skeletal muscle are augmented. Notice that the effects of cortisol and GH at the skeletal muscle are antagonistic. GH conserves amino acids within skeletal muscle, while cortisol promotes muscle breakdown. The amount of actual

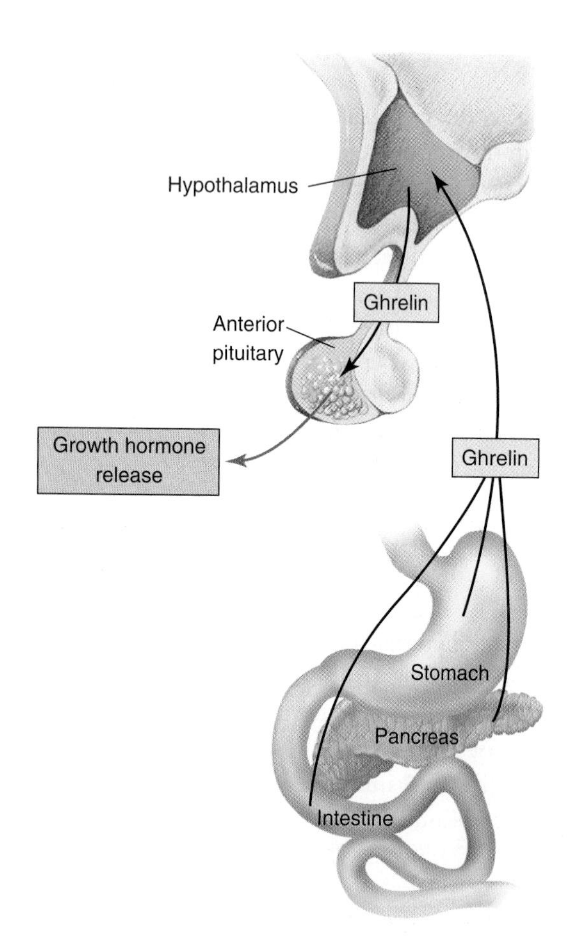

FIGURE 3.10 Ghrelin is produced peripherally by the stomach, pancreas, and intestine and centrally by the hypothalamus. It binds to G-protein coupled receptors in the hypothalamus to stimulate hunger. Ghrelin also stimulates the release of GH.

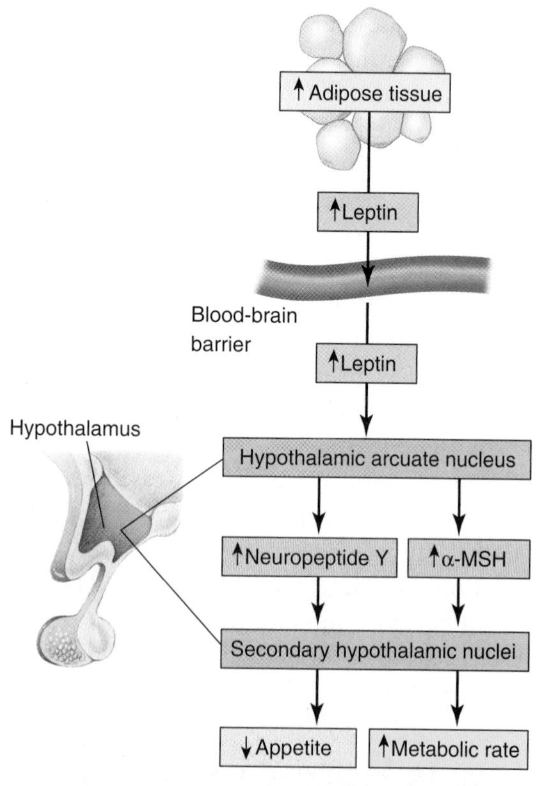

FIGURE 3.11 Leptin is released from adipose tissues and binds to receptors in the hypothalamus to signal satiety. This is done through neuropeptide Y, which suppresses feeding, and α-MSH, which increases cortisol release and thyroid hormone release.

proteolysis of skeletal muscle will be a sum of these two opposing actions. Glucagon release from the pancreas is stimulated, with all of its glycogenolytic actions at the liver. Thyroid hormone and insulin release may both be inhibited during starvation, reducing metabolic rate and glucose uptake by insulin-sensitive tissues.

What Is the Signal for Satiety?

How does the brain know when it's time to stop eating? Leptin is a peptide hormone released from adipocytes in response to an increase in energy stores. The exact signal for release is still unknown. Once in circulation, this hormone is transported across the blood-brain barrier by a specific transporter. Leptin then binds to its receptors within the hypothalamus. Leptin binding promotes the production of two additional hormones, neuropeptide Y and α-melanocyte-stimulating hormone (MSH). Neuropeptide Y suppresses feeding and promotes a sense of satiety; α-MSH stimulates CRH and therefore cortisol release (as well as thyrotropin-releasing hormone (TRH) and thyroid hormone release, discussed below). The net effects of leptin are to decrease appetite, increase fatty acid oxidation via cortisol, decrease blood glucose concentrations, and increase thermogenesis through thyroid hormone (**FIGURE 3.11**).

How is it possible, then, to become overweight? Leptin is very effective in decreasing appetite and reducing circulating glucose and fatty acids in the short term. Periodic feasts are well controlled by this hormone. However, in the face of a continuously plentiful food supply, high levels of leptin simply cause a downregulation of leptin receptors, and therefore less regulation of appetite. This makes sense, considering an abundant food supply has existed only for the past 60 years of our 200,000 year species history. While we have elaborate defenses against starvation, we have few adaptations for an excessive nutrient supply.

Is There a Modulator of ATP Production or Metabolic Rate?

Thyroid hormone, another of the hormones of the hypothalamic-anterior pituitary axis, is a general regulator of metabolic rate—i.e., how much ATP is hydrolyzed and produced. Where other hormones we have examined manipulate the available fuel substrates for metabolism, thyroid hormone increases metabolic rate and heat production, or thermogenesis. So, while stranded on the island off the coast of Maine, the low night-time temperatures will promote an increase in the actions of thyroid hormone, resulting in a higher metabolic rate.

Cold, sensed from the blood supply at the paraventricular nucleus of the hypothalamus, stimulates mRNA production

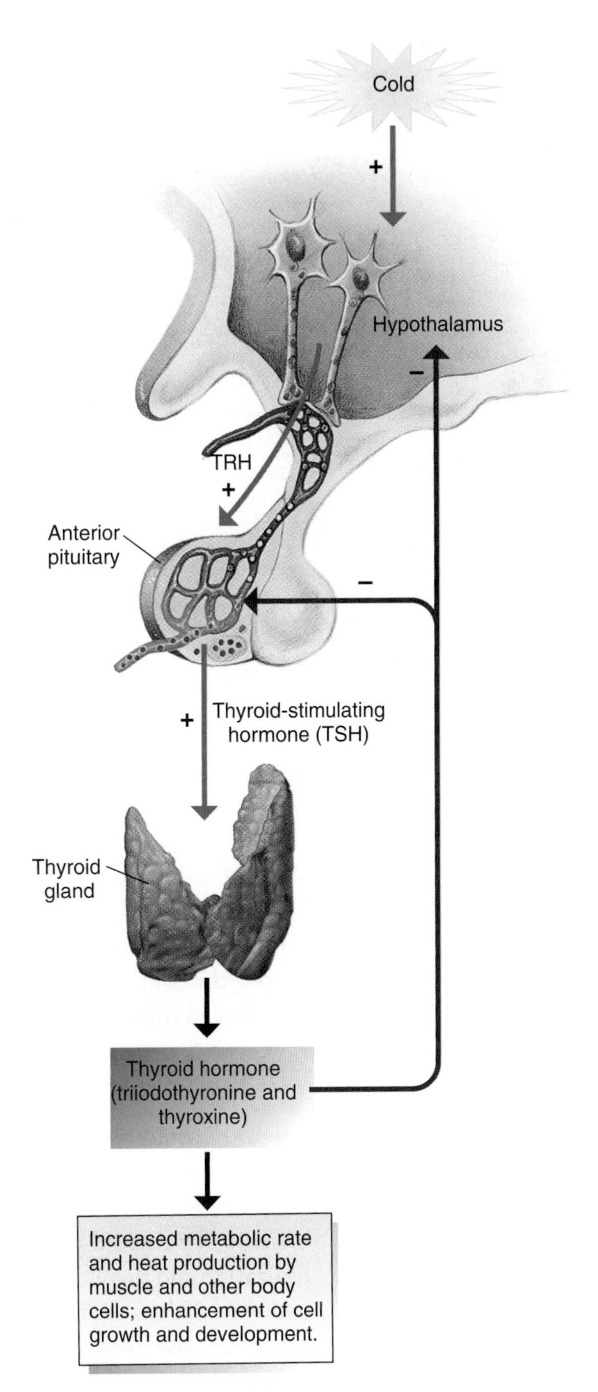

FIGURE 3.12 The tonic release of TH can be increased by cold temperatures. This raises metabolic rate and heat production. TH inhibits TRH release at the hypothalamus and TSH release at the anterior pituitary.

of TRH. This peptide hormone travels by the portal circulation to the anterior pituitary, where it signals the release of thyroid-stimulating hormone (TSH) from specialized cells. TSH has as its target follicular cells within the thyroid gland itself. TSH stimulates growth of these cells and increases blood supply to the thyroid (**FIGURE 3.12**). This effectively stimulates each part of thyroid hormone synthesis.

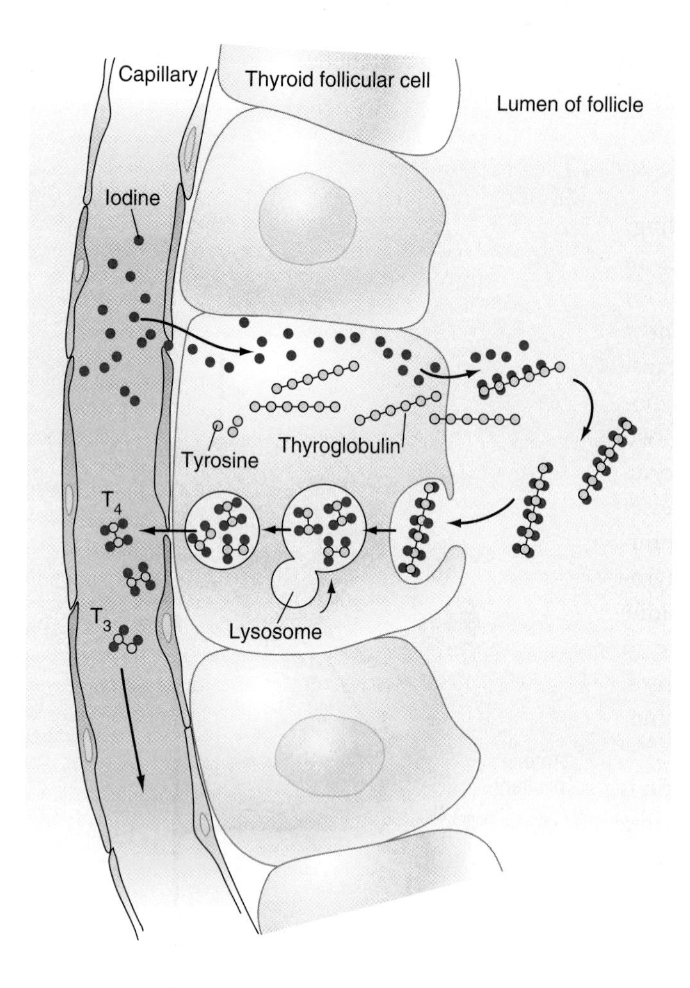

FIGURE 3.13 Synthesis of thyroid hormone.

How Is Thyroid Hormone Made?

Thyroid hormone comes in several forms, but the predominant form made by the thyroid gland is thyroxine (T_4), two tyrosine amino acids iodinated at four sites. The iodine, a trace element in our diets, is transported into the follicular cells by a Na^+/I^- symport. This symport protein is capable of transporting iodine against high concentration gradients, using the driving gradient of Na^+. Thus, iodine becomes concentrated in follicular cells and is subsequently transported into the thyroglobulin-containing lumen. Tyrosine residues on thyroglobulin proteins are iodinated within the lumen and then transported back into the follicular cell, where the active hormone is made by proteolysis of the iodinated tyrosines from the rest of the thyroglobin molecule (**FIGURE 3.13**). Each part of this enzymatic process is stimulated by TSH.

The predominant form of thyroid hormone released is T_4, which is converted to the more active form of triiodothyronine (T_3) in the peripheral tissues, particularly kidney and liver. T_3 and T_4, as end products of this pathway, feed back to the hypothalamus and the anterior pituitary to inhibit the release of TRH and TSH. When the nutrient iodine is in short supply, the consequences are less T_4 and T_3, and less inhibition of TRH and TSH, causing an enlarged thyroid gland, or goiter.

What Are the Actions of Thyroid Hormone?

Thyroid hormone, once in circulation, binds to plasma membrane transporters of target tissues, entering the cell. This transporter-hormone complex then binds to nuclear receptors and becomes a transcription factor, facilitating protein synthesis. The proteins that are increased by thyroid hormone include proteins such as Na^+/K^+ ATPase that increase ATP utilization and therefore, as a by-product, heat production. Many enzymes of the metabolic pathway, such as cytochrome oxidase in the mitochondria, are upregulated to increase the flux of glucose to energy. As a complement to this, glucose transporters are upregulated so that more glucose can enter intracellular metabolism. Hormone-sensitive lipase activity is stimulated, increasing lipolysis and the supply of circulating free fatty acids. Ca^{2+}-ATPase activity is increased in skeletal muscle, allowing for faster relaxation of skeletal muscle as calcium is resequestered in the sarcoplasmic reticulum. This facilitates faster muscle contraction. β-adrenergic receptors are also upregulated, increasing sympathetic tone. Finally, GH and antidiuretic hormone release are both increased by thyroid hormone.

What is the net effect of thyroid hormone? The increase in Na^+/K^+ ATPase activity alone increases basal metabolic rate. Because not all the energy of ATP hydrolysis results in work, the wasted energy becomes heat and helps us maintain our body temperature (thermoregulate). This is an essential mechanism for homeotherms, one that allows us to maintain constant body temperature. If metabolic rate and ATP hydrolysis are upregulated, movement of fuel substrate through the enzymes of metabolism must also be increased to supply the ATP for hydrolysis. Appropriately, this is another action of thyroid hormone. Increasing autonomic tone by upregulating β-receptors augments cardiac contractility, heart rate, and cardiac output, increasing blood flow to tissues for the delivery of nutrients (**FIGURE 3.14**).

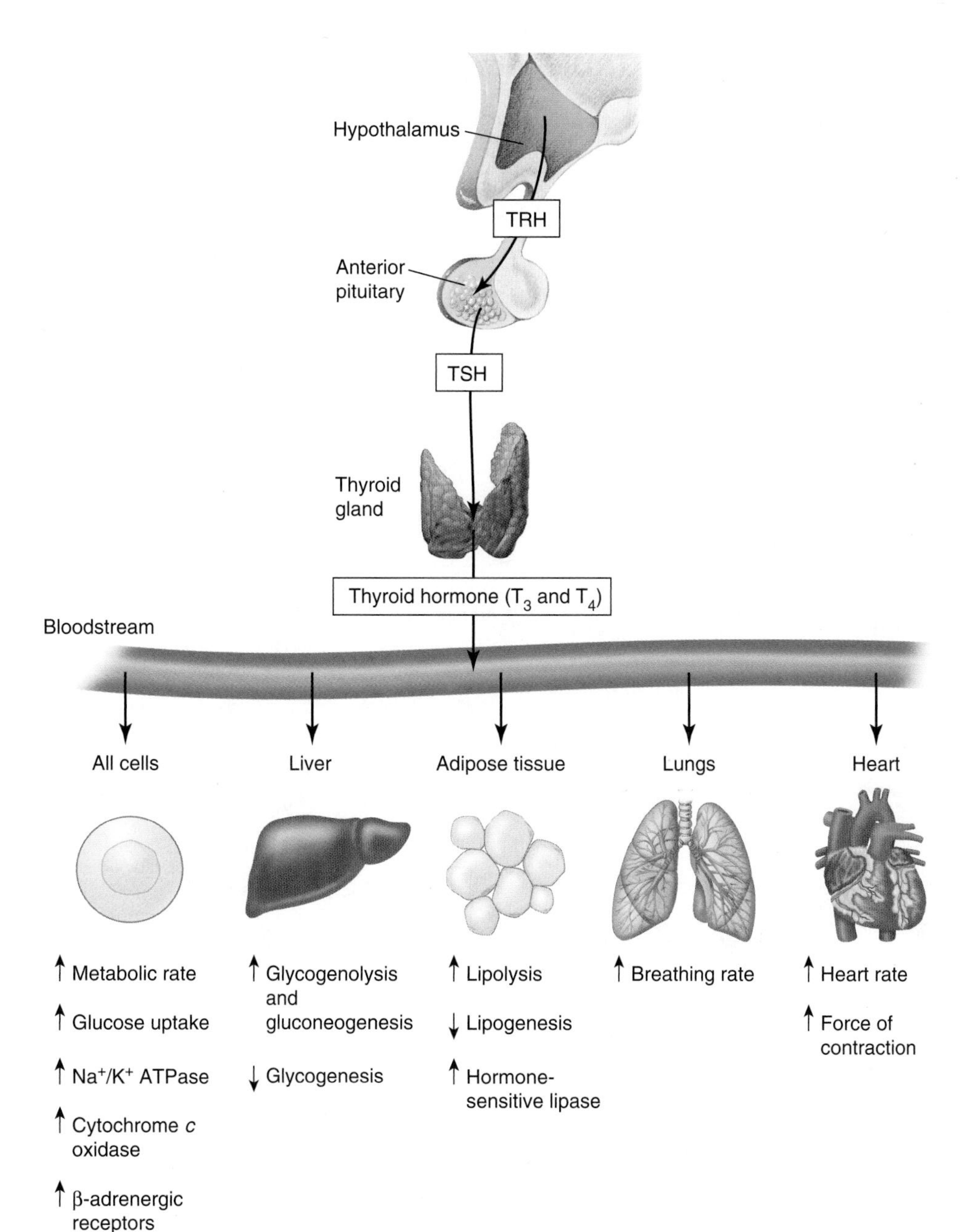

FIGURE 3.14 T3 and T4 stimulate the synthesis of proteins that increase ATP hydrolysis, such as the Na^+/K^+ ATPase and increase ATP production, such as the cytochrome *c* oxidase chain.

Starvation reduces thyroid hormone release, while cold stimulates it. Both of these conditions co-exist in our stranded-on-the-island example. Which stimulus is most important? Despite starvation, we must maintain our body temperature, so thyroid hormone release will increase. The result of this will be a more rapid weight loss in cold environments, as internal stores of glucose from glycogen reserves in muscle and liver, fatty acids from adipose tissue, and amino acids from protein in skeletal muscle are hydrolyzed to make ATP.

If There Is No Water, How Do We Regulate Our Fluid Balance?

Careful management of our total body water is essential for our survival. As land animals, we have elaborate hormonal mechanisms for conserving water, but only one for losing excess water. Let us begin with water conservation.

Antidiuretic Hormone (ADH) Acts Directly to Retain Water

ADH, also known as vasopressin, is released at the posterior pituitary from neuronal projections of hypothalamic neurons. These neurons in the hypothalamus receive information from nearby osmoreceptors, which monitor the osmolarity of your blood. When osmolarity increases, there are more ions per unit of water, so you need to conserve water—that is,

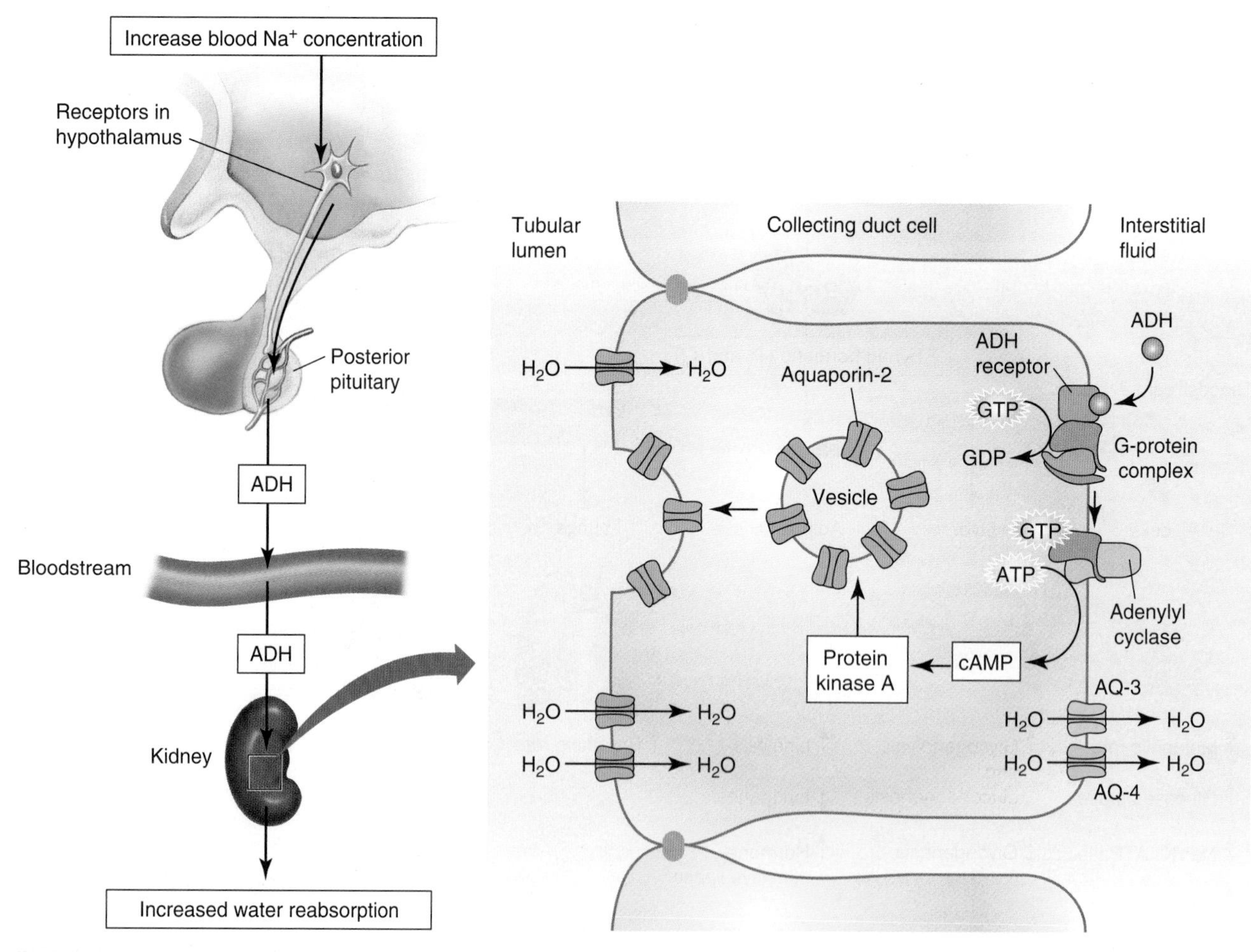

FIGURE 3.15 ADH is released from the posterior pituitary. Its target is cells of the kidney tubule. ADH binds to a G-protein-coupled receptor, signaling the exocytosis of vesicles within the tubule cells. The vesicles contain a water channel protein called aquaporin, which facilitates the uptake of water from the filtrate into the blood supply.

you are dehydrated. ADH is then released from the hypothalamic neuronal projections at the posterior pituitary and enters blood circulation. The target tissue for ADH is the collecting duct of the kidney tubule, where it binds to a G-protein-coupled receptor linked to cAMP-PKA (cyclic adenosine monophosphate–protein kinase A). Activation of PKA causes aquaporin-containing vesicles within the kidney tubules to fuse with the luminal membrane of the cells. Aquaporin proteins are water channels, multi-unit proteins that form a water-selective pore in the collecting duct membrane. Thus, as water moves through the kidney tubule, it can pass through these water channels and be taken back up into the blood supply (**FIGURE 3.15**). As we retain water (urinate less) and preserve a normal osmolarity, ADH secretion declines. This is a sensitive and rapid mechanism for the conservation of water within the vasculature.

Aldosterone Regulates Water Balance Indirectly

There is a second, slower-acting mechanism for water conservation involving the adrenal cortical hormone, aldosterone. Like cortisol, aldosterone is derived from cholesterol and is synthesized within the cortex of the adrenal gland. Unlike cortisol, aldosterone release is not triggered from the anterior pituitary. Instead, aldosterone release is stimulated by increased blood K^+ levels, which occurs during dehydration. An increase in extracellular K^+ depolarizes the membrane potential of the cortical cells of the adrenal gland, opening voltage-gated Ca^{2+} channels, and causing the release of aldosterone into circulation (**FIGURE 3.16**). Notice how similar this mechanism is to the one for insulin release from β cells of the pancreas. Aldosterone, being a steroid hormone, travels in the blood bound to transport

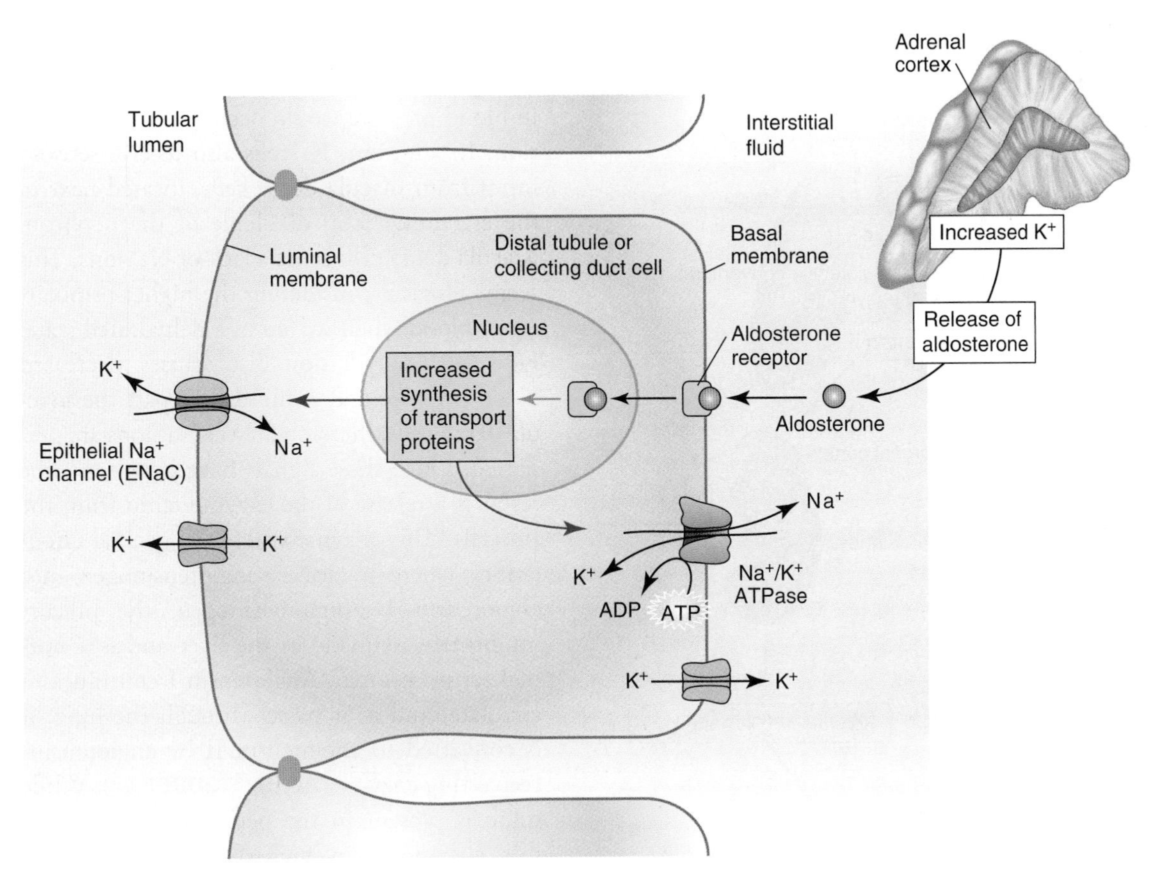

FIGURE 3.16 Increases in plasma K^+ concentration will cause the release of aldosterone from the adrenal glands. Aldosterone targets kidney tubular cells and causes the uptake of Na^+ from filtrate.

proteins such as albumin. Once at its target tissue, the distal tubule cells of the kidney, aldosterone binds to intracellular receptors, is translocated to the nucleus, and signals the synthesis of specific proteins. One of these proteins is the epithelial Na^+ channel, ENaC, which is inserted into the luminal membrane of the distal tubular and collecting duct cells. This protein is permeable to sodium and thus allows the salvage of Na^+ ions from the collecting duct fluid. Aldosterone also upregulates synthesis of a specific kinase that increases the activity of the Na^+/K^+ ATPase in the basal membrane of these same cells. The Na^+/K^+ ATPase transports the saved Na^+ back into the blood supply. The K^+ brought into the cell through the Na^+/K^+ ATPase is excreted into the luminal fluid via a K^+ channel. Because protein synthesis takes time, this method of water conservation is slower and manipulates not water, but sodium and potassium balance. Remember that Na^+ is osmotically active, so conservation of salts is essential for the kidney tubule to recover water.

There is another important hormone that contributes to the conservation of body water and also regulates the release of both ADH and aldosterone. This hormone is angiotensin II.

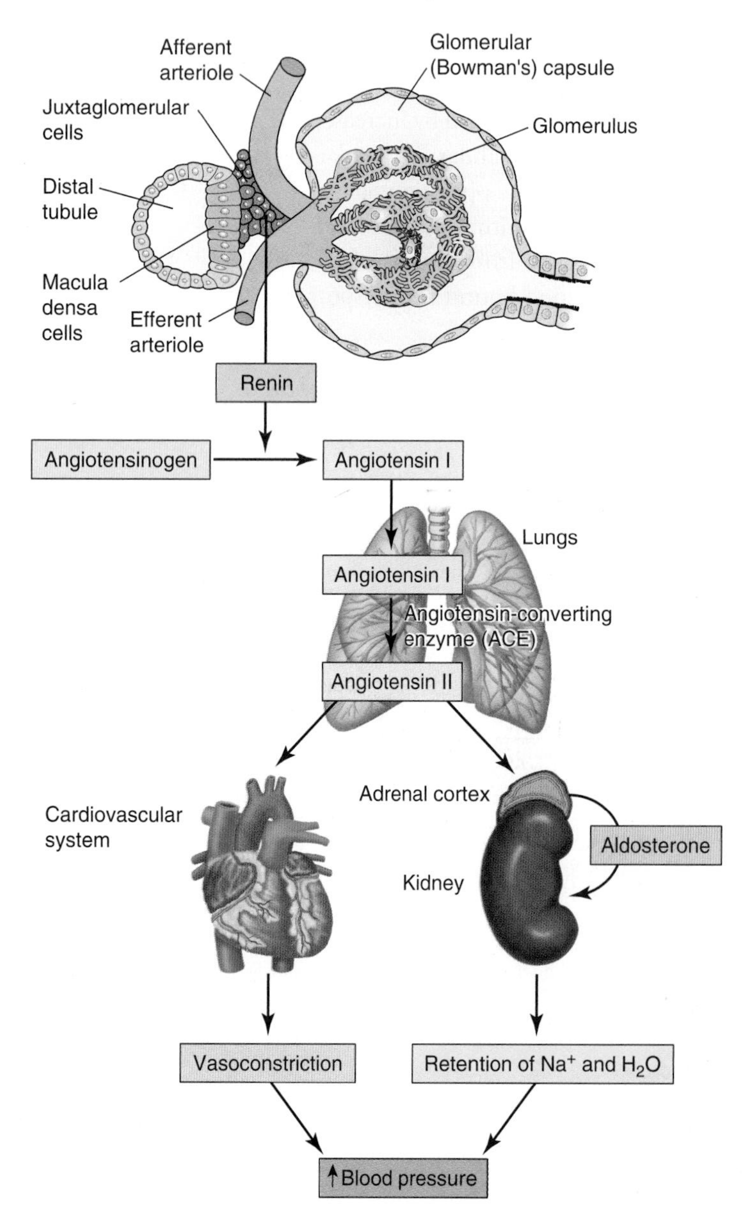

FIGURE 3.17 The renin-angiotensin-aldosterone system requires multiple organs for angiotensin II synthesis.

The Formation of Angiotensin II Requires Enzymes in Several Tissues

The sensor for initiating angiotensin II synthesis lies within the kidney. Juxtaglomerular (JG) cells, located at the afferent arteriole near the glomerulus, can sense a change in blood pressure going into the nephron. Such a decrease in blood pressure would occur during dehydration. In addition, JG cells also receive sensory input from macula densa cells, located next to the ascending loop of Henle of the nephron. Macula densa cells sense a lack of Na^+ ions. This seems illogical considering the higher osmolarity of blood when we are in a dehydrated state. However, dehydration also causes decreased blood pressure and reduced flow past the macula densa cells; therefore, fewer Na^+ ions are presented. Once these signals have been sent, the result is a release of the enzyme renin from the JG cells. This enzyme enters the general circulation, where it cleaves angiotensinogen into angiotensin I. Angiotensinogen, the precursor protein, is made in the liver and is a normal serum protein. Angiotensin I continues to circulate, and as it passes through the lung, it is converted to angiotensin II by angiotensin-converting enzyme (ACE) (**FIGURE 3.17**). While some is present in the peripheral vasculature, this enzyme is in highest concentration on the endothelial surface of the pulmonary vasculature. Its high concentration and the vast network of pulmonary circulation ensure effective conversion of angiotensin I to angiotensin II.

Angiotensin II has several target organs and effects. Binding to a Gaq-coupled, G-protein-coupled receptor in the adrenal cortex, it stimulates aldosterone release, thus potentiating the K^+ effect. Angiotensin II binds to the same receptor on the vasculature to cause vasoconstriction, thus maintaining blood pressure in a potentially dehydrated individual. (Although we are not currently discussing cardiovascular physiology, it is worth pointing out here that the effects of angiotensin II also have key significance for heart function.)

Notice that a potent aldosterone effect requires the participation of cells of the adrenal cortex, JG and macula densa cells of the kidney, liver protein (angiotensinogen), and enzymes of the endothelial layer of the blood vessels. Should any one of these cells be compromised, our ability to conserve water would be seriously hindered.

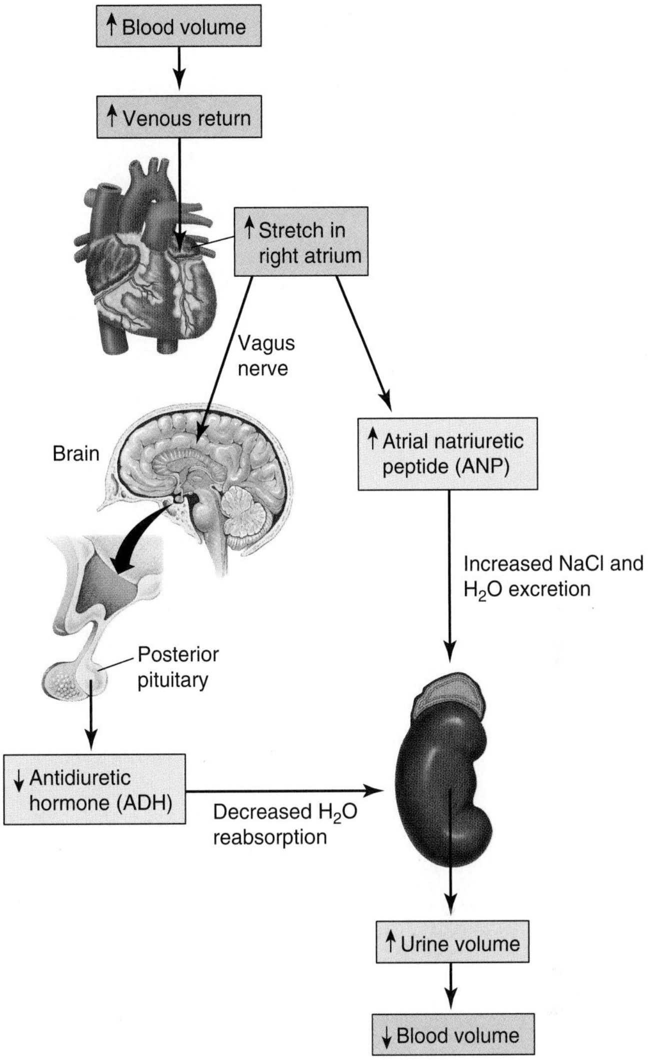

What Happens If We Drink Too Much Water?

There are several powerful hormones that protect us from water loss. However, there is only one hormone performing the inverse function—protecting us from excess water—and that is atrial natriuretic peptide (ANP). This peptide is made in cells of the cardiac atria in response to stretch, an effect of a greater blood volume. ANP binds to guanosine triphosphate (GTP)-linked receptors in the kidney to promote blood flow and filtration rates through the nephron. This hemodynamic change increases Na^+ loss, and therefore water loss, from the nephron. ANP also inhibits aldosterone synthesis, ADH, and renin release (**FIGURE 3.18**). Through all these mechanisms, more water is lost, and our normal fluid volume and osmolarity are restored.

FIGURE 3.18 ANP regulates blood volume when it increases.

To maintain the proper fluid balance involves many sensory cells: the brain senses a change in osmolarity, the kidney nephron senses flow and Na^+, the adrenal cortex senses K^+, and finally, the cells of the atria sense mechanical stretch.

How Is Blood Calcium Regulated and How Is Lost Calcium Regained?

There is one final issue to be resolved: how do we regulate plasma Ca^{2+}, and how do we restore Ca^{2+} to our bones after it has been released by cortisol? Constant blood calcium concentrations are vital for cell signaling and cardiac function, so blood Ca^{2+} levels are maintained between 2.1 and 2.6 mM—a very narrow range indeed! How is this regulated? Parathyroid hormone (PTH), produced by two small parathyroid glands lying next to the

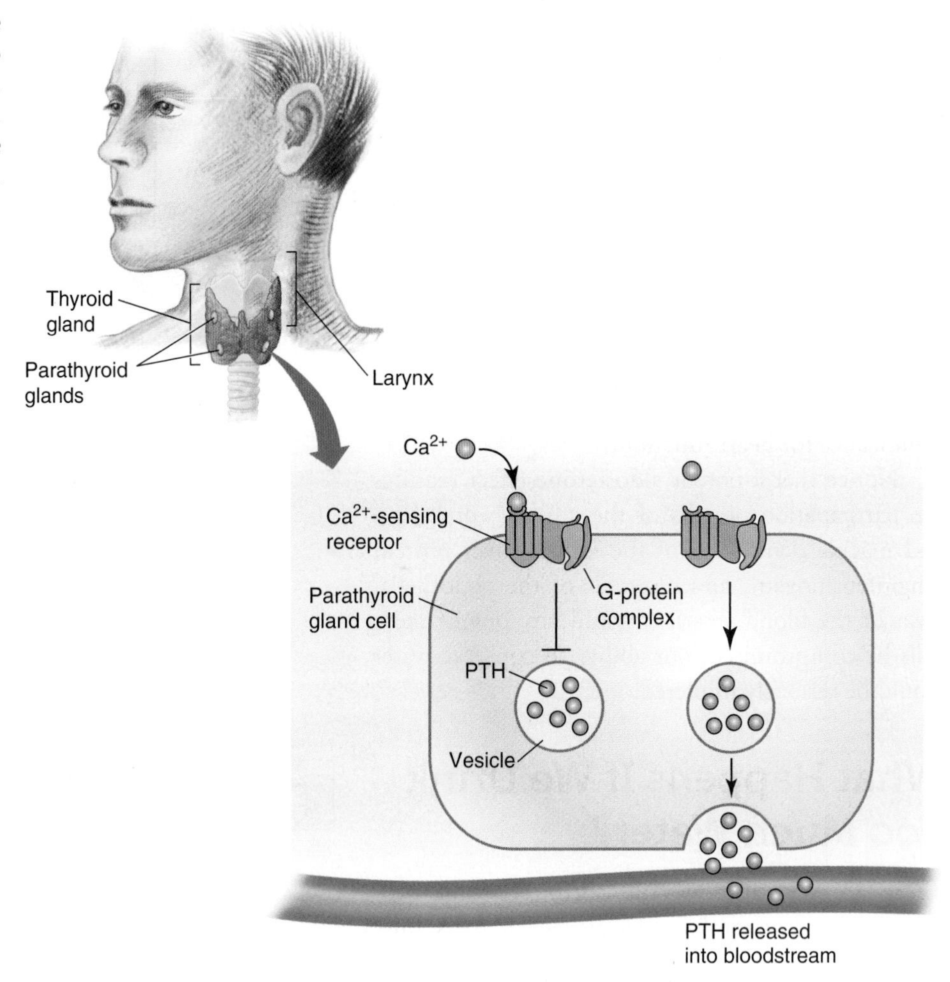

FIGURE 3.19 Parathyroid hormone is released whenever plasma Ca^{2+} is too low. It targets bone for Ca^{2+} resorption, intestine for increased Ca^{2+} uptake, and kidney for Ca^{2+} uptake from the urine filtrate.

thyroid gland, is the primary regulator of blood Ca^{2+}. Cells of the parathyroid glands have a G-protein-coupled receptor known as the Ca^{2+}-sensing receptor in their membranes. These receptors bind Ca^{2+} with low affinity, from the blood that nourishes the parathyroid glands. The receptors are sensitive to a reduction in the rate of Ca^{2+} binding, so when blood Ca^{2+} levels fall, even slightly, and the Ca^{2+} sensing receptor binds less Ca^{2+}, that signals the release of vesicles of parathyroid hormone into circulation (**FIGURE 3.19**). Notice how similar this cellular mechanism is to that of insulin release from β cells when glucose concentrations increase!

Once in circulation, PTH acts at the bone, an enormous reservoir of the mineral calcium, to release Ca^{2+} into circulation, thus adding to the effects of cortisol. At the kidney, PTH promotes Ca^{2+} reabsorption by the nephron. The net result of these actions is an increase in blood calcium to normal levels.

What replaces Ca^{2+} into our bones? Vitamin D, created in precursor form in the skin upon exposure to sunlight, is converted by the liver into a different precursor form, and finally, is converted into an active vitamin D by the kidney. Vitamin D stimulates calcium uptake at the small intestine, allowing the calcium you ingest to be efficiently absorbed. It also promotes Ca^{2+} deposition into the blood and calcium reabsorption in the kidney (**FIGURE 3.20**). This is why vitamin D is added to milk in the United States. Having the vitamin that promotes calcium absorption present in a food that is a rich source of calcium

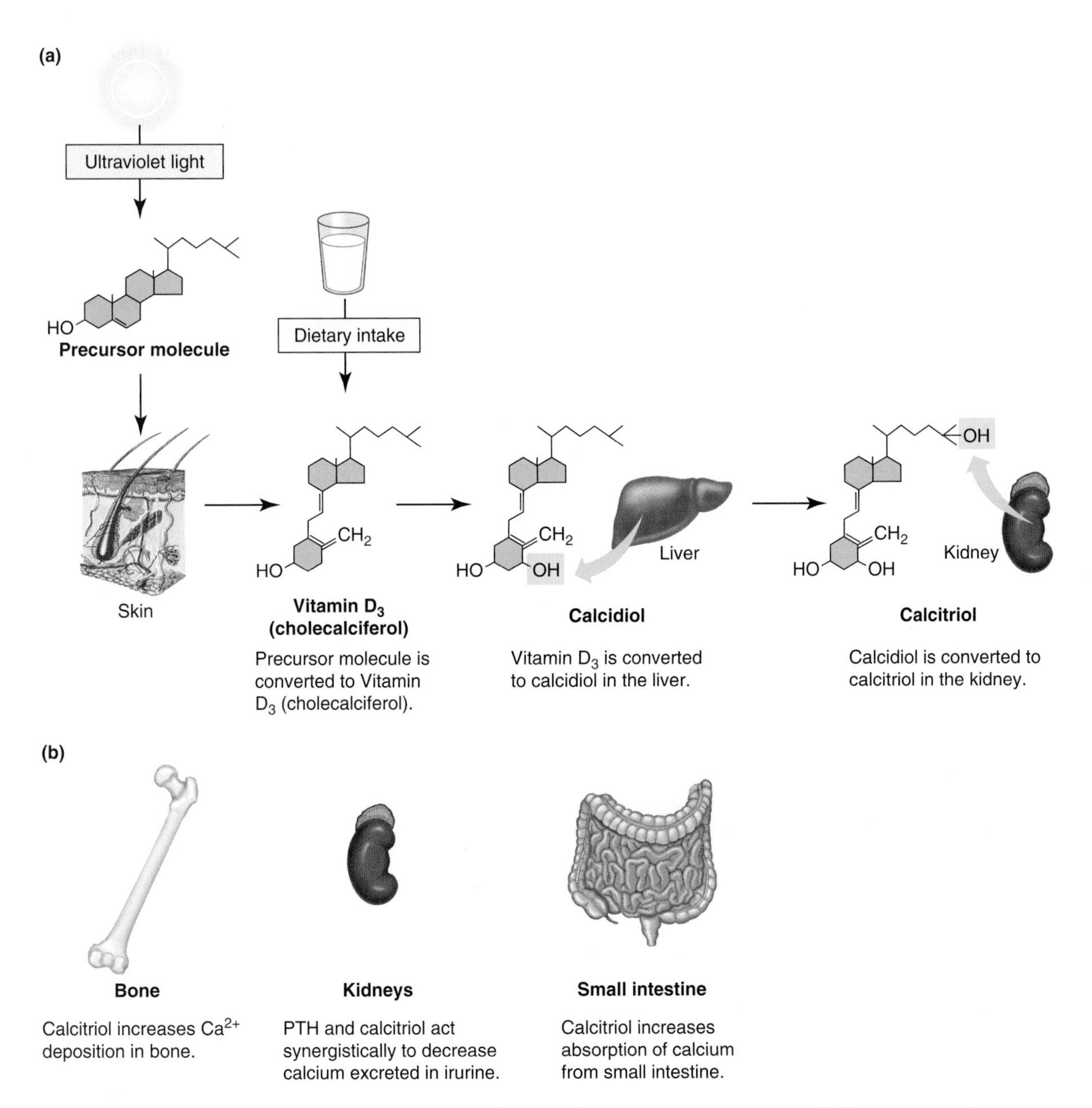

FIGURE 3.20 Vitamin D requires multiple tissues for its synthesis. It targets intestine to facilitate Ca^{2+} uptake.

promotes bone growth, even when there is no sunlight for the production of natural vitamin D by the skin!

Summary

During our daily lives, as we eat and drink irregularly, starving and feasting by turns, there is a host of hormones that manage our energy supply, water balance, and mineral balance. Some of these are locally controlled and some are centrally released. Although there are many more hormones than the ones we have discussed—hormones of the digestive system as well as reproductive hormones, for example—those discussed here have important effects on metabolism, thermogenesis, and energy transfer that impact many bodily systems and functions.

Key Concepts

Glucose regulation
Mechanism of insulin release
Cortisol action
Hypothalmic-anterior pituitary axis
Growth hormone action
Appetite control
Regulation of metabolic rate
Regulation of fluid balance

Key Terms

Glucagon
Insulin
Glucose transporter type 4 (GLUT4)
Cortisol
Suprachiasmatic nucleus (SCN)
Corticotropin-releasing hormone (CRH)
Adrenocorticotrophic hormone (ACTH)
Growth hormone (GH)
Growth hormone–releasing hormone (GHRH)
Insulin-like growth factor (IGF-1)
Ghrelin
Leptin
Thyroid hormone
Antidiuretic hormone (ADH)
Aldosterone
Angiotensin II angiotensin converting enzyme (ACE)
Atrial natriuretic peptide (ANP)

Application: Pharmacology

1. Type 2 diabetes mellitus (T2DM) is a common condition in which the ability of cells to respond to insulin becomes impaired. As a result, there are a variety of drugs that have been developed to treat it. One of the most effective is metformin. This drug acts by increasing insulin sensitivity, so the circulating insulin signals more effectively for glucose and fatty acid uptake. It also inhibits gluconeogenesis and glycogenolysis at the liver while aiding in the oxidation of fatty acids.
Given what you know about glucose balance in T2DM, explain how this drug could improve the key, interrelated dysfunctions associated with the condition:

- Insulin resistance/diabetes
- Hyperinsulinemia
- Hyperlipidemia
- Obesity
- Hypertension

- Atherosclerosis
- Endothelial dysfunction

2. The thyroid gland sets the basal metabolic rate. Iodine is essential for the creation of thyroid hormones thyroxine (T_4) and triiodothyronine (T_3). When there is insufficient dietary iodine available, the thyroid gland hypertrophies, producing a swelling of the neck known as a goiter. Iodized salt is sold to ensure that sufficient quantities of iodine are available to people, regardless of diet. Even with adequate iodine, some thyroid glands fail to produce enough thyroid hormone. What would you expect the symptoms would be of low thyroid hormone? Levothyroxine, an analog of T_4, can be given to people with too little thyroid hormone. What would you expect to change, once a person with an underperforming thyroid gland began to take levothyroxine? What changes might you expect if the person took too much levothyroxine?
3. ADH is released in response to an increase in plasma osmolarity, e.g., when you become dehydrated. What are the actions of ADH? Alcohol inhibits the release of ADH. Describe what might happen to your fluid volume and plasma osmolarity if you failed to drink anything for an entire day, and then sat down and drank three beers.

Clinical Case Study

Type 2 Diabetes Mellitus, Part 2

BACKGROUND

Insulin stimulates fatty acid uptake by adipose tissue through its stimulatory effects on lipoprotein lipase. Fatty acids, carried in the blood supply by various lipoproteins, are removed from the protein complex by lipoprotein lipase, an enzyme present on the extracellular surface of adipose cells. Thus, digestion of a fatty meal will increase the blood fatty acid content until insulin stimulates lipoprotein lipase, causing the fatty acids to be stored in adipose tissue. Insulin is as important for storage of fats as it is for storage of glucose. Excess fat storage into adipose tissue increases the hormonal output of adipose tissue, including the hormone resistin, which as its name suggests increases insulin resistance. Obesity, particularly visceral or belly fat, is thus linked with insulin resistance and the beginning of type 2 diabetes mellitus (T2DM).

In T2DM, lipoprotein lipase activity is reduced, causing fatty acids to remain in circulation rather than being stored. This results in hyperlipidemia, high circulating plasma lipids. Lipids are a fuel source for many tissues, including skeletal muscle and liver, so these tissues, faced with a lavish supply of lipids and with little access to glucose, take up lipids. The lipids are used for fuel, but they are also stored in these tissues. Abnormal amounts of fats stored in liver and skeletal muscle can cause inflammation and increase insulin resistance, aggravating the original problem. High circulating fatty acids also contribute to dysregulation of whole body lipid handling, which we will look at in more detail following the chapter on digestion.

THE CASE

Your aunt, recently diagnosed with T2DM, has changed her diet to reduce her carbohydrate intake as her doctor suggested. Unfortunately, while she has been careful to limit glucose and carbohydrates in her diet, she has continued enjoying her "triple-stackers" from the local fast food restaurant (after removing the bread, of course!) and deep-fried mushrooms at home. She has substituted a high-fat diet for her previously high-carbohydrate diet. Always plump, she has gained a bit more weight despite the new low-carb diet, adding 2 inches to her waist. At her next physical exam, her blood triglyceride levels were found to be extraordinarily high. The doctor warned her of the dangers of this new behavior, and gave her 6 weeks to reduce her triglyceride levels by diet. Failure to do so would mean one more medication that she would need to take to control triglyceride levels. Devastated by the knowledge that she can no longer take refuge in her high-fat treats, she comes to you for help.

THE QUESTIONS

1. If T2DM is a disease defined by high blood sugar, why is high fat a problem?
2. What advice will you give your aunt concerning her diet?
3. What would be the advantage if your aunt lost all her excess weight?

Clinical Case Study

Heart Failure, Part 2

BACKGROUND

Management of fluid balance is an important part of managing heart failure. Excess blood volume can "back up" into tissues, particularly the lungs. Indeed, one of the first signs of heart failure may be shortness of breath on exertion.

How is the excess blood volume generated? When the stroke volume (beat by beat output from the heart) decreases, the kidney senses a reduction in flow. The sensors interpret this as a reduced blood volume and release renin into the blood supply. Renin, an enzyme, cleaves circulating angiotensinogen into angiotensin I and angiotensin converting enzyme in the lung cleaves angiotensin I into the active hormone angiotensin II. Angiotensin II has a variety of effects, including stimulating the release of ADH and aldosterone, each of which has their effects at the kidney, and function to conserve plasma water. This increases extracellular fluid volume. Notice that this happens even when there is sufficient extracellular fluid volume because of the nature of the kidney sensor itself.

Increases in plasma volume burden the already compromised heart with additional blood to pump. The solution to this problem is to reduce extracellular fluid volume, which is generally done with two types of drugs. First, ACE inhibitors prevent the formation of angiotensin II and its suite of actions. This prevents water salvage by the kidney. Diuretics are the second line of defense, causing an increase in urine production and a decrease in fluid volume.

Use of these two classes of medications reduces the volume of blood and therefore the amount of blood that must be pumped by the heart, requiring the heart to do less physiological work.

THE CASE

Your grandfather has never been a sickly person, has taken few drugs in his lifetime, and is rebelling against taking his heart failure medications now. He especially dislikes taking the diuretics, which make him arise from sleep during the night in order to void his bladder. Once awakened, he has trouble going back to sleep, and the lack of sleep makes him grumpy. You discover while talking to him that he has decided on his own to cut the dosage on the ACE inhibitor and has abandoned the diuretic altogether. Since changing his medication, he can sleep through the night and notices no other difference except for being winded when he climbs the stairs at night. His conclusion is that the drugs are more of a hindrance than a help.

THE QUESTIONS

1. What is the ACE inhibitor doing?
2. What specifically does angiotensin II do?
3. What are the actions of aldosterone?
4. What are the actions of ADH?
5. What might you say to your grandfather to impress upon him the importance of taking his medications as prescribed?

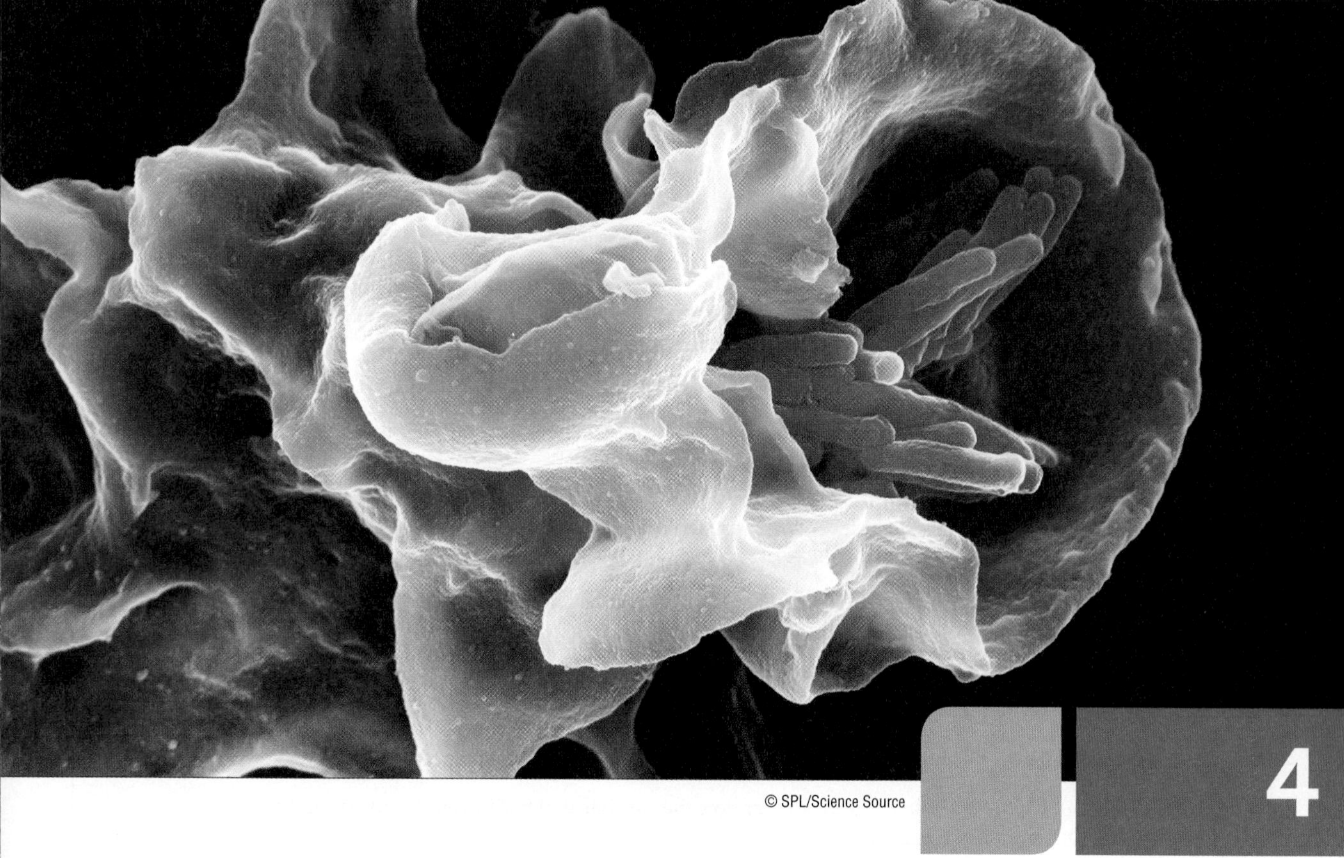

© SPL/Science Source

4

Immune System Physiology

Case 1

It is early in the morning, and you are ready for your morning run. You sit on a bench next to the track to re-tie your shoes, which have mysteriously come undone. As your hand sweeps across the grass, you get a deep cut in your hand from an unseen piece of glass. Annoyed with the situation, you wrap a tissue around your hand and go out for your normal 5-mile run. After the run, you realize that you will be late for your calculus exam unless you really hurry. As a result, you take the fastest possible shower and do nothing to disinfect your cut hand. It has stopped bleeding, but there is a significant opening in your skin. Unfortunately, the day is a busy one, and you fall asleep that night with the hand still untreated. By morning, your hand is inflamed and swollen, and you have a slight fever. What caused these tissue changes overnight? What defense mechanisms have been launched by the immune system?

Case 2

A month after your incident with the infected hand, your physician recommends that you get a flu vaccine to protect you during the upcoming flu season. You take her advice and wonder how the vaccine works with your immune system.

Introduction

We know that the autonomic nervous system and the endocrine system regulate our internal environment, including our heart rate, vascular diameter, body temperature, gut motility, and metabolism. The sympathetic nervous system responds when we are in physical danger—if we are being chased by a Bengal tiger, for example, our fear is sufficient to begin the "fight or flight" response. But the world outside of our bodies contains less obvious dangers than a Bengal tiger. We are exposed to bacteria, viruses, and parasites, all of which can threaten our lives as seriously as any mammalian predator. The immune system has defenses in place to prevent us from falling prey to the microbes around us. Like the endocrine system, the immune system has organs throughout the body, and specialized cells of the immune system travel in the blood, as do some of the molecules that are unique to the immune system.

The immune system is complex, but it is important to understand the basics of this system and how it influences other organ systems of the body.

Physical Barriers—Our First Line of Defense

Skin provides an excellent barrier to bacteria, viruses, and parasites. The epidermis is nonvascularized, and the most external layer of skin cells is composed of dehydrated cells filled with keratin protein. This layer is continuously replaced from deeper layers, so we slough off our outermost layer. The anatomy of our epidermal layer prevents microbes from gaining access to moisture or the blood supply, thus effectively confining them to our exterior (**FIGURE 4.1**). We grant microbes access to our bodies only when we break the skin or contaminate the few entry points (mouth, eyes, nose, etc.) by various means—rubbing our eyes or nose with dirty hands or eating uncooked, unwashed fruits, to name just two examples.

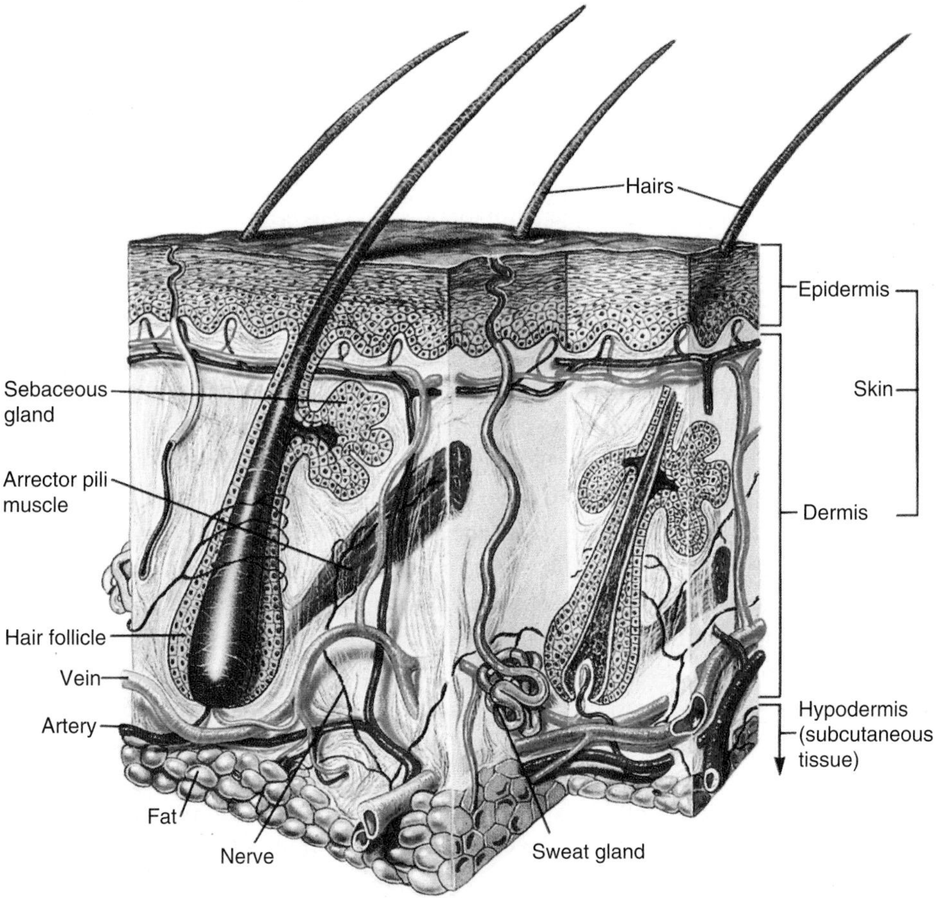

FIGURE 4.1 The epidermal layer of the skin is not vascularized.

What about all of our openings to the outside world? Our mouth, nose, eyes, and ears all provide potential entry points from the exterior. However, in each case, there is fluid movement from the inside to the outside acting as a protective barrier. In the mouth, we have a continuous production of saliva that washes the mouth. Saliva contains enzymes that can kill microbes, and once the saliva is swallowed, microbes end up in the stomach, where the pH is approximately 2, about the same acidity as battery acid. This low pH effectively kills most microbes. Lacrimal glands secrete tears to the lateral surface of the eyes that wash over the eye and drain into the nasal cavity. This continual flow of fluid washes the eyes. Tears also contain the enzyme lysozyme, which cleaves sugars within the cell walls of bacteria, killing them.

Our nasal passages contain hairs that trap particulate matter, including bacteria. Active mucus-producing cells in the nasal passages also ensnare microbes in their thick, sticky mucus. The airways from the nose to the lungs also contain numerous mucus-producing cells that serve the same purpose. In addition, cilia on the epithelial cells beat rhythmically, moving the mucus from the airways to the mouth, where trapped bacteria and viruses can be expelled with the mucus. Ears produce a waxy secretion that serves a similar purpose. Thus, our own secretions provide important physical barriers against the microbial world.

The Healing Process: Tissue Repair and Immune Response

White Blood Cells Are Cells of the Immune System

Most of us are aware that we have immune cells in our blood. White blood cells and red blood cells all come from common hematopoietic stem cells within bone marrow. These stem cells differentiate into two major stem cell lines. One, the myeloid stem cells, differentiate into reticulocytes, which give rise to red blood cells. Red blood cells are vital for carrying oxygen to tissues, but they are not immune system cells. However, these same myeloid stem cells also differentiate into progenitor cells that give rise to leukocytes: neutrophils, monocytes, eosinophils, and basophils (**FIGURE 4.2**). Throughout life, these cells are made within the bone marrow, travel in the blood, and spend much of their time within tissues. Each of these mature white blood cells has its own functions, but all are capable of ameboid movement and can leave the blood supply to enter tissue.

Neutrophils are the most common leukocyte and are very mobile, phagocytic cells that target and engulf bacteria (**FIGURE 4.3**). This is a lethal process for neutrophils, and they die after consuming one or two dozen bacteria. Neutrophils can also secrete proteins called cytokines that will increase inflammation. Their role in the inflammation process will be discussed shortly.

Monocytes are the next most common leukocyte. These large cells travel in the blood but can enter tissue and transform into mobile macrophages, which engulf bacteria. Monocytes are slower to reach damaged tissue, but once they enter tissue as macrophages, they have a greater bacterial killing capacity than neutrophils. In addition to the macrophages that arise transiently from monocytes, there are tissue-resident macrophages that permanently reside in specific tissues. Kupffer cells in the liver, alveolar macrophages in the lungs, macrophages within connective tissue, and microglia in the brain are all examples of tissue-resident macrophages. The tissue-resident macrophages can release a class of cytokines called chemokines that attract neutrophils and monocytes to damaged tissue.

Basophils and eosinophils are the least common white blood cells. Basophils store histamine granules, which can be released in response to specific antigens (foreign particles). Eosinophils also contain granules of enzymes that, when released, will destroy parasites and some infectious diseases. Unfortunately, these enzymes are also lethal to normal cells in the same area.

One final cell type is produced by the same stem cell line: the nonmotile megakaryocyte. These cells reside in the bone marrow and release noncellular cytoplasmic fragments known as platelets into the blood supply, where they play an essential role in blood clotting.

How do these cells protect us? How do they travel between blood and nearby tissue? The answers to these questions are most easily answered by examining our first case study, where you cut your finger. Please notice that by cutting your hand, you have broken the physical defense barrier, the skin, and have exposed internal tissue to the outside world. You have also severed blood vessels, allowing blood to be lost. Maintaining blood supply to the tissues requires that the blood loss be stopped, and protecting those tissues from bacteria now requires an immune system response that mobilizes leukocytes to the injured area and initiates a systemic immune response as well.

How Does Blood Clot and Stop the Flow of Blood?

As soon as you cut your hand, many processes begin simultaneously. In the interest of clarity, we will go through them separately, but remember that each of these signaling cascades

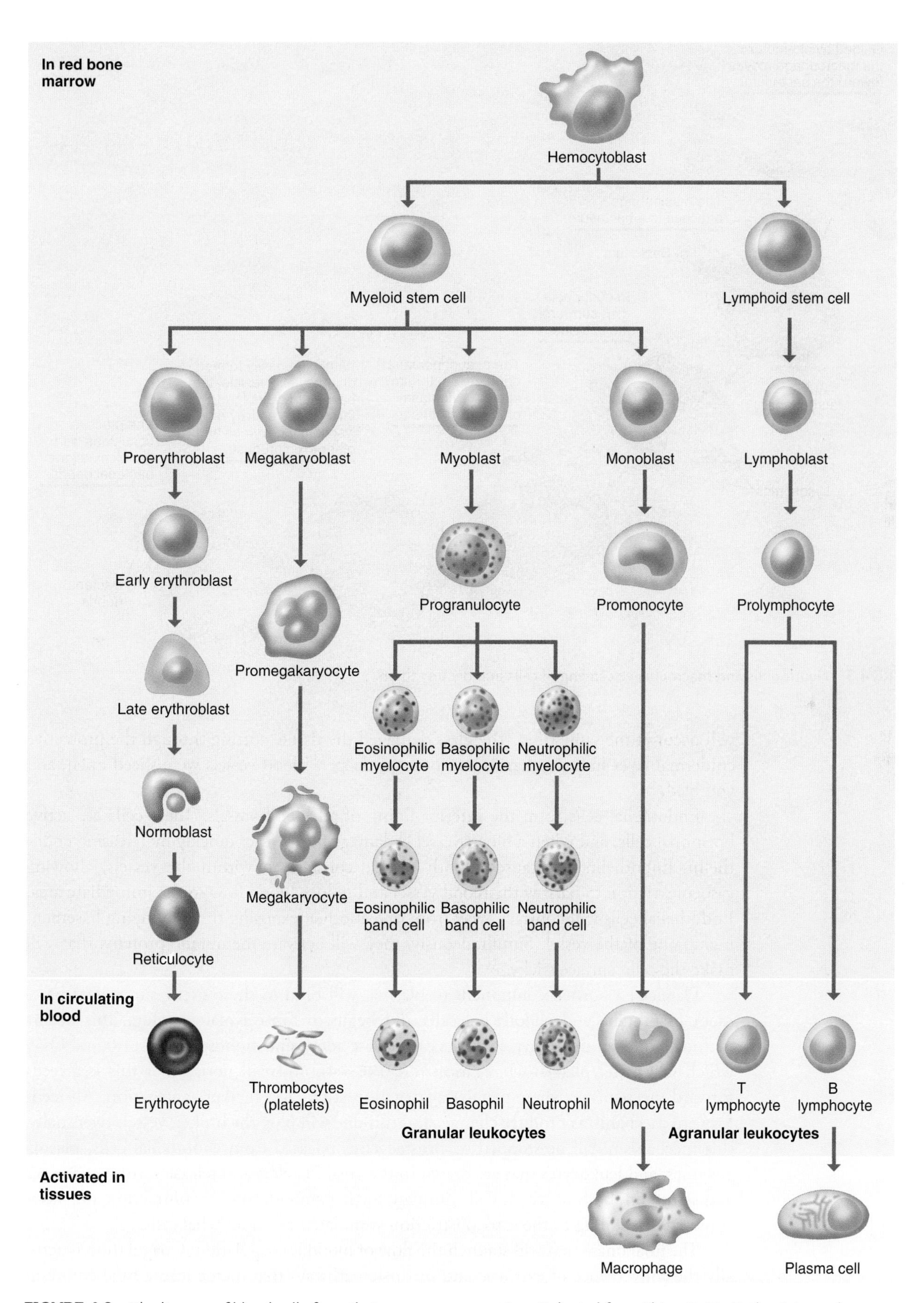

FIGURE 4.2 The lineage of blood cells from their common progenitor. (Adapted from Shier, D. N., Butler, J. L., and Lewis, R. *Hole's Essentials of Human Anatomy & Physiology*, Tenth edition. McGraw-Hill Higher Education, 2009.)

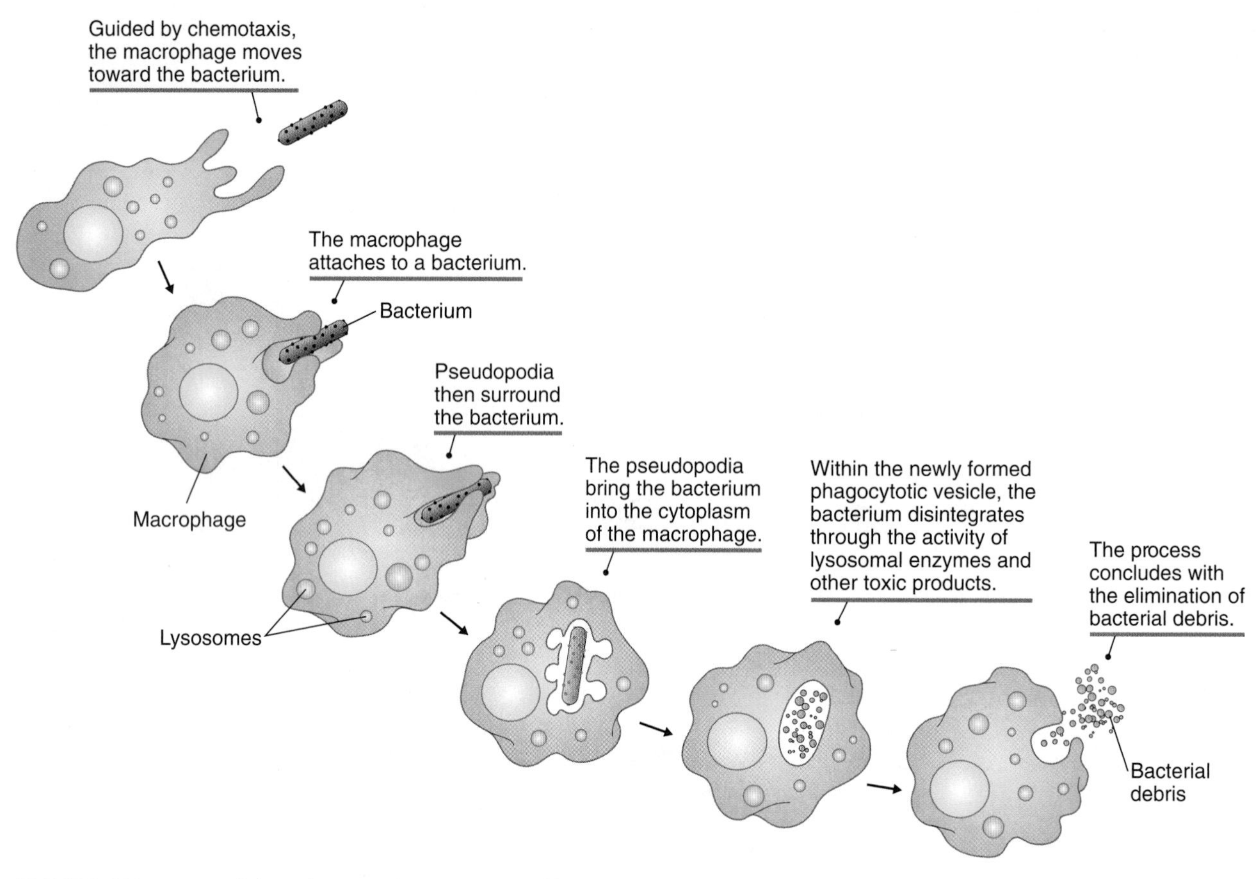

FIGURE 4.3 Neutrophils and macrophages can engulf cells and destroy them.

will occur at the same time. The glass damaged the tissue, cutting through the protective epidermal layer into the dermis and perhaps deeper. Blood vessels were sliced open, and you bled.

Endothelial cells form the interior lining of all blood vessels. These cells are active hormonal cells, and when a blood vessel is damaged they react quickly by releasing endothelin. Endothelin will cause smooth muscle contraction within the vessels, allowing vasoconstriction to narrow the blood vessel and reduce blood flow to the immediate area. Endothelial cells will contract away from one another, exposing the underlying basement membrane of the vessel. Simultaneously, they will activate membrane proteins that will make the cells' surface sticky.

Platelets, a normal component of plasma, will bind to these sticky surfaces of basement membrane and endothelial cells and begin to form a platelet plug. This occurs within seconds of the injury. Platelets can release adenosine diphosphate (ADP) and Ca^{2+}, which will foster platelet aggregation (**FIGURE 4.4**). You will notice that this is a feed-forward mechanism, where platelet aggregation stimulates further aggregation. All feed-forward mechanisms produce change, and this one will plug the broken vessel. Eventually, this mechanism will be stopped by prostacyclins released from endothelial cells, plasma enzymes, and leukocytes that are drawn to the area. Platelets also release thromboxane A2 and serotonin, both of which will stimulate further smooth muscle contraction and vasoconstriction, adding to the vasoconstriction stimulated by endothelial cells.

The final phase that will staunch the flow of blood is coagulation. Coagulation is actually the convergence of extrinsic and intrinsic pathways that merge into a final common path. The extrinsic pathway is initiated by tissue factor III released from endothelial cells.

In combination with Ca^{2+}, tissue factor III combines with another clotting factor protein to form Factor X. Factor X activates a proenzyme, which in turn activates another proenzyme, ultimately creating fibrin, a fibrous protein that will form a net that entraps cells and platelets into a stable plug. The intrinsic pathway begins with platelet factor 3, released from platelets in the wound, and in combination with Ca^{2+}, and some additional clotting factors, will cause the formation of Factor X, again leading to the formation of fibrin (**FIGURE 4.5**). Endothelial cells and platelets are important in each phase of blood clotting, using redundant mechanisms to seal the wound.

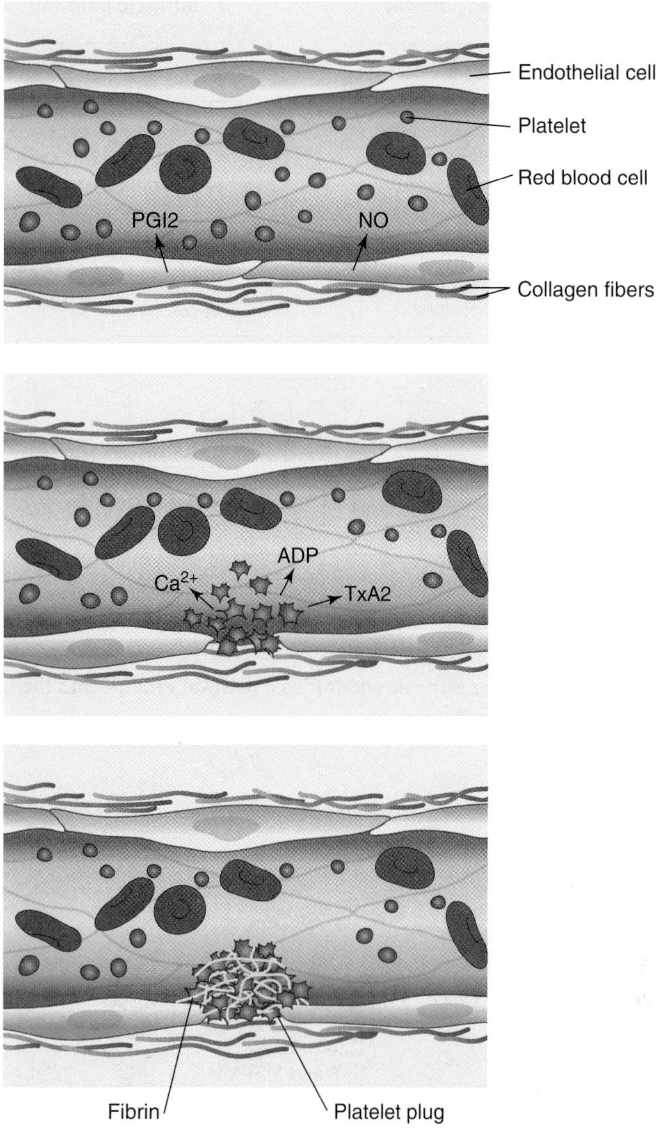

FIGURE 4.4 Before the injury, endothelial cells release nitric oxide (NO), which prevents platelet aggregation. After injury, these same endothelial cells will release endothelin, which will cause vasoconstriction. Collagen exposed by vascular damage plus newly activated endothelial proteins will start platelet aggregation. Platelets contribute to their own clumping by secreting ADP and Ca^{2+}.

What Happens When I Cut My Hand?—Recruitment of Leukocytes and Local Inflammation

At the same time that endothelial cells and platelets are initiating the blood clotting cascade, mast cells resident within connective tissue release histamine in response to the injury. Histamine will cause vasodilation, which will increase blood flow to the region and make the injured area appear red. Histamine will also increase capillary permeability, i.e., widen spaces between the endothelial cells that comprise the capillaries. This allows fluid within capillaries to leak into the tissue space and will facilitate movement of neutrophils out of circulation and into the damaged tissue. Resident tissue macrophages within connective tissue will secrete a special class of cytokines called chemokines, which will attract neutrophils within the blood supply to the area. Leukocytes release a wide variety of small signaling molecules, cytokines, known as interleukins (ILs), that provide a communication link between these widespread cells of the immune system. Two of the chemokines released by resident tissue macrophages are IL-8 and macrophage inflammatory protein 1β (MIP-1β). These proteins will attract neutrophils to the site of injury, facilitate their movement out of circulation (extravasation), and promote their conversion into active phagocytic cells. Phospholipids, released from damaged endothelial cells, are also a trigger for activation of neutrophils (**FIGURE 4.6**).

Neutrophils are the first leukocytes recruited to the site of injury, arriving in an hour or less, traveling through the swollen, fluid-rich tissue by ameboid movement to engulf bacteria. Monocytes are slower-moving cells that arrive much later, 12 to 24 hours later, attracted by fragments of bacterial wall or complement proteins (we will discuss complement proteins shortly). Monocytes, once they leave the circulation and enter tissue,

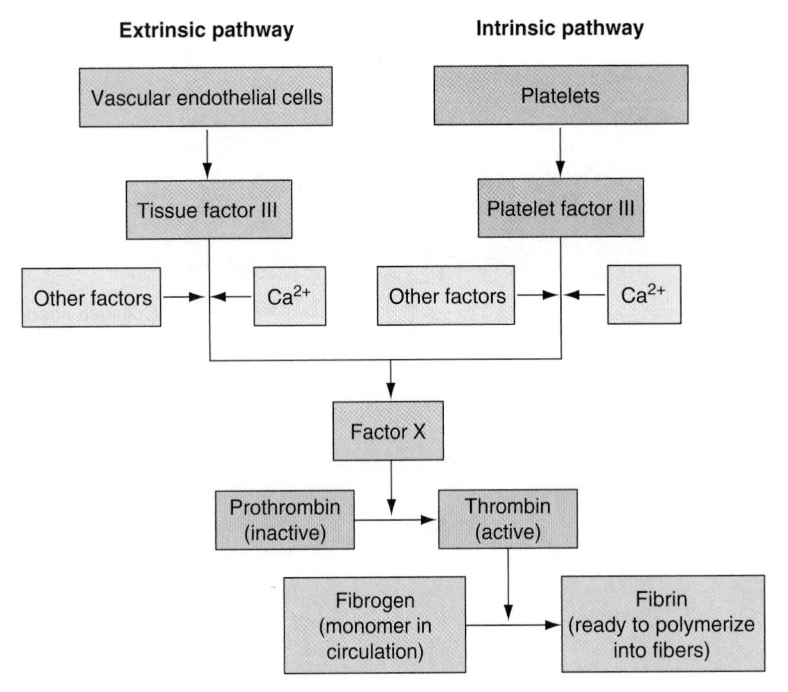

FIGURE 4.5 The extrinsic and intrinsic pathways merge into the final common pathway, forming fibrin.

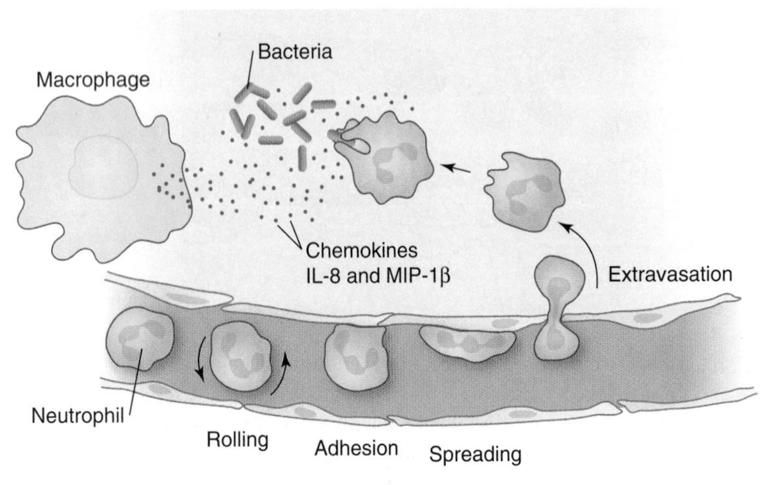

FIGURE 4.6 During extravasation, neutrophils move out of the vascular space into the tissues to phagocytose bacterial cells.

transform into macrophages. While slower to arrive, macrophages can kill hundreds of bacteria before they die. The act of killing bacteria is ultimately lethal to both neutrophils and macrophages, but they are replaced as bone marrow stem cell division leads to increased production of leukocytes. A cytokine named colony-stimulating factor increases the rate of stem cell production of leukocytes.

Several hours after the injury, you are experiencing all the symptoms of local inflammation: (1) redness from the vasodilation, (2) swelling, from the vasodilation and increased capillary permeability, (3) heat from the increased blood flow to the area that will occur after the clot has formed, and (4) pain, which is a result of pressure of the swelling caused by increased capillary permeability.

One of the blood-clotting proteins is a precursor for a peptide called bradykinin. Bradykinin causes vasodilation and stimulates pain receptors. Thus, the blood-clotting and inflammatory pathways intersect. By the next morning, dead macrophages and neutophils, along with cells that were destroyed by the cut itself and bacteria, will produce the pus that is present in your wound.

The Complement System of Plasma Proteins Targets Bacteria

Continuously circulating in your plasma are the 30 or so proteins of the complement system. This system complements the rest of the immune system, hence its name. There are several pathways for activating the complement proteins, each starting with a different trigger molecule. The classical pathway recognizes antibodies linked to bacteria (how the antibodies are attached to bacteria will be discussed later). The lectin pathway recognizes the carbohydrate mannose on bacterial walls, and the alternate pathway recognizes other carbohydrates common to bacteria. Each of these pathways involves a cascade of activation, much like we saw with coagulation, but this cascade occurs on the bacterial wall. The final portion of the cascade forms a membrane attack complex (MAC), which forms a pore within the bacteria, allowing fluid to pass inside the bacteria and eventually causing it to burst (**FIGURE 4.7**). While the MAC requires the entire cascade of complement proteins, some of the individual proteins have effects of their own. The complement molecule C3b, for example, binds to bacteria and to phagocytic cells to ensure that the bacteria is recognized by the phagocytic cell.

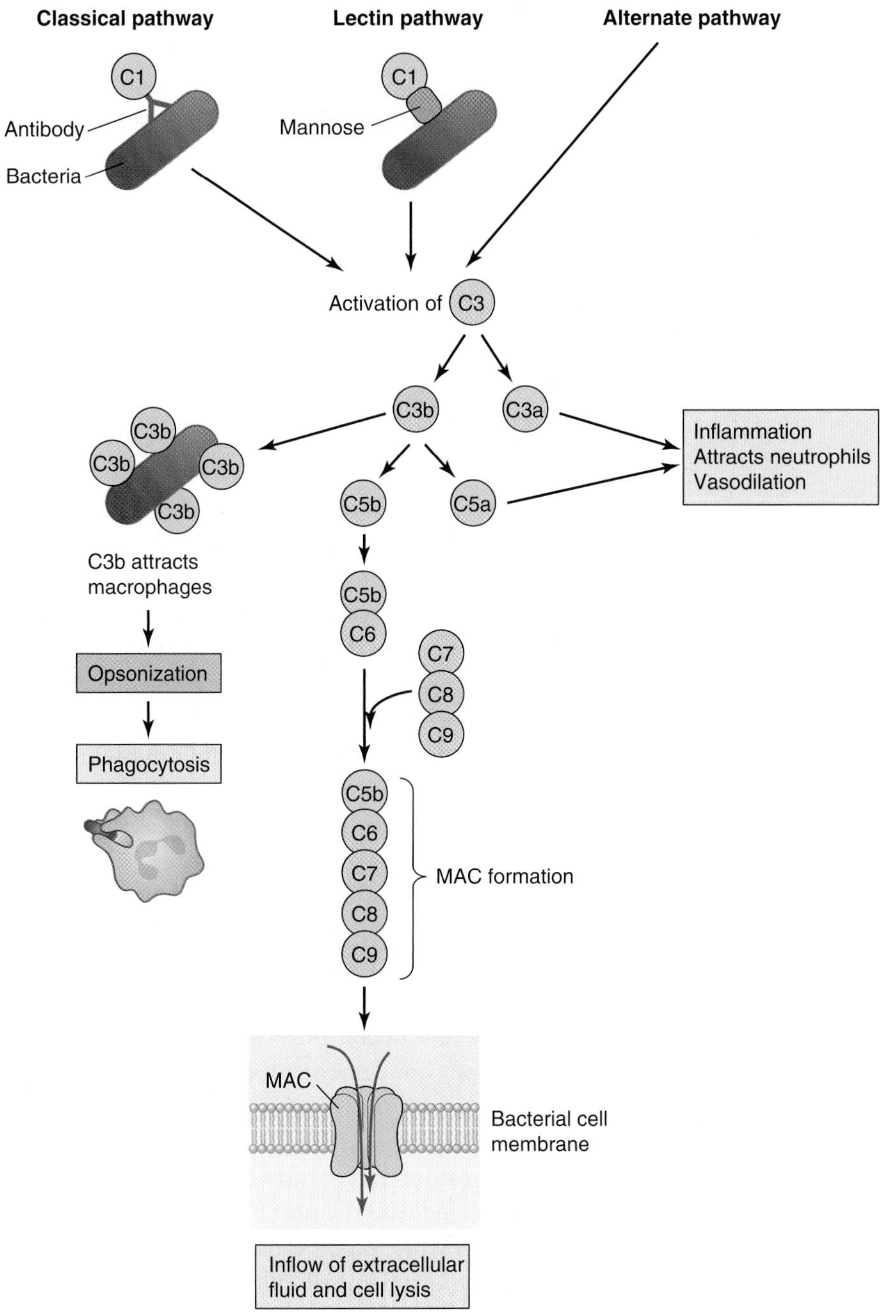

FIGURE 4.7 The complement system attacks bacterial cells in several different ways.

This is known as opsonization. Once physically joined, the phagocytic cell engulfs the bacteria. Other complement proteins, C3a, C4a and C5a, promote inflammation, activate mast cells, cause vasodilation, and activate neutrophils. Thus, the complement proteins can work immediately to increase local inflammation and recruit neutrophils to the site of injury.

How Does Local Inflammation Progress to Systemic Inflammation?

Resident tissue macrophages are a rich source of cytokines. In addition to the chemokines discussed above, which attract neutrophils to the site of injury, tissue macrophages also release two ILs, IL-1 and IL-6, as well as tumor necrosis factor α (TNF-α). Locally, these

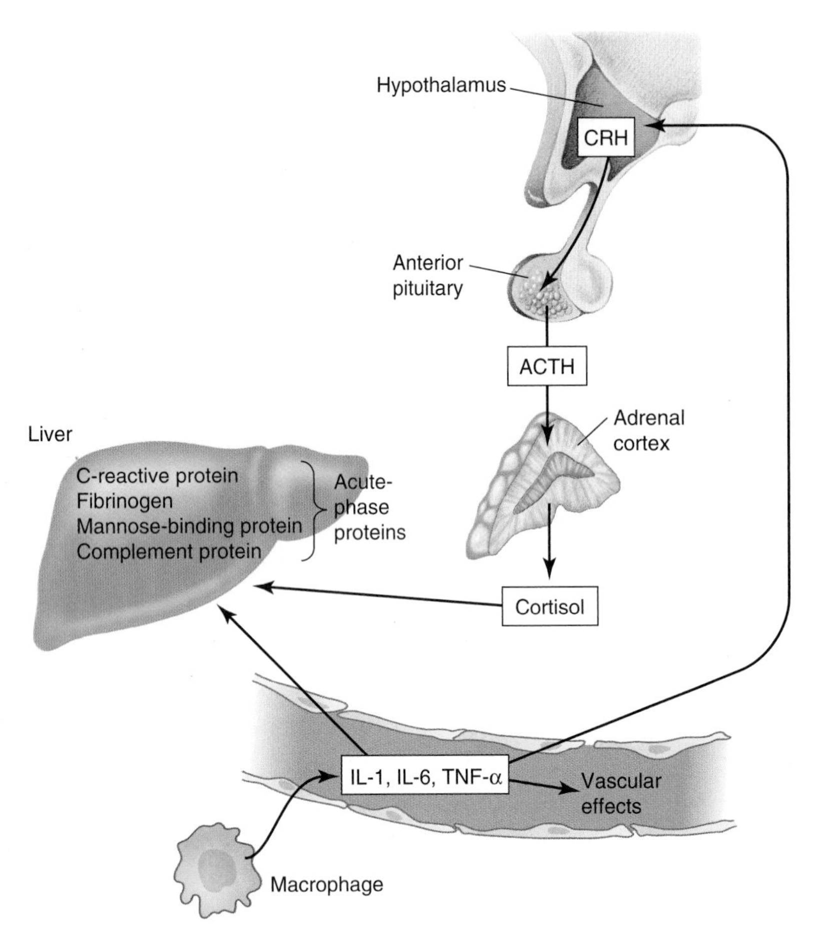

FIGURE 4.8 Cytokines released from macrophages help a local inflammation progress to systemic inflammation.

three cytokines increase blood coagulation and promote vascular permeability, thus helping your blood to clot and allowing more circulating neutrophils to enter the injured area. However, these same three cytokines also exert systemic effects by entering the blood supply and functioning as hormones.

At the hypothalamus, IL-1, IL-6, and TNF-α stimulate the release of adrenocorticotrophic hormone (ACTH) which acts at the adrenal cortex to promote cortisol release. As always, cortisol stimulates the synthesis of proteins, and its target here is the liver. The liver produces C-reactive protein, fibrinogen, mannose-binding protein, and pieces of the complement protein. C-reactive protein binds to microorganisms and tags them for destruction by phagocytic cells, macrophages, and neutrophils. Mannose-binding protein binds to carbohydrates present on the cell walls of bacteria, fungi, yeasts, and viral envelopes. It also activates complement via the lectin pathway. Thus, mannose-binding protein also tags microbes for destruction. Fibrinogen is the precursor to fibrin, so its production improves blood clotting. C-reactive protein, mannose-binding protein, and the components of complement produced by the liver are known collectively as acute-phase proteins, because they are released soon after injury. IL-1, IL-6, and TNF-α can stimulate the production of acute-phase proteins by the liver directly, without the mediation of the hypothalamus and cortisol (**FIGURE 4.8**). IL-6 and TNF-α travel to the bone marrow, where they increase the production of colony-stimulating factor, which increases the production of white blood cells, thus, replenishing the supply of neutrophils and monocytes necessary for microbial killing.

This system-wide mobilization of antimicrobial defenses includes one more response—fever.

How Is Fever Generated and Maintained?

When you awaken in the morning with a fever, you know immediately that you are ill. Fever is a hallmark of a systemic immune response. In addition to their effects on bone marrow and liver, IL-1, IL-6, and TNF-α have an effect at the hypothalamus, which is to increase body temperature, or put another way, to cause a fever. IL-1, the first of the cytokines to be linked with fever, was originally described as an endogenous pyrogen, meaning a molecule that was always present that could induce heat production. With further research, it was found that all three of these cytokines function as endogenous pyrogens, capable of stimulating the production of prostaglandins within the hypothalamus. Prostaglandins, acting through a G-protein coupled receptor, activate protein kinase A, which inhibits warm temperature-sensing neurons within the hypothalamus. This increases the firing of cold temperature-sensing hypothalamic neurons, which are responsible for raising body temperature to a higher value (**FIGURE 4.9**). A higher body temperature inhibits microbial replication and increases the activity of immune system cells. Fever will continue as long as the circulating cytokines continue to stimulate prostaglandin synthesis within the hypothalamus. Perhaps you took an aspirin when you awoke. Aspirin interferes with prostaglandin synthesis, thus reducing your fever.

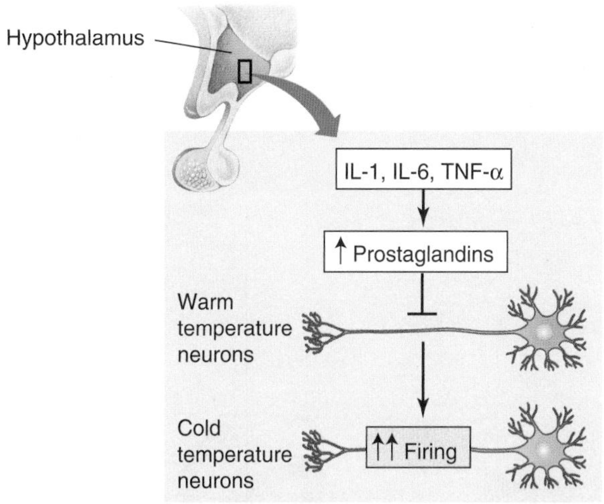

FIGURE 4.9 The generation of fever.

Lymphocytes Are the Cells of the Adaptive Immune System

All of the responses we've discussed so far are elements of the innate immune system, derived from myeloid stem cells. The original hematopoietic stem cell also gives rise to another stem cell line—the lymphoid line. This line produces three cell types: B lymphocytes, T lymphocytes, and natural killer (NK) cells. These cells are not fully functional at birth but "learn" or adapt to microbial challenges we encounter in our lives. Each of these cells begins in the bone marrow, as do the cells of the innate immune system. However, the B cells and NK cells mature within bone marrow, while T cells mature within the thymus gland. How do these cells travel within the body? What do they do? How do they interact with the innate immune system? In order to answer these questions, we need to look at the organs and circulation within the lymphatic system.

Anatomy of the Lymphatic System

The major tissues of the lymphatic system include the tonsils, thymus, spleen, localized patches of tissue in the intestine, the appendix, and the lymph nodes, located throughout the body. Connecting these tissues is the lymphatic circulation. Unlike the blood circulatory system, the lymphatic system has no pump, and it is not a continuous, closed system. Lymphatic flow begins at the lymphatic capillaries, which lie near the capillaries of the blood circulatory system. Lymphatic capillaries are also thin-walled vessels, but with one significant anatomical difference—the endothelial cells that form the capillary walls overlap one another. This allows the cells themselves to function as one-way valves, allowing water, nutrients, proteins, and even cells to enter the lymphatic capillary, but not to exit again. In this way, viruses, bacteria, damaged cells, and lymphocytes can leave the blood space and enter the lymphatic space, where they remain trapped for a time. Lymphatic capillaries join with lymphatic vessels, and lymphatic fluid flows into lymph nodes (**FIGURE 4.10**). Ultimately, lymphatic fluid empties into the thoracic duct, where the fluid flows into venous circulation, reentering the blood circulation. However, the flow of lymphatic fluid is slow relative to blood flow, and there is time in the lymph nodes for B cells and T cells to confront foreign particles.

T Cells Engage in Cell-Mediated Immunity

T cells mature within the thymus gland, and by the time they leave the thymus for residence in other lymphoid tissues, they have differentiated into four types of T cells: NK cells, cytotoxic T cells, helper T cells, and regulatory T cells.

FIGURE 4.10 The organization of the lymphatic system.

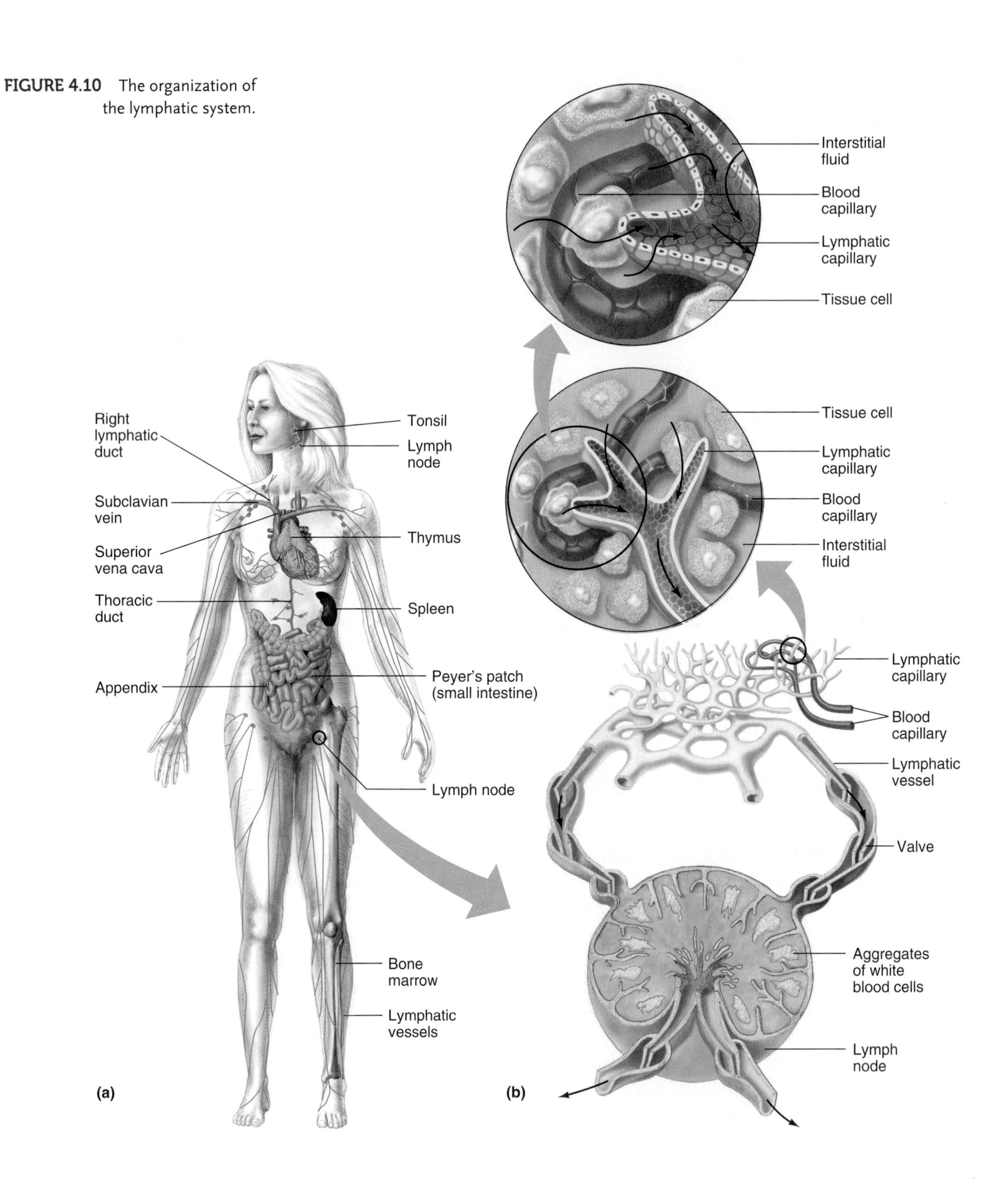

NK cells bind to plasma membranes of cells that contain abnormal antigens. That means that NK cells can bind to bacteria, to a normal cell that has been invaded by a virus, or to a cell that has become cancerous. Regardless of what cell they bind to, NK cells will release perforins via exocytosis. Perforins will form pores in the affected cell and prevent it from controlling its intracellular environment, thus causing the death of that cell. In order to encounter these cells in tissues, as in your cut hand, NK cells will need to leave blood circulation through the capillary endothelium and enter the injured tissue. NK cell activity is enhanced by cytokines released from tissue-resident macrophages, so they work in conjunction with the innate immune system to eliminate foreign organisms or cells infected by them.

Cytotoxic T cells recognize infected cells in a more specific way. T cell receptors bind simultaneously to foreign antigen on a cell and to a major histocompatibility (MHC) protein. MHC proteins are different in each person, so a specific MHC marks your cells as "self," effectively allowing the immune system to distinguish friend from foe. Cytotoxic T cells specifically hunt down infected "self" cells. Thus, they would not recognize a bacterium as foreign, but they would recognize a virally infected tissue cell. Once bound to this cell, the cytotoxic T cell can kill it in one of several ways: via perforin release, as the NK cells do, by releasing toxins to kill the cell, or by stimulating apoptosis, which is the programmed cell death of the infected cell. In any case, the infected cell is killed directly by the cytotoxic T cell (**FIGURE 4.11**). Cytotoxic T cells are specific for certain antigens and form clonal colonies that are designed to target that one antigen. Memory cytotoxic T cells form a reserve of these cells that can proliferate quickly if this antigen is encountered again.

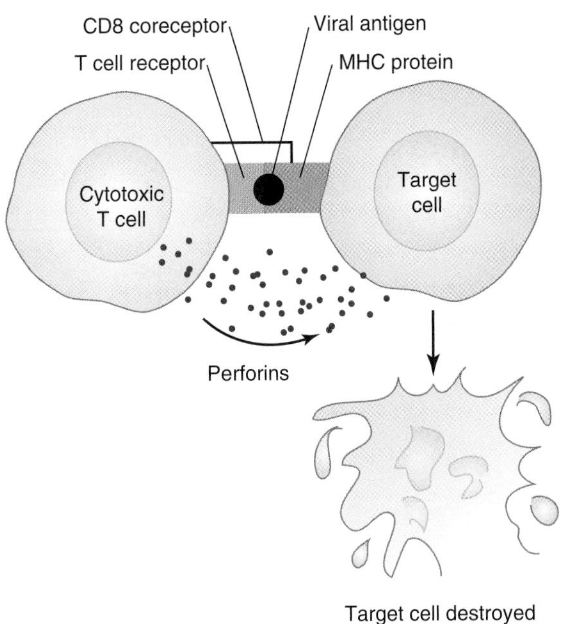

FIGURE 4.11 Killer T-cells can destroy infected cells of the body.

So far, it appears that the T cells are not much help with the bacterial invasion of your cut. However, all of the phagocytic cells of the innate immune system—monocytes that transform into macrophages, neutrophils, and tissue-resident macrophages—can function as antigen-presenting cells. Once they ingest bacteria, they will "present" a portion of the bacterial protein on their plasma membrane. Helper T cells can bind to these antigen/MHC complexes. Once bound, helper T cells release cytokines that activate cytotoxic T cells, attract macrophages to the site, and finally activate B lymphocytes.

B Lymphocytes Produce Antibodies

Like T lymphocytes, B lymphocytes originate in the bone marrow, and they mature there. Once mature, they leave the bone marrow and circulate in blood and lymph, spending most of their time in lymph nodes. B lymphocytes have MHC molecules on their surface and can bind foreign antigens on these MHC molecules. When a helper T cell encounters a B lymphocyte with a bound antigen, it secretes cytokines that cause B lymphocytes to transform into plasma cells and produce antibodies. The antibodies are small molecules that can bind to the antigen, cross-linking them so that they agglutinate, or stick together. This can cause the bacteria to clump and precipitate out of solution, inhibiting their function. Antibodies binding to bacteria will activate complement, causing bacterial death by MAC. Finally, antibody binding to bacteria will target these cells for phagocytosis by macrophages or for lysis by NK cells (**FIGURE 4.12**).

Some of B lymphocyte plasma cells will become memory cells, a clonal population that will be ready for activation and proliferation if you encounter that antigen again. Now we come to your flu vaccine. Vaccines are made from killed viruses or bacteria, or from a portion of a protein from that organism. This antigen is recognized as foreign by your adaptive immune system, specifically the B lymphocytes. It takes several weeks for sufficient "immunity" to be built up after the vaccination, because it takes time for the B lymphocytes to encounter the antigen, be activated, create the unique antibody to that

FIGURE 4.12 Helper T cells can activate B cells to produce antibodies.

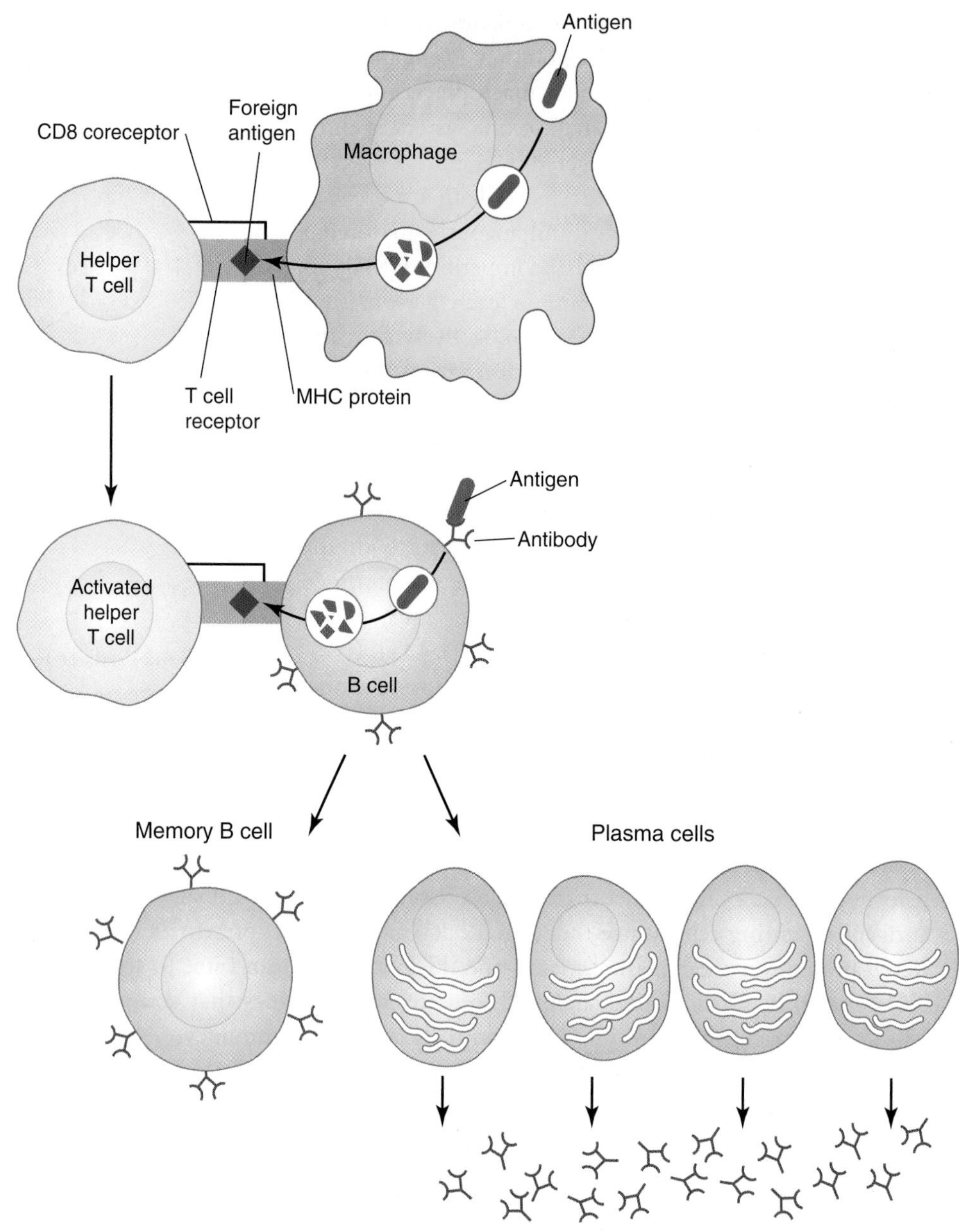

antigen, and produce memory cells against that particular antigen. Later in the season, should you be infected with the flu, the proliferation of B memory cells and the production of specific antibodies will protect you from illness.

Summary

Exterior defenses, the innate immune system, and the adaptive immune system all work in a coordinated way to protect you from invasion by external organisms. There are special cases of immune system activity in many of the organ systems of the body.

Key Concepts

Physical barriers to bacteria
Development of blood cells

Inflammation
Fever
Blood clotting
Innate immunity
Adaptive immunity
Lymphatic system
Complement

Key Terms

Neutrophils
Basophils
Monocytes
Eosinophils
Platelets
Megakaryocytes
Cytokines
T cells
B cells
Complement proteins
Antibodies

Application: Pharmacology

1. You wake up one morning feeling warm, and after taking your temperature, you discover that you have a fever of 102°F. You immediately take an aspirin. How does aspirin reduce your fever?
2. After your experience with an infected hand, you are now more careful when you cut yourself. Of course, it has happened again, but this time you immediately wash the wound and apply a topical antibiotic cream. How will this affect the physiological response to injury? Be certain to think about the environment the neutrophils will experience.
3. Your father takes a baby aspirin every morning with his breakfast for his health. Indeed, this is often recommended for cardiac health. You know from your study of biochemistry that prostacyclins and prostaglandins come from the same synthetic pathway. What do you think that the aspirin is doing in your father's circulatory system, and why might this be beneficial?

Clinical Case Study

Type 2 Diabetes Mellitus and Chronic Inflammation

BACKGROUND

We already know that type 2 diabetes mellitus (T2DM) is associated with obesity, but why? At least part of the answer lies with the interaction between adipose cells and macrophages, the same cells that came to your rescue when you cut your hand. The cut in your hand began a series of signaling events that transformed neutrophils and monocytes into phagocytic cells, i.e., macrophages. Lipopolysaccharide from bacterial cell walls potentiates the activation of macrophages. Macrophages then release factors that reduce insulin sensitivity in tissues. Why? Because a temporary insulin resistance may be beneficial for fighting infection. Neutrophils and macrophages can use the increased supplies of glucose in circulation to produce the ATP they need for bacterial killing.

However, what is beneficial in the short term can be problematic in the long term. As adipose tissue increases in volume, adipose tissue begins to release inflammatory cytokines like TNF-α. These cytokines attract macrophages to adipose tissue, where they release more inflammatory cytokines, such as more TNF-α, IL-6, and IL-1β. These cytokines have two effects: First, they recruit even more activated macrophages to the tissue, in a positive feedback mechanism. Second, the cytokines released by macrophages begin a signaling cascade that interferes with insulin receptor substrate, causing insulin resistance. As fat stores increase, lipids are deposited into liver and skeletal muscle, and macrophages join them in those tissues as well. Insulin resistance in liver and skeletal muscle increases, and blood glucose levels increase along with them.

Inflammatory cytokines produced by macrophages will enter circulation and affect tissues throughout the body. Thus, a simple excess of weight can begin a chronic, long-term systemic inflammatory response. As you may already know, inflammation forms part of the link between diabetes and atherosclerosis and heart disease.

THE CASE

Your aunt, who has diabetes, is struggling with her diet. She is not supposed to indulge in the desserts she has loved all her life, and recently you informed her that the big greasy hamburgers she eats are hardly better. She knows she should be eating whole grains, fruits, and vegetables, but they hold little appeal for her. She tests her blood sugar regularly, and if it is not extravagantly high, she is happy to eat to her heart's content. You happen to drop by one day while she is eating a plate of fried chicken and fries. You have just finished studying the immune system, and you are concerned for your aunt's health, given her habits.

THE QUESTIONS

1. What effect might this diet have on her skeletal muscle? Liver?
2. How might this diet affect her immune system?
3. How will her weight and diet influence the insulin sensitivity of her tissues? How does that happen?

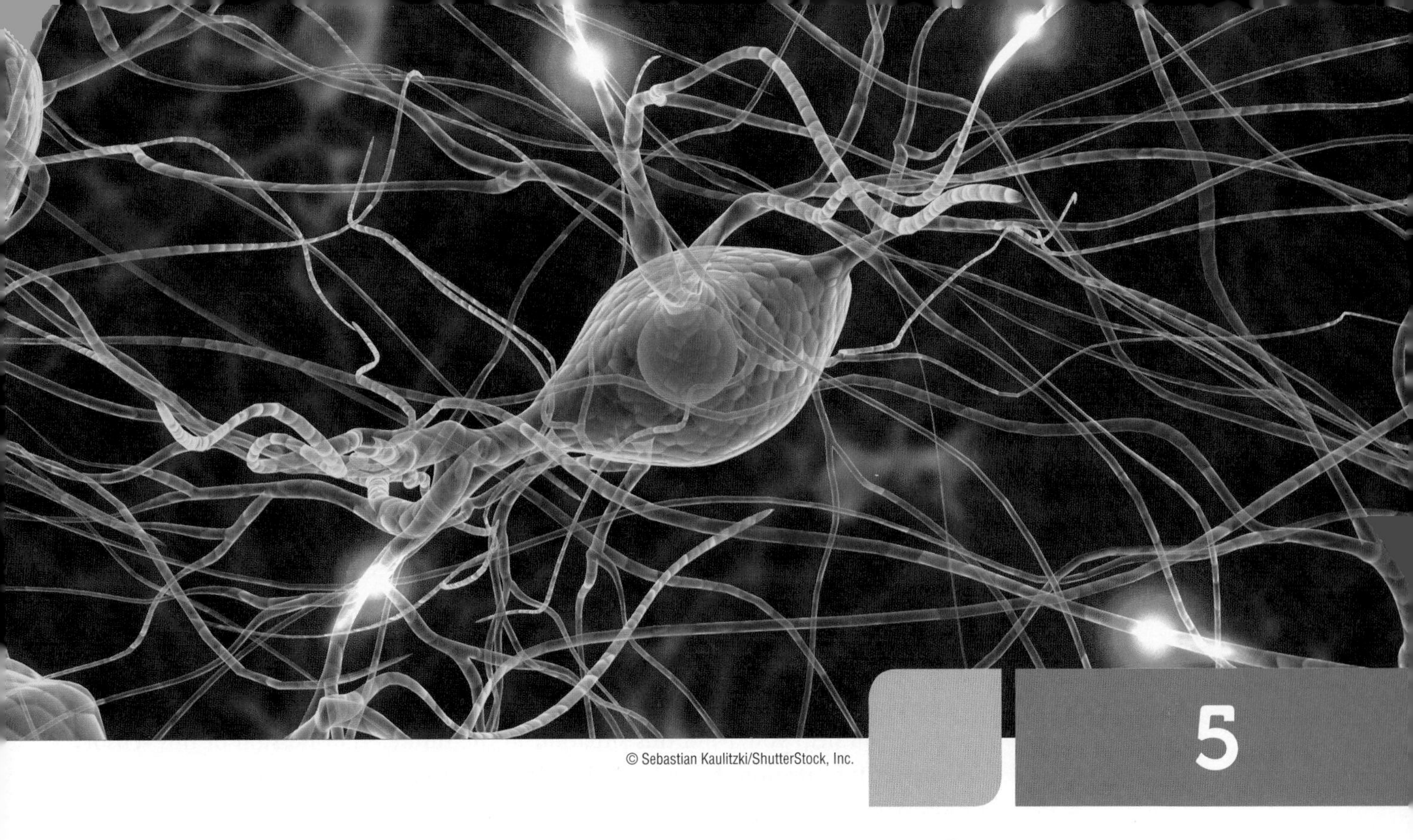

© Sebastian Kaulitzki/ShutterStock, Inc.

5

Somatic Nervous System and Special Senses

Case 1

You are walking barefoot on the beach, talking to a friend, when you step on something sharp. You recoil in pain, hopping on one foot while rubbing the other. Finally, you walk back to the car and put on your shoes, remembering that your mother always told you not to walk barefoot. Describe the sensory pathways that were stimulated as you walked on the warm sand. Which receptor and pathway allowed you to feel pain? How do analgesic pathways alleviate pain? How did you maintain your balance on one foot? What motor pathways allow you to walk? What cellular mechanisms are involved in memory?

Introduction

As an organism, we must gather information about the outside world—locating danger or food, for example—and then be able to respond to these external stimuli. Our sensory systems obtain information for us, which must be processed so we may use our motor systems to make the appropriate response. In our opening case study, pain warns you of potential danger, the withdrawal reflex moves your leg away from harm, and memory will help you avoid this situation in the future. Transmission of this sensory and motor information throughout the nervous system involves a wordless communication system made up of changes in membrane potential, action potentials, and a host of neurotransmitters. In this way, a variety of signals can be constructed from a simple action potential. We cannot appreciate the number of neural connections that are made each second of our life; the networks are too far reaching and the synapses involved are too vast in number. However, in this chapter, we will learn the major neural pathways and strategies of neural communication.

General Organization of the Somatic Nervous System: The Neuron

The basic cellular unit of the brain is the neuron and, very often, it is a multipolar neuron. The multipolar neuron has a cell body, notable for the dendrites that protrude from it, and an axon, ending in one or multiple synaptic bulbs. In most cases, the neuronal axon is myelinated, covered with a sheath of fatty myelin. Within the central nervous system, the brain and spinal cord, this myelin is produced by oligodendrocytes, a supporting cell associated with neurons. Myelin serves to seal the plasma membrane of the neuronal axon, covering it completely except for the internodal spaces, which are often called the nodes of Ranvier.

If the neuron is wrapped with myelin, how does it conduct an action potential? In a myelinated neuron, the only portions of the membrane that are exposed to the extracellular space are these internodal spaces, and ion channels are concentrated in these spaces. So, unlike the model action potential, where ion channels are located all along the axon, most neurons are myelinated, and their action potentials "skip" between internodal spaces in a saltatory conduction (**FIGURE 5.1**). This is a much faster version of the model action potential with fewer ions carrying the current! In fact, without the insulating properties of myelin, action potentials cannot be maintained over distances within the nervous system, which is why demyelinating diseases such as multiple sclerosis are so devastating.

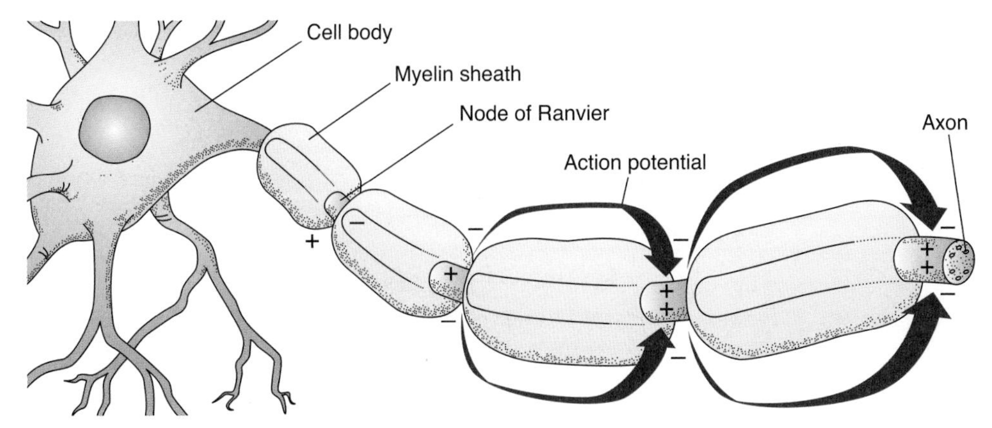

FIGURE 5.1 In myelinated neurons, voltage-gated Na^+ channels are located at the internodes, and the action potential "skips" between the nodes. This increases the speed of action potential conduction.

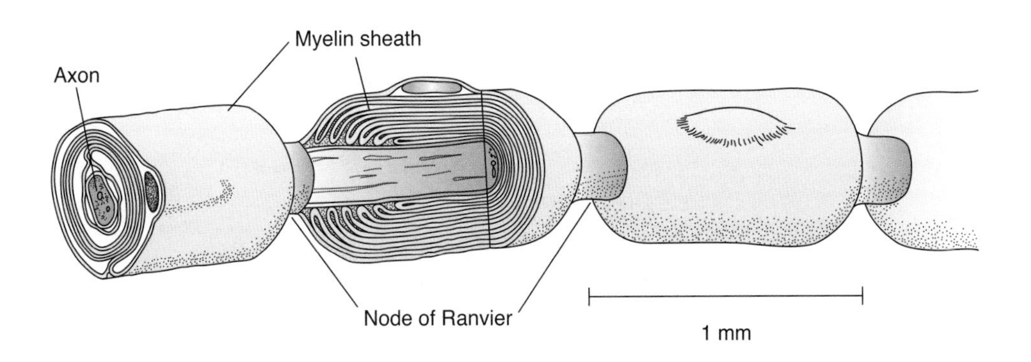

At the end of the axon lies the synaptic bulb, or multiple synaptic bulbs, as we will see in many instances. It is common for the action potential to be spread across many synaptic bulbs, all of which will connect to different neurons. So, one signal can be sent to many discrete neurons.

Neurons can be exceptionally long cells, extending from toes to spinal cord, or from spinal cord to fingers. At the same time, these long cells contain one nucleus, so that proteins created within the nucleus, in the cell body, may need to be transported to the active synaptic bulb, several feet away! Waste products at the synaptic bulb must be transported back to the cell body for processing at the lysosomes. Axoplasmic transport is necessary for moving molecules within these extremely long cells, and the movement must be directional. Anterograde transport from the cell body to the synapse moves along microtubules with an associated kinesin protein. This molecular motor moves vesicles of neurotransmitters or enzymes and mitochondria to the periphery, using adenosine triphosphate (ATP) as an energy source. Returning transport, or retrograde transport, uses microtubules and a different associated molecular motor, dynein, to move waste products back to the cell body. This process also requires ATP, but it is slower than anterograde transport.

Glial Cells

While we tend to focus on the neurons as the most important cells of the nervous system, they could not operate without the various glial cells, each of which has a specific function. Astrocytes are in very close contact with neurons, storing glycogen and providing neurons with lactate for ATP production. Astrocytes regulate ion balances, particularly extracellular K^+. Finally, astrocytes produce (and can take up) neurotransmitters released by neurons. The best example of this is glutamate, which is released by neurons, taken up by astrocytes,

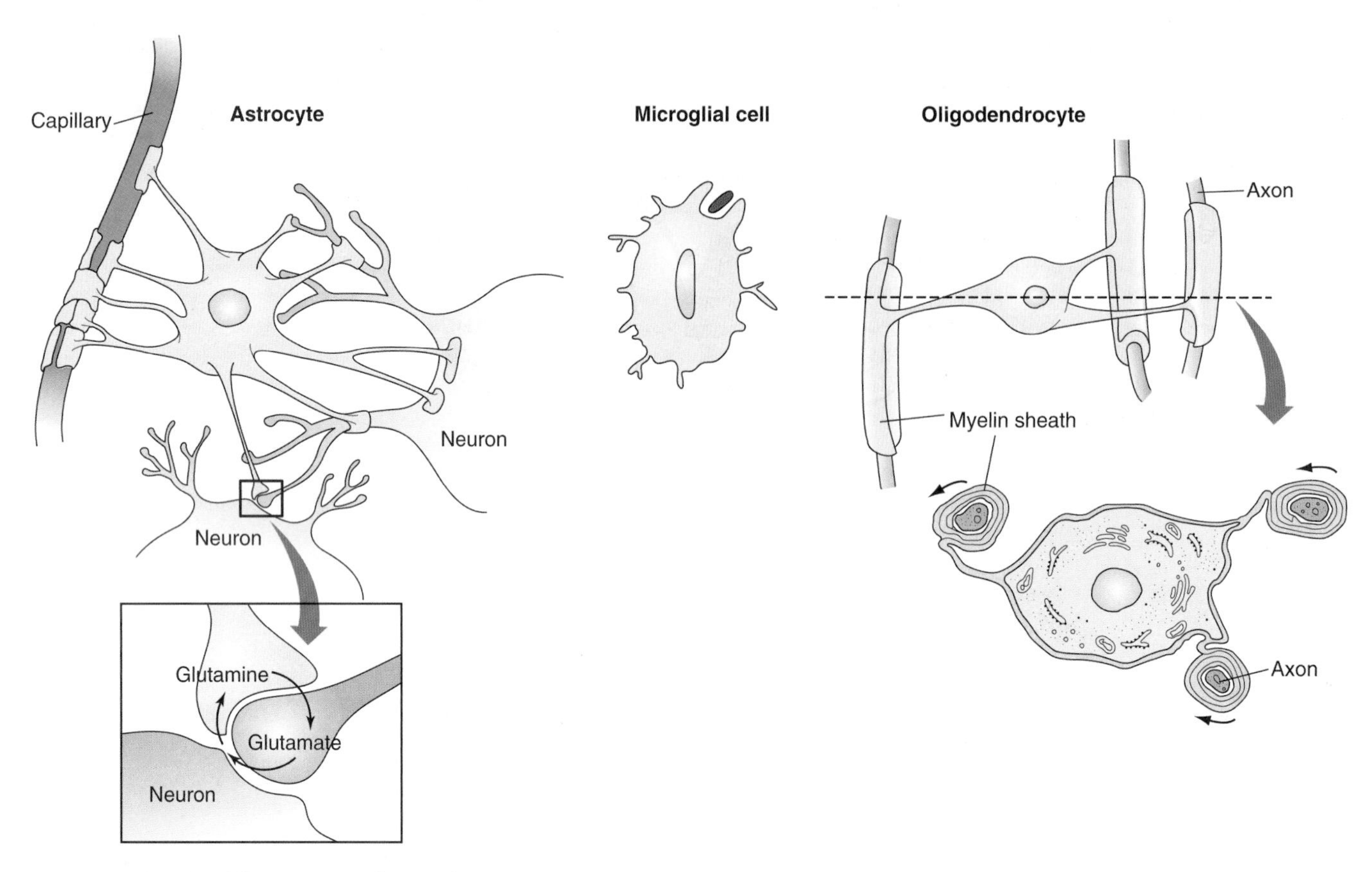

FIGURE 5.2 The different types of microglial cells. Astrocytes take up glucose from circulation and glutamate from the synapse and return them to the neuron.

metabolized to glutamine, and released back into the extracellular fluid, where neurons can take it up and enzymatically convert it back to glutamate. The blood-brain barrier, which we will examine in more detail later in this chapter, is largely maintained by astrocytes, the foot projections of which form a barrier around brain capillaries. Astrocytes are so important that changes in astrocyte health have been implicated in neuronal diseases.

Oligodendrocytes are the glial cells that produce myelin within the central nervous system. Unlike the Schwann cells, which are associated with a single neuron, oligodendrocytes myelinate several neurons, thus providing a support bridge between two neurons and giving them structural strength as well as the insulation of myelin.

Microglial cells function as macrophages within the brain and spinal cord. They scavenge cellular debris, phagocytosing damaged neurons in cases of injury. Microglia are also capable of having immune functions. They can become antigen-presenting cells should infection reach the brain through the normally sealed blood-brain barrier (**FIGURE 5.2**).

Finally, ependymal cells create cerebrospinal fluid (CSF) in which the brain floats. This fluid is a unique filtrate of blood that is made continually and circulates around the brain, removing waste products from the brain and carrying them to venous circulation. We will investigate the production and flow of CSF in more detail later in the chapter.

Organization of the Brain: Brain Areas Have Specialized Functions

As a tissue, the brain appears soft and unstructured, but in fact it is very well organized, and areas of the brain are devoted to specific functions. If we were to remove the cranium and expose the brain, we would reveal the cerebral hemispheres, the portion of the brain with which we are most familiar. There are two cerebral hemispheres divided into matching lobes. The frontal lobe lies anterior to the central sulcus (deep invagination) and is the area where we make decisions, exercise judgment, and store memories. It also contains the premotor cortex and the motor cortex, so the frontal lobe is important in our voluntary motion. Posterior to the central sulcus lies the parietal lobe, which includes the sensory cortex and sensory association cortex. The sensory cortex allows us to recognize and localize sensory input, such as touch, temperature, or pain, while the sensory association cortex puts those sensations into the context of memory and experience. The parietal-occipital sulcus separates the parietal lobe from the occipital lobe, the most posterior of the cerebral lobes. The occipital lobe is dominated by the primary visual cortex and the visual association cortex. The primary visual cortex receives neural input from the eyes, and the visual association cortex translates these neural signals into images we can recognize.

If we look at the brain from a lateral view, the final cerebral lobe becomes obvious—the temporal lobe. The temporal lobe contains the auditory cortex and lies near the olfactory cortex, which lies interior to the lobe proper. The hippocampus, essential for memory formation, lies interior to the temporal lobe. Because of its proximity to the hippocampus, the temporal lobe is also involved in memory, particularly of names and word memory.

If we view the brain from the posterior, we will see a structure discrete enough that it is not classified as a lobe of the cerebrum but is given its own name: cerebellum, or little brain. The cerebellum has two lobes joined by a vermis (**FIGURE 5.3**). As we will see in more detail later, the cerebellum is primarily involved in motion. It does not initiate motion but rather refines our motions so that they are smooth and coordinated. Also the time and distance calculations we take for granted when we run at speed toward the net on a tennis court, and then slow down before hitting it, are all done by the cerebellum. Unlike the cerebrum, where we can access sensory information and memories, or decide to walk down a beach, we have no conscious access or control over the cerebellum.

If we were to remove the cerebrum, we would find even more brain area. We lack conscious control over these hidden portions of the brain. The diencephalon contains the thalamus and the hypothalamus. The thalamus is important in both sensory and motor pathways as we will see later. The hypothalamus is the source of releasing factors for hormonal control, and it monitors and regulates body temperature, satiety, and hunger,

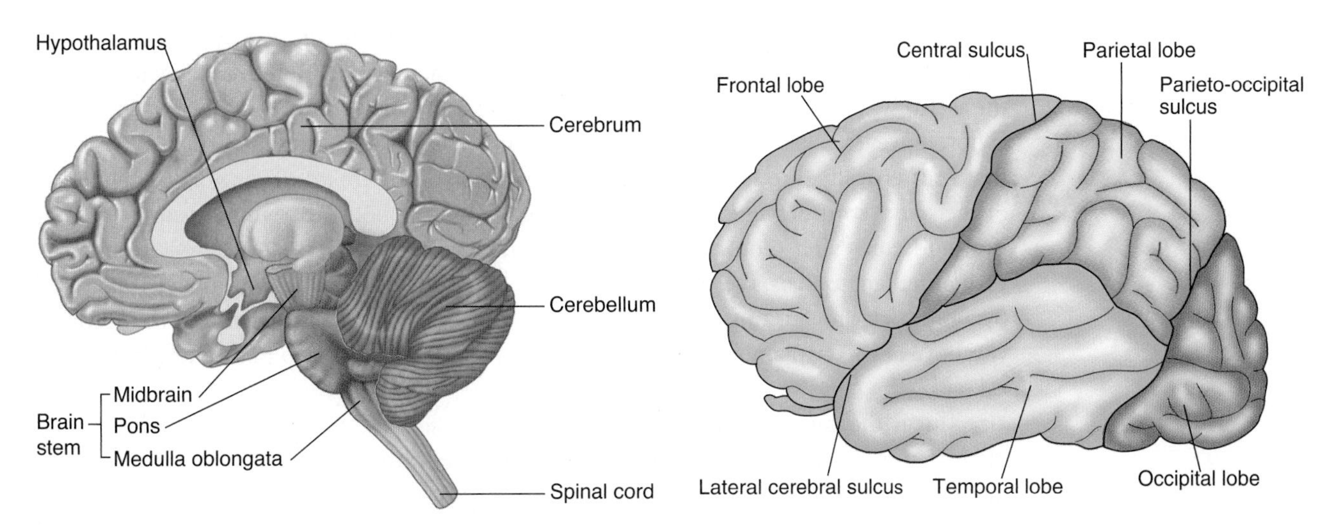

FIGURE 5.3 Major internal structures (left) and lobes (right) of the brain.

as well as osmolarity of the blood, and contains nuclei for the parasympathetic and sympathetic nervous systems.

Below the hypothalamus lies the midbrain. The corpus quadrigemini of the midbrain coordinate with the thalamus and visual and auditory pathways to allow central reflex action. We will investigate these in more detail during our discussion of sensory systems. The midbrain, despite its anatomical structures, contains nuclei that encompass several of the anatomical structures, running through them apparently randomly. This is unfortunately confusing for students of the brain, but true nonetheless. The substantia nigra and red nucleus each contribute to voluntary motion.

The reticular formation, a long structure that goes from midbrain to medulla, is responsible for our awareness. It is, for example, the reticular formation that will wake you in the night when you hear the sound of the refrigerator opening, especially if you live alone! You may have slept through road noise, sirens, and the sound of music from next door, but this simple sound, out of place, awakened you. The reticular formation is always active and provides background information, having neural connections to hearing, vision, and sensory systems. It is the reticular formation, connected to pain pathways, that awakens you in the night when you have pain.

Below the midbrain lies the pons, an important site of cranial nerve connections and the home of nuclei that contribute to our respiratory rhythms. The pons is also a network of nerve fibers that connect the cerebellum to the motor cortex. This mass of fibers gives the pons its cumberbund look. The medulla also contains nuclei for cranial nerves. However, it also contains the chemosensors that regulate respiration and cardiovascular centers that regulate heart rate and vascular tone. The medulla is the junction between the brain and spinal cord, so all neurons ascending or descending must also pass through the medulla.

While each brain area has discrete functions, the areas are connected to one another by myelinated neural pathways, bundles of neurons. Some fibers connect the two hemispheres, some connect the frontal lobe to the occipital lobe, and some are shorter paths that connect adjacent gyri to one another. These myelinated neuronal pathways allow you to see a face (occipital lobe), recognize the face (temporal lobe), and remember the identity of the person from the party you went to last Friday (frontal lobe), all within seconds. It also has the anatomical effect of making the white matter of the brain interior to the gray matter, which lies on the surface of the cerebrum.

Spinal Cord and Cranial Nerve Organization

In order for the brain to receive information from the body, or give instructions to the body, there must be neural pathways that convey these action potentials. Eventually, all these pathways must converge on the spinal cord for transit to the brain. The spinal cord is not, however, a disorganized cluster of neurons, but carefully arranged tracts of neurons that run vertically through the spinal cord. Further, each dermatome, or horizontal section of the spinal cord, serves a particular cross-section of the body, receiving sensory information and conveying motor information through spinal nerves (**FIGURE 5.4**). If we were very gifted anatomists, we could physically follow the neuronal action potential initiated by a pinprick on your finger to the brain, where you actually feel the pain, to the memory you have of this happening before! This is a difficult and multi-step path, but it is a physical pathway.

This is the general physical organization of the somatic nervous system. The cellular organization of this system is also complex, and we must also understand the pattern of neuron-to-neuron communication before we appreciate the plasticity of this system.

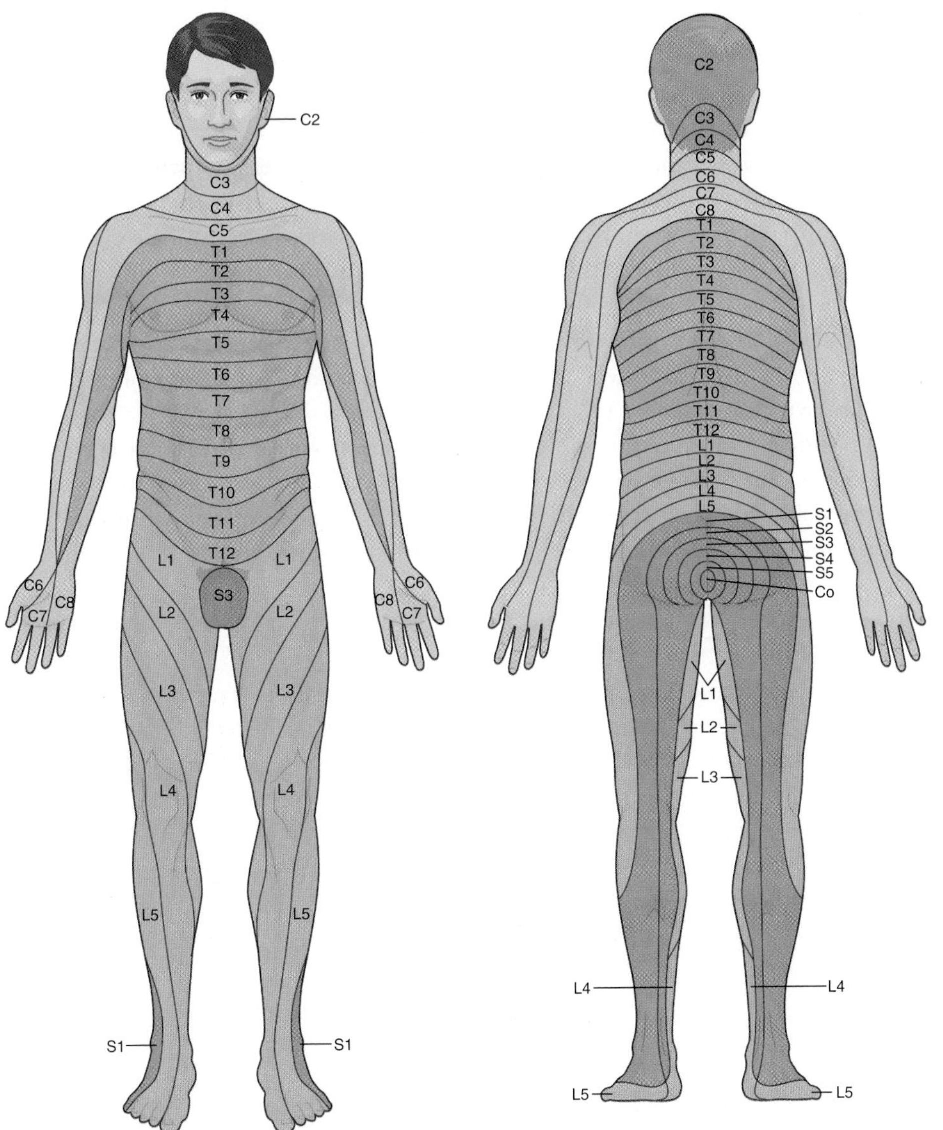

FIGURE 5.4 Dermatomes are the area served by each of the spinal nerves. Spinal nerves carry sensory information to the central nervous system and motor information from the spinal cord to the skeletal muscle.

Neuronal Patterns of Communication

Let's return to a multi-polar neuron to see how neurons receive and integrate information. The dendrites projecting from the cell body of the neuron receive information from other neurons, in the form of changes in membrane potential. A synaptic bulb of a neuron forms a synapse with the plasma membrane of the dendrite. Neurotransmitters will be released into the synaptic cleft, bind to a receptor, and cause a change in membrane potential at the dendrite. Dendrites contain large numbers of Ca^{2+} channels, so the change in dendritic membrane potential is generally caused by calcium influx—a different positive ion than we see in the neuromuscular junction. However, the neuronal action potential is still caused by Na^+ channels, so depolarization at the dendrites must be sufficiently positive to open Na^+ channels in the initial segment of the axon, an area with a high concentration of Na^+ channels. Depolarization at the dendrite frequently results in a graded potential. Graded potentials degrade over distance, so a small depolarization at a single dendrite is unlikely to cause an action potential. As you may know, the action potential occurs in a motor neuron when the depolarization is sufficient to reach threshold, the membrane voltage at which Na^+ channels would open. A graded potential remains below threshold, causing a change in membrane voltage without generating an action potential. If all this is true, how does a neuron synapsing at a dendrite ever cause an action potential? There are several ways.

Each neuron possesses many dendrites and dendritic spines, each of which is a potential site for neuronal synapses. While we generally show you one synapse between cells, for simplicity, this is a time when you should imagine five synapses on a single multi-polar neuron. At each of these synapses there will be neurotransmitter released, but only some will release excitatory neurotransmitter, in an excitatory postsynaptic potential (EPSP). These EPSPs will cause a depolarization of membrane potential. Some of the synapses may release inhibitory neurotransmitters, causing a hyperpolarization of the membrane or an inhibitory postsynaptic potential (IPSP).

Let's imagine a moderately complex synaptic pattern where three of five synapses caused depolarization of the membrane through EPSPs, and the remaining two synapses contain hyperpolarizing neurotransmitters causing IPSPs. The multipolar neuron will sum all of these signals and if the change in membrane potential reaches threshold for Na^+ channels, an action potential will ensue (**FIGURE 5.5**). This is spatial summation, and occurs whenever there are multiple inputs to a single cell, even if all of the input is excitatory, or all is inhibitory.

Summation can also occur over time (temporally). Using our example above, if the two inhibitory synapses are firing quickly, releasing neurotransmitter more rapidly than the excitatory synapses, they may prevent threshold from being reached. Temporal summation changes the amount of neurotransmitter delivered, and therefore the amount of signal that will change the membrane potential.

Neuronal connections within the somatic nervous system are more complex than the simple connection we see in the neuromuscular junction. However, the complexity allows

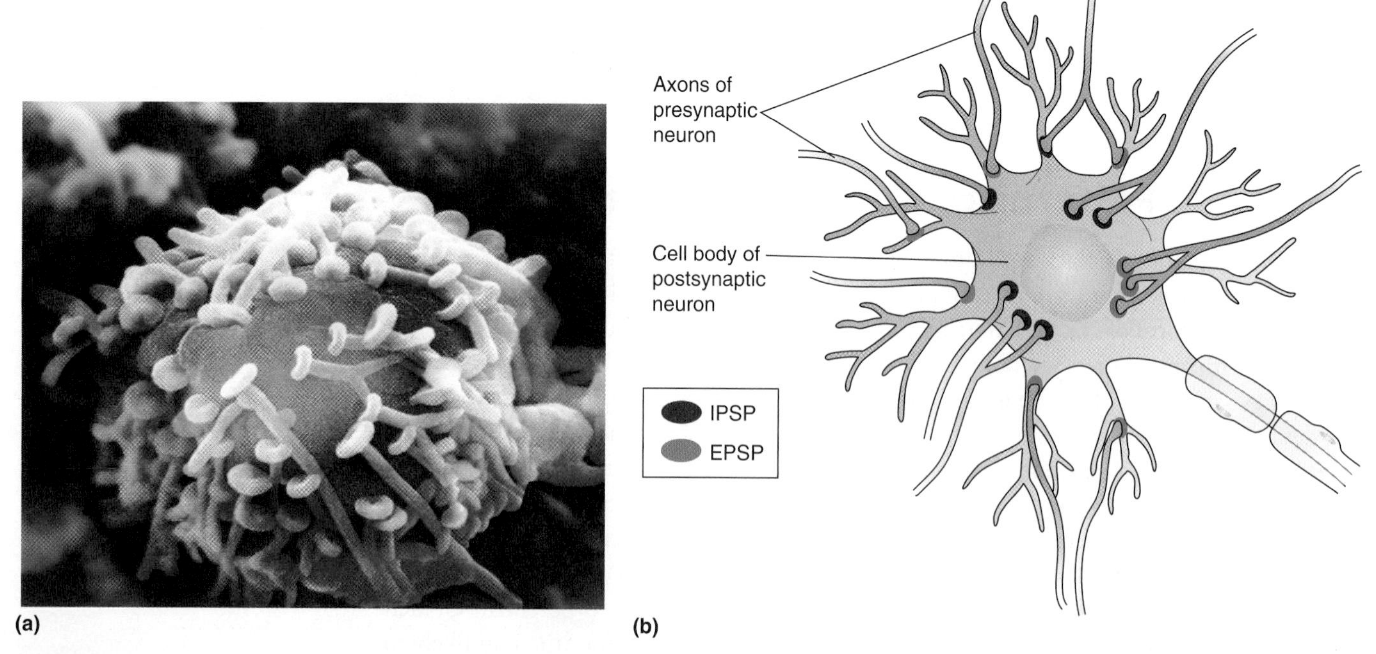

FIGURE 5.5 (a) Scanning electron micrograph of a neuron cell body (purple) with numerous synapses (blue) from other neurons (x 80,000). (b) Presynaptic neurons can synapse with the dendrites or the soma of the neuron. Excitatory signals will cause EPSPs, and inhibitory signals will cause IPSPs. (© Science VU/Lewis-Everhart-Zeevi/Visuals Unlimited, Inc.)

for input from many areas of the brain as we make a decision, for example, and a great deal of plasticity in our reactions. A simple "all or nothing" action potential thus gains refinement.

Neurotransmitters of the Somatic Nervous System and Their Receptors

While skeletal muscle is activated by a single neurotransmitter, there are many neurotransmitters at work within the brain, with each neurotransmitter binding to more than one receptor type. This is another way that signaling in the brain is more complex and allows more variety than signaling in other tissues.

Some neurotransmitters are made in discrete brain areas and then widely distributed throughout the brain by neurons. The neurotransmitter norepinephrine, for example, is made in the locus coeruleus, a nucleus within the pons, and it is distributed by neurons to all areas of the brain and spinal cord. Another neurotransmitter, serotonin, is made in the raphe nuclei located in the pons and medulla, and from there is delivered to all parts of the brain and spinal cord. Acetylcholine is made in three separate areas of the diencephalon and midbrain and then widely released within the brain. Finally, dopamine is made in the substantial nigra and vental tegmental area. Dopamine exerts its effects at the frontal lobe, the limbic system, and nuclei of the midbrain associated with motion (**FIGURE 5.6**). These four neurotransmitters have orderly pathways from origin to destination.

The most common neurotransmitters are simply made by the neurons releasing them; these neurons may be located in any part of the brain. Excitatory stimuli within the brain are generally caused by either glutamate or aspartate, binding to ligand-gated cation channels. So, they function in a similar fashion to acetylcholine, allowing Na^+ into the postsynaptic cell to generate an action potential. However, there are 14 different subtypes of glutamate ligand-gated cation channels, each of which possesses slightly different characteristics. Like acetylcholine, glutamate can also bind to a G-protein-coupled receptor, of

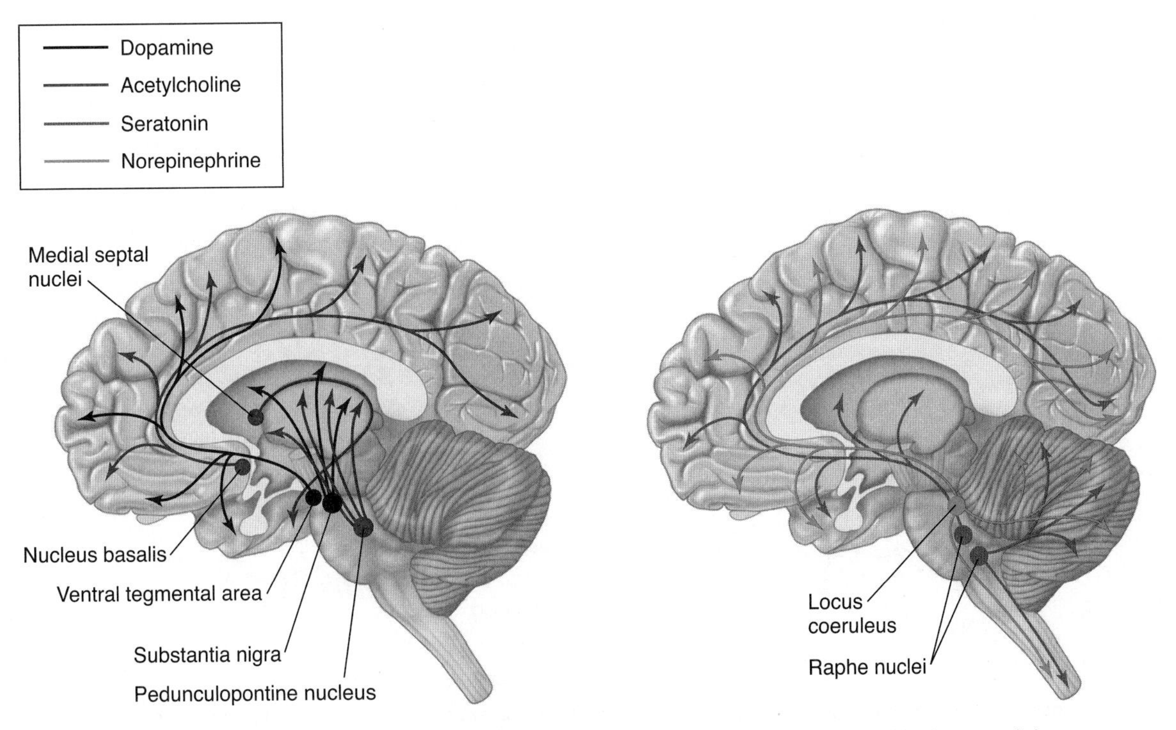

FIGURE 5.6 Some neurotransmitters are made by neurons clustered in specific areas of the brain and then dispersed around the brain by axonal projections.

which there are eight separate types. While we will not go into the details of each receptor and its functioning, it is important to remember that the variety of receptors allows many intercellular messages from a single neurotransmitter.

Inhibitory signals are simpler. Gamma-aminobutyric acid (GABA) and glycine are the neurotransmitters that inhibit, and these open ligand-gated chloride channels (two subtypes) or signal through one type of G-protein-coupled receptor. It is easy to envision how an open Cl^- channel, allowing negative ions into the cell, will hyperpolarize the membrane, moving it further from threshold and an action potential, thus causing inhibition.

Each of the neurotransmitters mentioned above may also be packaged within the presynaptic neuron with one or more other peptides, which will also influence the open times of ion channels or the binding at G-protein-coupled receptors. Packaging of these "partner" peptides may differ between neurons or by portion of the brain. Finally, nitric oxide and carbon monoxide, two gaseous molecules, are also short-lived but active neurotransmitters.

Everything we discussed in the last section pertained to neurotransmitters released from presynaptic neurons into a synaptic space and affecting a postsynaptic neuron. You could easily presume that all of those action potentials began at the dendrites of the presynaptic neuron, and indeed they probably did. However, there is another mode of signaling we must consider. Sometimes a third neuron is involved, and it forms a synapse with the presynaptic neuron, at the point of the synaptic bulb. Now we have a presynaptic neuron forming a synapse with our other presynaptic neuron! This type of connection generally affects the quantity of neurotransmitter release from the second presynaptic neuron. If neurotransmitter release is augmented, it is called presynaptic facilitation, because is facilitates the release of neurotransmitter. If neurotransmitter release is reduced, it is called presynaptic inhibition. For example, if the third neuron releases serotonin, which binds to a serotonin receptor on the presynaptic neuron, this will increase Ca^{2+} entry into the synaptic bulb, increasing neurotransmitter release, or presynaptic facilitation. If that neuron released GABA, it would open Cl^- channels, hyperpolarize the presynaptic membrane, and reduce neurotransmitter release, thus initiating presynaptic inhibition (**FIGURE 5.7**).

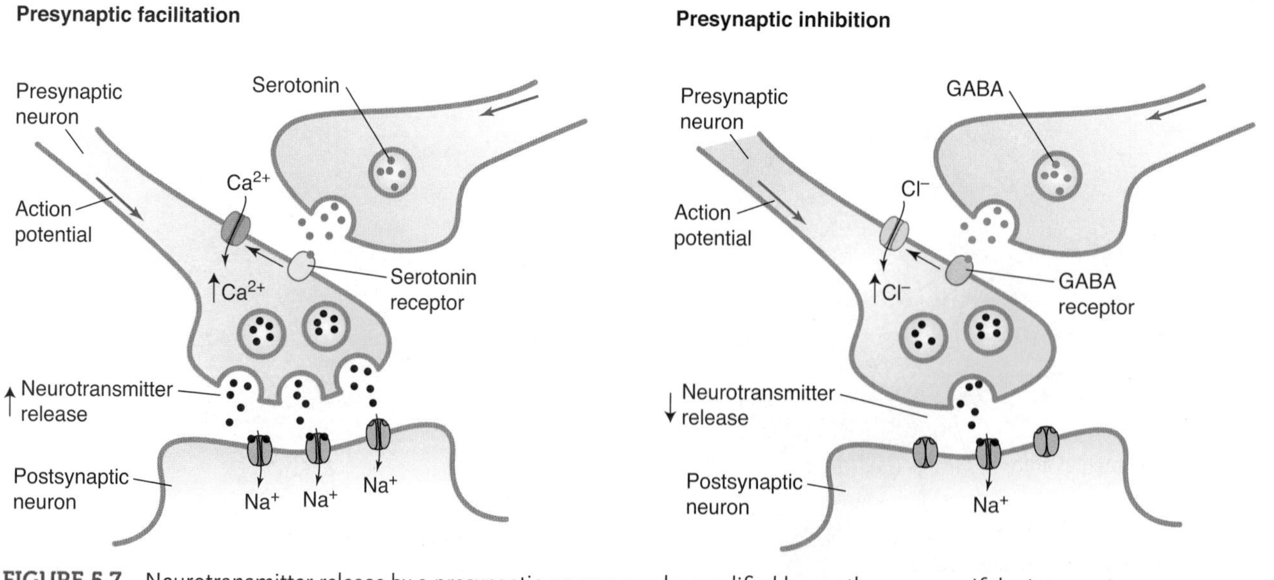

FIGURE 5.7 Neurotransmitter release by a presynaptic neuron can be modified by another neuron. If the intervening neuron releases serotonin, it can increase the amount of neurotransmitter released. If GABA is released, it will inhibit or decrease the amount of neurotransmitter released to the postsynaptic neuron.

If all these patterns of communication between neurons seem confusing, imagine describing the variety of forms of communication on Facebook and the number of possible interconnections. I think we can make a credible argument that the brain was the original Facebook!

How Do We Get Information from the Outside World?

All of our knowledge of the outside world comes in through our sensory pathways, and most of these are in the skin. There are, of course, a variety of special senses—taste, hearing, sight, and smell, which will be detailed later—but the tactile sensory inputs from the skin comprise the majority. We feel the warmth of the sun on our skin, or the coolness of a breeze. We feel a fly landing on our arm or the touch of a coat as we put it on. We feel pain when a staple left in the sleeve scrapes our skin, or itching from a label rubbing against the back of our neck. These are all our general tactile senses—fine touch, crude touch, pain, warm, cold, pressure, or vibration—that tell us about our environment. This sensory information comes through receptors—but this time the receptors are not membrane proteins, but small sensory organs located within skin layers. Without these receptors, we would likely damage ourselves by placing our hands or feet on hot surfaces, failing to notice damaging cold, or being unaware of the location of a touch. The simple act of swatting a mosquito would become impossible, because you would never feel its presence! In this section, we will learn about the receptors for each of these sensory types, or modalities, and the pathways they use to reach the brain.

Several Skin Sensory Organs Respond to Mechanical Stress

As you walk down the beach, in our opening case, you can feel the grains of sand on the soles of your feet and between your toes. When you step on a small pebble, you feel the difference in its shape and touch, and you know which part of your foot touched that pebble. You also feel the ocean breezes on your face and arms. How is this sensory information transduced?

Within and just below the dermal layer of skin lie a variety of small sensory organs (receptors) that can respond to mechanical deformation. These organs were named for the anatomists and physiologists who discovered them: Ruffini's corpuscle, Pacinian corpuscle, Merkel's disc, and Meissner's corpuscle all contain ion channels that can be opened mechanically, allowing an action potential to begin when the receptor has been stimulated. These types of receptors containing mechanically gated ion channels are anatomically unique and lie at various depths within the skin, which allows them to respond to different types of touch. The Pacinian corpuscle, for example, is most sensitive to rapid touch, like vibration. The nerve ending is encased in layers of gelatinous fluid that deforms rapidly when touched, but just as rapidly recovers its original shape. These receptors are deep in the subcutaneous layer, so are not in a position to sense a light touch. Contrast this to Meissner's and Ruffini's corpuscles, located just below the dermal layer. These corpuscles, along with Merkel's discs, are located most superficially and are deformed by light touch. Action potentials from a single receptor will result in an action potential that progresses down a nervelet. Our knowledge of touch will come from this action potential.

The breeze on our face is sensed by a different type of receptor; the small hairs projecting from the epidermis are connected to small nervelets within the dermis. Movement of these hairs will cause a mechanical opening of ion channels in this nervelet, again allowing us to sense air motion. This is the same mechanism that allows us to swat a mosquito, knowing

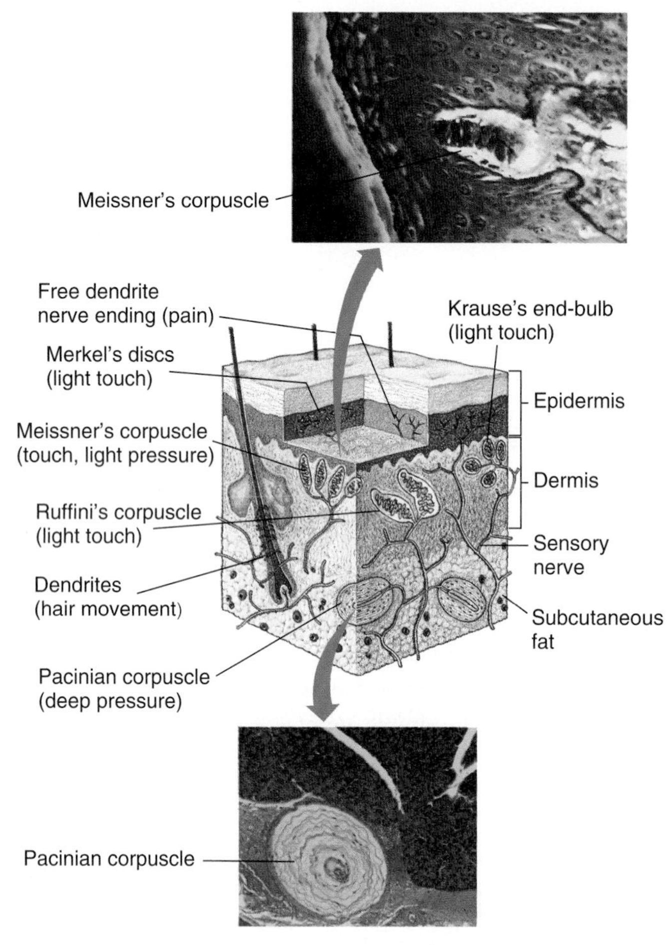

FIGURE 5.8 Sensory receptors are located under the epidermal layer of the skin. Each receptor responds to different types of sensory stimulation, allowing us to differentiate between light touch, pressure, and hair movement. (Top photo © Biophoto Associates/Science Source. Bottom photo © Donna Beer Stolz, Ph.D., Center for Biologic Imaging, University of Pittsburgh Medical School.)

precisely where it is located, without even looking! While we are not very hairy creatures, the hairs we have are important mechanosensory receptors.

Receptor Fields and Tactile Localization

These receptor organs are not evenly distributed within your skin but are more concentrated in some areas and widely dispersed in others. Each receptor will serve a specific area of skin. Some areas are large and some are small, and this is called a receptor field. A receptor is connected to a tiny nerve, which will join into a larger nerve, ultimately ending in the brain. This is not an amorphous path, but a very specific one. Even with your eyes closed, a pinprick on your finger can be localized within a few millimeters because the receptor density is quite high on the hands, and the receptor field is very small. However, you may mislocate this same pinprick by several centimeters in the center of your back because the receptors are much less concentrated there (**FIGURE 5.8**).

Receptors Adapt to Stimuli

In our opening case study, you put your shoes on after stepping on a sharp object. Undoubtedly, when you first placed the shoes on your feet, you could sense the weight of the shoes, the strap that rubbed on your left foot, and their touch on the soles of your feet. However, moments later, all of these sensations disappear! The mechanoreceptors are all adapting receptors: some adapt quickly and some more slowly, but all of them will fail to respond when stimulated over a long period. The mechanoreceptors transduce and localize information about a change in touch on any part of our skin.

Receptors Also Sense Temperature and Pain

As you walk down the beach, you can feel the warmth of the sand. This perception comes from temperature receptors. We possess separate warm and cold receptors that transduce information about peripheral temperature. Warm receptors are stimulated at temperatures in excess of 37 °C, and cold receptors are stimulated when temperatures fall below 37 °C. At extremes of temperature on either end of the scale, these receptors become less responsive, and the signal transduction is done by pain receptors. Touching a hot stove is more painful than hot.

When you step on a sharp object, pain receptors are triggered, inspiring you to remove your foot from that area. While heat and cold receptors are specific, pain receptors can be stimulated by any strong stimulus, whether it is touch—in the guise of a sharp rock—or extreme temperature, or even chemical stimulus, like a strong acid or base that burns the skin. Pain receptors are essential for our safety and are the only receptor that is never adapting—the sensations continue as long as the stimulus is present. This means that pain can be chronic, lasting days to years.

Sensory Stimuli Travel to the Brain by Organized Pathways

Mechanoreceptors, thermoreceptors, and pain receptors are all associated with a neuron and send their signals to the brain by action potential. We can discern the difference between a touch and warmth because these signals travel by different pathways, routes traveled by large numbers of neurons all carrying the same modality of sensation. Thus pain pathways are distinct from touch pathways in the route they take to the brain. Within each pathway, the origin of the sensory neuron is maintained, and the point of origin will determine where it ends within the brain. This is how we are able to localize a sensation. Let's follow some sensations so we can understand how this system works.

The breezes on your skin, sensed by nerves wrapped around skin hairs, produce an action potential that travels via the fine touch pathway—the dorsal column pathway. The dorsal column pathway begins with the first-order sensory neuron, the one that begins at the receptor. The first-order neuron travels to the midbrain, where it crosses the centerline of the brain and synapses at the medial lemniscus. The second-order neuron begins at the medial lemniscus and synapses at the thalamus. The third-order neuron begins at the thalamus and ends at the postcentral gyrus of the parietal lobe, the sensory cortex. The exact location where the third-order neuron terminates in the sensory cortex depends on the origin of the first-order neuron. In other words, we could, if we were talented anatomists, trace a labeled line from your forearm, where you felt the breeze, to the exact spot on the sensory cortex that relates to that portion of the skin.

The sensory cortex has been mapped, and the map is called the sensory homunculus. You will notice that some areas are larger than others. The face is enormous compared to the thigh, for example. This relates to the density of sensory receptors in each of those areas, and the ability we have to localize a fine touch. In addition to fine touch, the dorsal column pathway also carries sensations of vibration and proprioception.

Pain and temperature follow the anterolateral pathway, also called the spinothalamic tract. Pain and temperature receptors in the skin connect to nerves that form the first-order neuron, which travels to the spinal cord. The first-order neuron crosses the midline of the spinal cord and synapses there with the second-order neuron. The second-order neuron ascends to the thalamus where it synapses with the third-order neuron. The third-order neuron terminates in the sensory cortex, at the site appropriate to the body part stimulated (**FIGURE 5.9**). The sensory

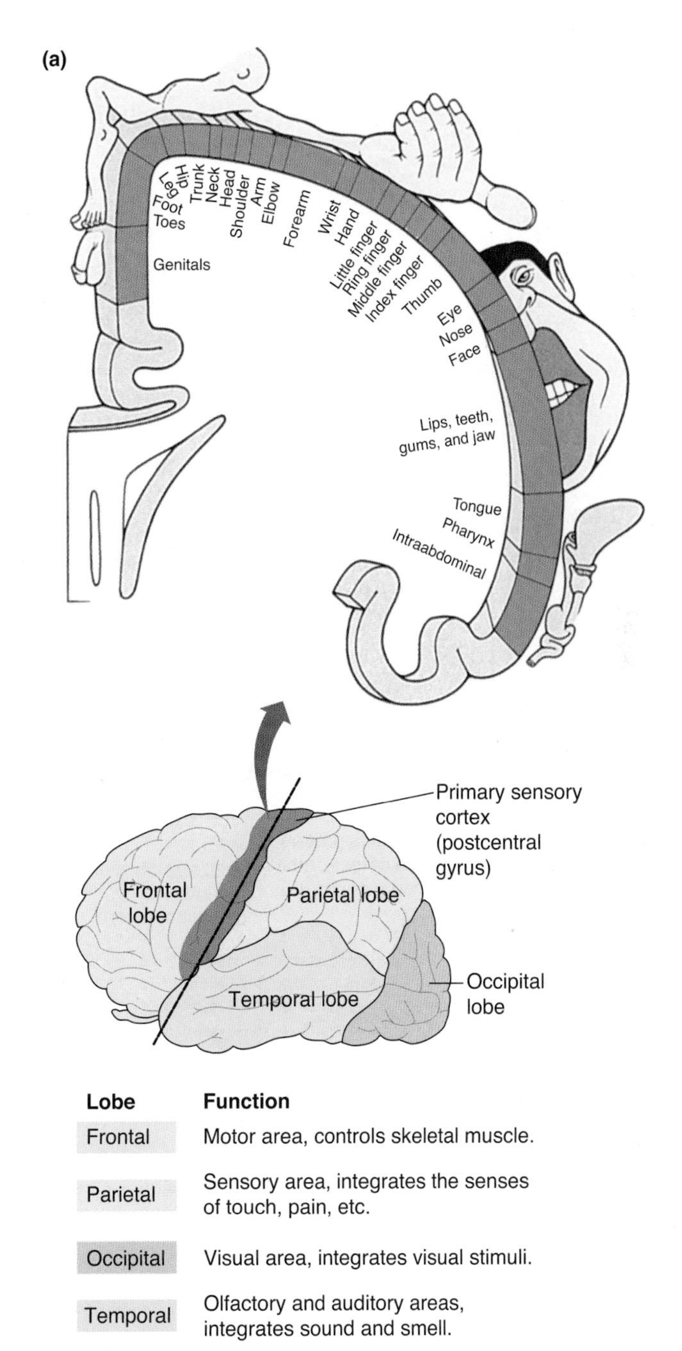

Lobe	Function
Frontal	Motor area, controls skeletal muscle.
Parietal	Sensory area, integrates the senses of touch, pain, etc.
Occipital	Visual area, integrates visual stimuli.
Temporal	Olfactory and auditory areas, integrates sound and smell.

FIGURE 5.9 The neuronal pathways for the anterior lateral and dorsal column pathways ending at the sensory cortex.

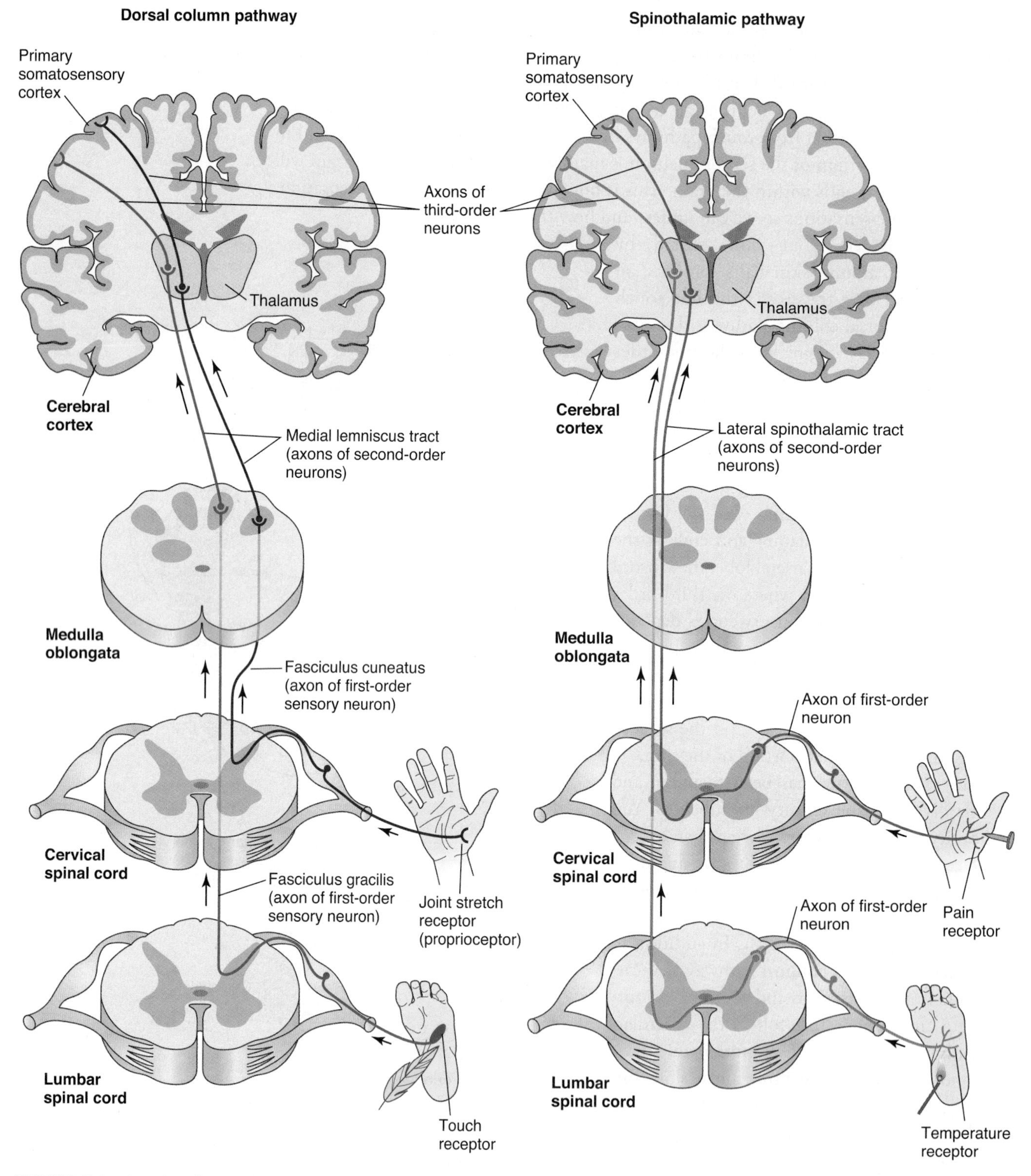

FIGURE 5.9 *(continued)*

homunculus remains the same for all sensory pathways.

Notice that these two pathways, dorsal column and anterolateral, are in separate tracts, or nerve bundles, so specific neurons are carrying specific sensory modalities. This is the concept of the labeled line: given sufficient skill, we could trace a neural pathway from receptor to brain. The point at which these tracts decussate or cross is important clinically. A loss of pain sensation in one leg may give information regarding a spinal cord lesion and can be used in differential diagnosis of neural injury.

All of these sensory pathways are ascending pathways, with information traveling from peripheral receptors to the central nervous system, ultimately reaching the brain. Most of the action potentials relaying this information are in myelinated neurons of fairly large diameter, so the signals travel quickly. Pain is carried this way. However, pain is an ancient sensation, a signal of danger. The receptors are nonadapting, so we are forced to acknowledge the pain. In addition to the large, fast, myelinated neurons that carry pain as a modality, we also have smaller, unmyelinated fibers that give us the poorly localized dull ache after the initial sharp pain. These fibers are thought to have evolved earlier than the myelinated pain fibers, and a lack of myelination makes this signal slower, which is why it is always later.

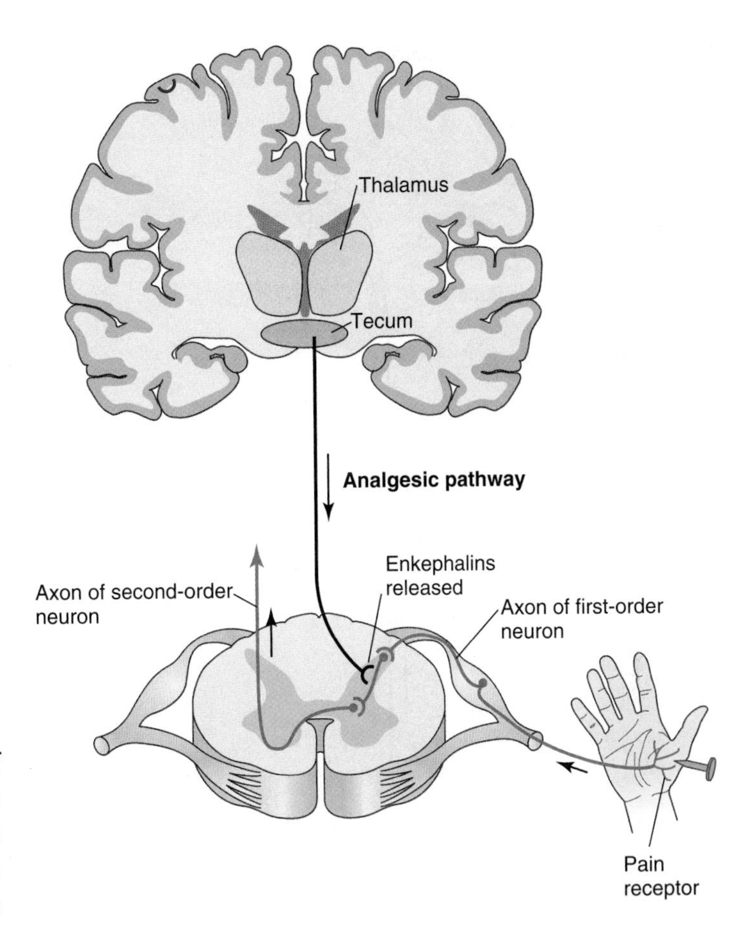

FIGURE 5.10 The analgesic pathway interferes with the transmission of pain signals from nociceptors (pain receptors) and reduces the sensation of pain.

Our perception of pain is influenced by another pathway—the analgesic pathway. The analgesic center lies in the tectum, just below the thalamus. Neurons descend to the spinal cord and an interneuron releases enkephalins, inhibiting the transmission of an action potential to the second-order neuron. The interference with communication of the pain signal may not be total but will dampen the sensation of pain (**FIGURE 5.10**). The activity of the analgesic pathway is thought to account for the wide range of pain perception in people. What feels like a mild prick on the finger to one person may feel like a stabbing pain to another. Clearly, being able to control the analgesic pathway and relieve chronic pain is an area of intense investigation.

The Spinocerebellar Tract Senses Muscle Tension

This final sensory pathway is one of which you may be completely unaware. The receptor is a muscle spindle organ, located deep within skeletal muscle. A nervelet wrapped around non-contractile muscle fibers is stimulated by stretch and sends action potentials from the muscle to the cerebellum. The cerebellum, which is responsible for calculating time and distance of our motions, gets sensory information about the state of contraction of each of our muscles (**FIGURE 5.11**). Think of the simple act of picking up a pebble from the beach. You bend down, move your arm and hand toward the pebble, decelerate the speed of arm

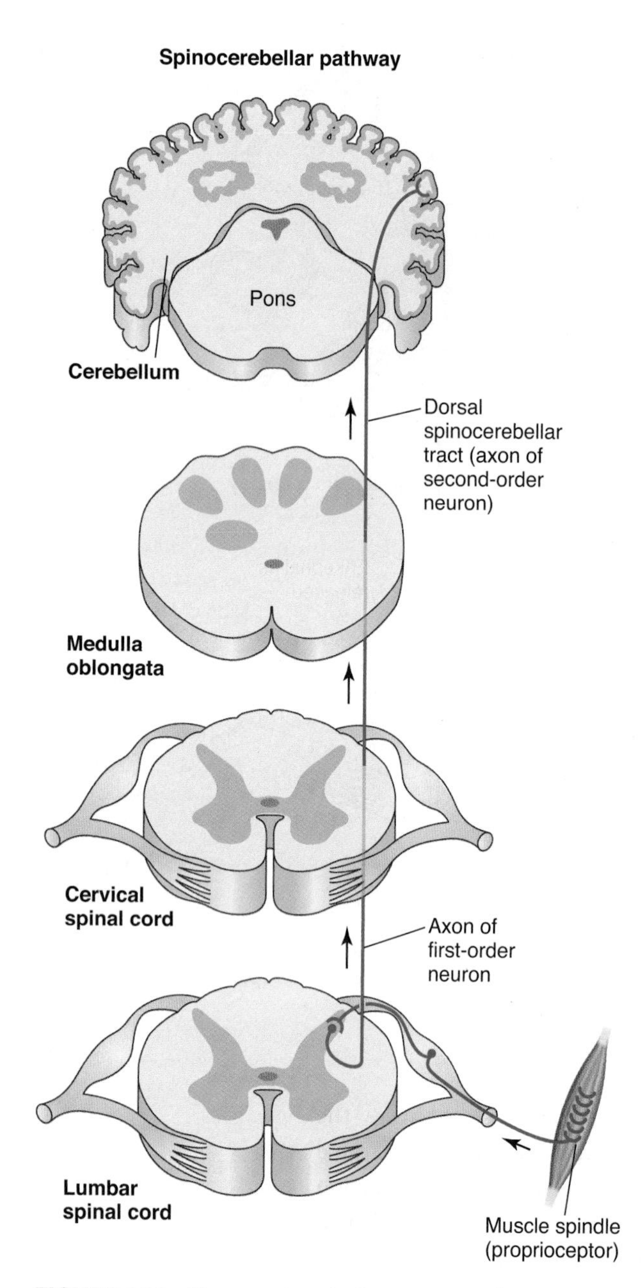

FIGURE 5.11 The muscle spindle organ provides sensory input to the cerebellum via the spinocerebellar tract.

motion, pick up the pebble, accelerate the contraction of your arm, and then decelerate it again as you draw the pebble to your face to take a good look at it. All this acceleration and deceleration of motion is managed by the cerebellum using the sensory information provided by the spinocerebellar tract to calculate accurately your motions. The cerebellum does not command the motion but coordinates it. All of the thousands of motions each day, from brushing your teeth, to typing on a keyboard, to scratching your head are made smooth and accurate by the sensory input from the spinocerebellar tract. Notice that this tract is a simple two-neuron track, going from the sensory receptor directly to the cerebellum. Because it does not go to the cerebral cortex, we have no conscious awareness of these sensations.

Special Senses: More Information Collected from the Outside World

Taste, smell, vision, hearing, and balance are additional senses that provide us with a wealth of information about the world. Like the general senses, there is a receptor that senses the physical input and, via action potentials, that information is relayed to a specific sensory cortex for that modality.

Taste

Taste buds are located primarily on the tongue, but there are some taste cells in the pharynx as well. The five recognized taste sensations are salty, sour, sweet, umami (savory), and bitter. Each of these tastes has a unique receptor. Salty taste is transmitted by Na^+, which goes through Na^+ channels causing a depolarization; the resulting action potentials travel along cranial nerves X, IX, and VII to the thalamus, where they synapse. Neurons from the thalamus travel to the sensory cortex, where they terminate at the deepest portion of the postcentral gyrus, the gustatory cortex. This pathway is the same regardless of the specific tastant (**FIGURE 5.12**). H^+ ions trigger the taste of sour by closing K^+ channels and causing a depolarization. Again, the resulting action potentials end up at the sensory cortex. Sweet, umami, and bitter tastants bind to separate receptor cells but transmit their signal by the same signaling mechanism. Each of these tastants binds to G-protein-coupled receptors that open channels via an increase in IP_3 and intracellular Ca^{2+}. The action potentials, again, end on the sensory cortex.

The number of each type of receptor or the sensitivities of the receptors change with age. As children, we are very sensitive to bitter tastes. Because many toxins are bitter, this is thought to be a protective mechanism. As we age, we are less sensitive to bitter and actually

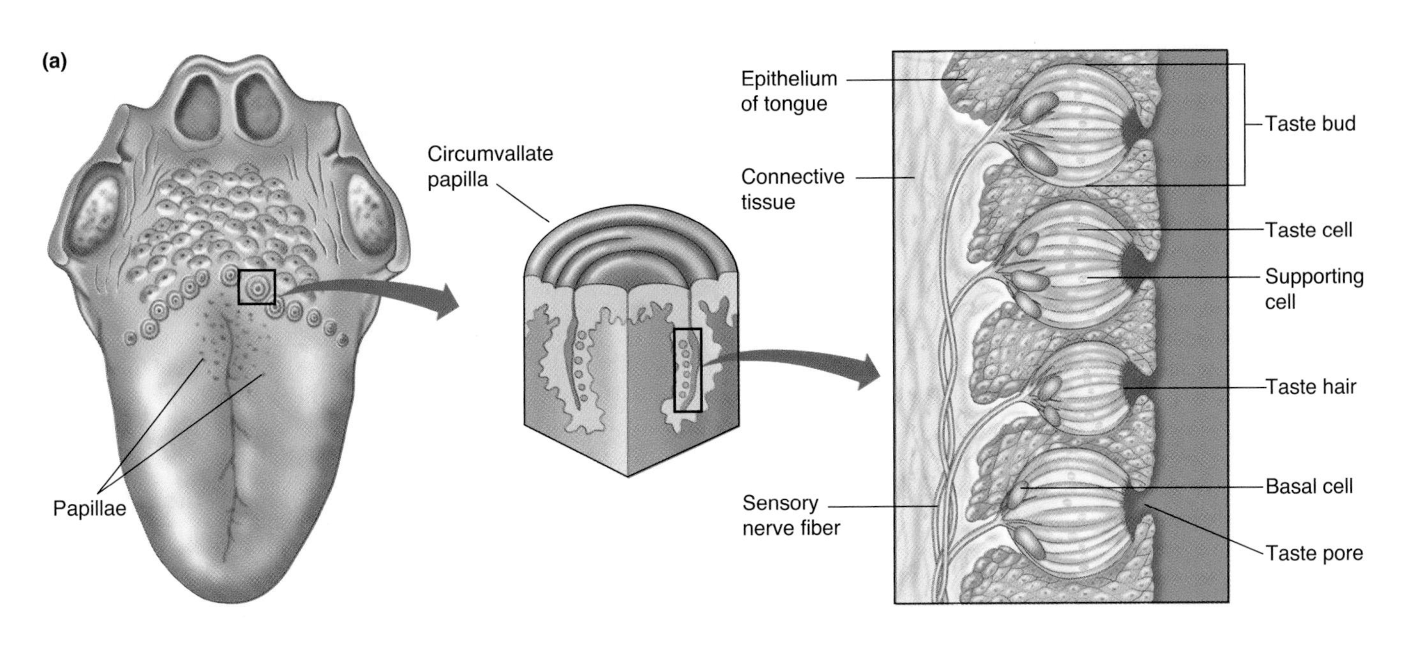

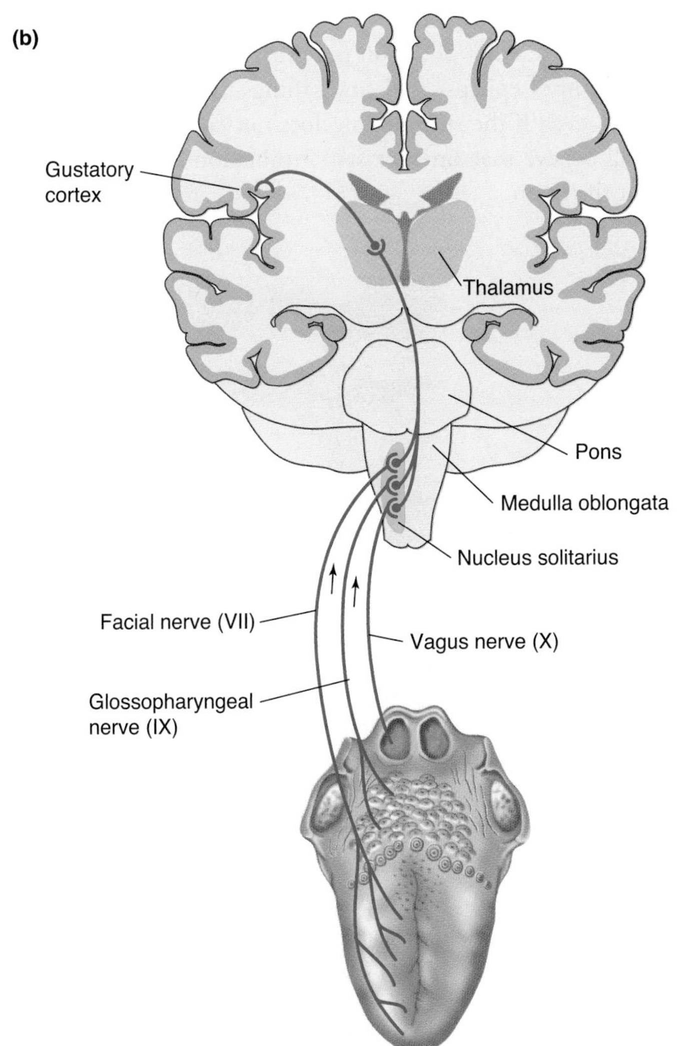

FIGURE 5.12 Taste buds contain receptor cells that react to specific ions within the food we eat. Action potentials from these receptor cells travel to the thalamus via sensory portions of the cranial nerves and then to the gustatory cortex.

begin to enjoy foods with a bitter taste. We also lose our sensitivity to salt with age and can tolerate much saltier foods than we could as children. Despite what we know about the individual taste receptors, it is unclear how these five tastes can encode the subtleties of taste we enjoy even at a simple meal.

Olfaction

The olfactory epithelium, located within the nose, contains more than a thousand different G-protein coupled receptors that react to specific odorants. The signals are transduced through cyclic adenosine monophosphate (cAMP) and adenylate cyclase, which open Na^+ channels and begin the action potential that will travel along the olfactory nerve to the olfactory cortex (**FIGURE 5.13**). Collateral fibers also travel to the limbic system, which is the primary way that smells influence our mood.

Notice that the olfactory nerves must travel from the nose through the cribriform plate of the skull to reach the brain. The opening for these nerves is small, and head injuries that damage the cribriform plate may damage the olfactory nerve, causing a loss of smell. While not life-threatening, the loss of the sense of smell can seriously affect appreciation of food and contribute to depression.

The olfactory cortex lies within the temporal lobe of the brain. In instances of temporal lobe epilepsy, there is an inappropriate burst of action potentials in the temporal lobe of the brain. People suffering this form of epilepsy will smell things that are not present. If the olfactory cortex is stimulated, even if the stimulation does not come from odorant proteins at the nose, the brain will believe that smell exists. Stimulation anywhere along the pathway has the same effect at the brain.

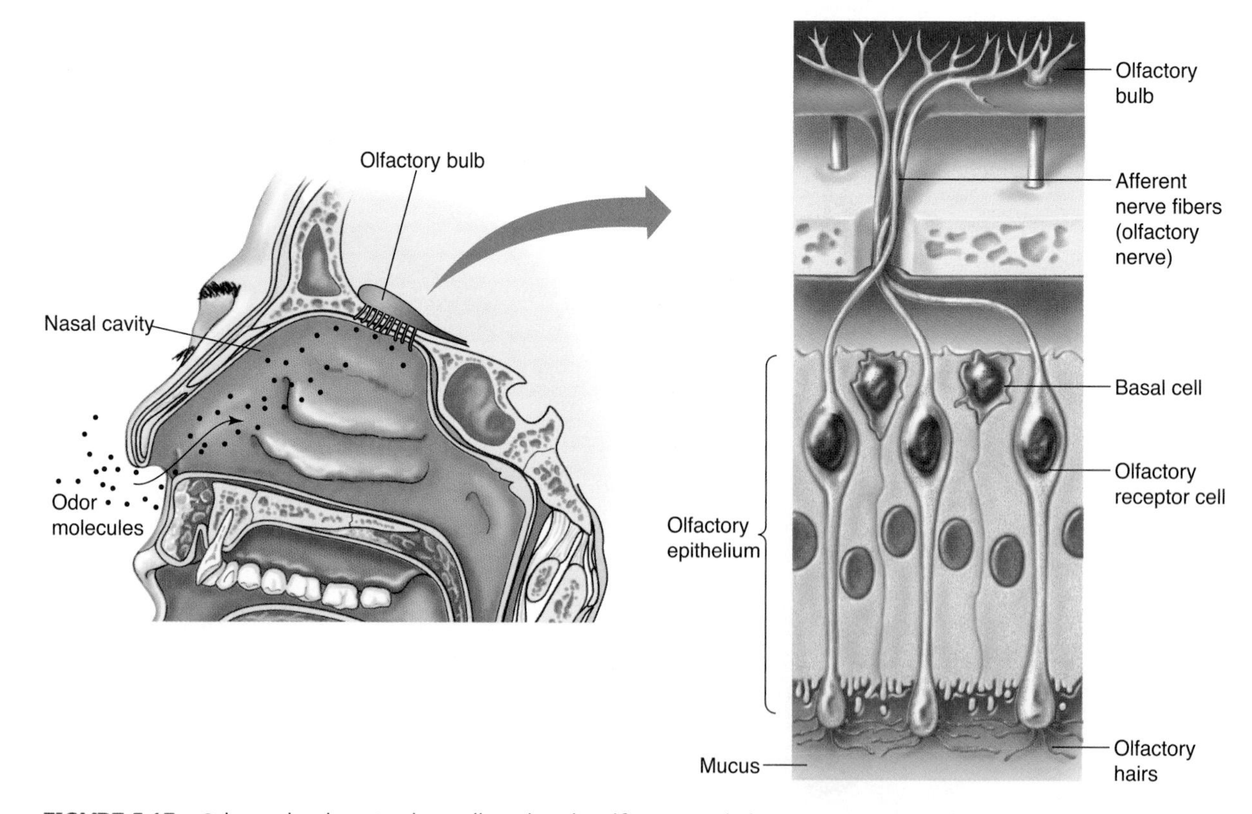

FIGURE 5.13 Odor molecules stimulate cells within the olfactory epithelium. Action potentials travel through the olfactory bulb directly to the olfactory cortex of the cerebrum.

Vision

The visual sensory mechanism and pathway is more complex and the visual cortex occupies a larger portion of our brain than the other senses. Unlike taste or smell, the receptor of the eye doesn't react to a molecule, but to a wave of light. How can a wave of light be converted into an action potential?

If we follow the path of light as it enters the eye, we can trace this fascinating process. Light waves pass through the cornea, through the pupil, and then the lens, which will focus the light waves onto the retina. The lens is made of crystalline proteins, making the lens clear and compliant. With age, these crystalline proteins cross-link, making the lens stiffer and more opaque. Stiffness in the lens causes presbyopia in older adults, and opaqueness causes cataracts in some elders. In the young, with normal eyesight, the rays will end at the retina, where the receptors are located, in the rod and cone cells (**FIGURE 5.14**).

Rod cells contain disc of rhodopsin, a pigment derived from vitamin A. When light hits rhodopsin, it causes the 11-*cis* conformation to switch to an 11-*trans* conformation. Light energy is responsible for this conformational change in rhodopsin. The 11-*trans* rhodopsin then activates a G-protein that stimulates a phosphodiesterase that converts cyclic guanosine monophosphate (cGMP) into 5′ GMP. Na^+ channels, dependent upon cGMP to remain open, close. This hyperpolarizes the membrane and less neurotransmitter is released (**FIGURE 5.15**). So light interrupts a current that normally flows in the dark! This is a variation of the action potential scenario we've seen to this point, but it works in a similar way, with the signal creating a change of action potential generation. Rhodopsin must be enzymatically reconverted to 11-*cis* rhodopsin before it can be used again. Rod cells are very sensitive to light and will react to a single photon of light. Cones work in a similar fashion but are color-specific and much less sensitive to light, requiring hundreds of photons of light before they can send a signal. This is why, in low light conditions, colors are less vibrant.

You will notice that the rods and cones connect to layers of bipolar cells, horizontal cells, and amacrine cells. These cells are processors that will convert raw light into images

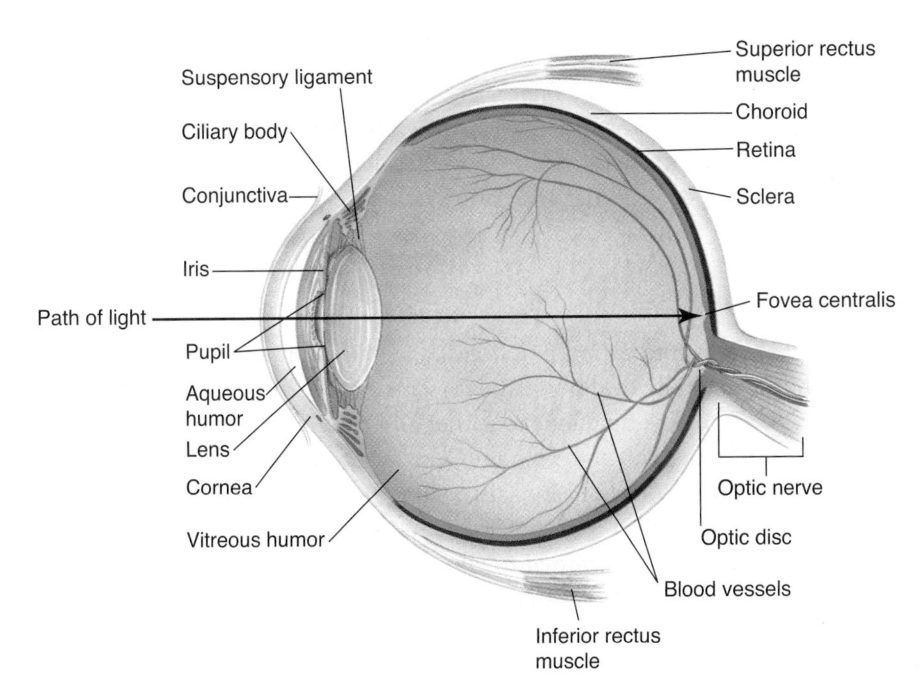

FIGURE 5.14 The path of light to the retina.

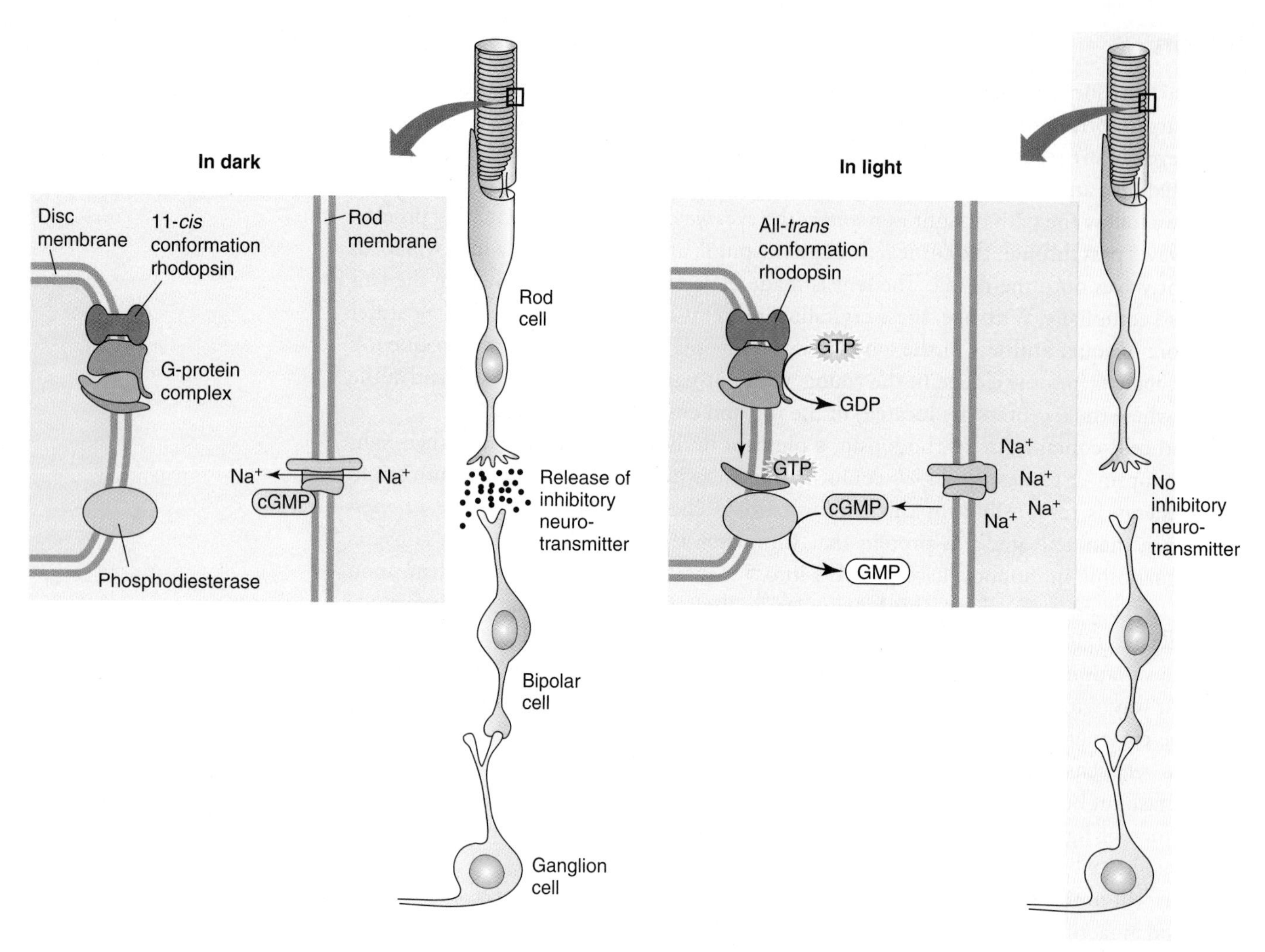

FIGURE 5.15 Light converts *cis*-retinal to *trans*-retinal, starting the G-protein-coupled activation of phosphodiesterase that will close the cGMP regulated Na^+ channel. The resulting hyperpolarization allows the bipolar cells to release neurotransmitters, and light is seen.

based on on-off (light-dark) patterns of light (**FIGURE 5.16**). Once processed, the action potentials travel through the optic nerve to the optic chiasma and back to the visual cortex. The primary visual cortex and the visual association cortex take up most of the occipital lobe of the brain (**FIGURE 5.17**).

Hearing

The eye converts light waves into action potentials, and the ear must do the same with sound waves. Again, the anatomy of the system is essential for its function. The pinna of the ear, the part we see externally, functions to gather and direct the sound waves toward internal structures. Sound waves hit the tympanum, or eardrum, and cause it to vibrate; thus, sound waves are initially converted into tissue vibration. In the middle ear, just on the other side of the tympanum, lie the auditory ossicles—the hammer, anvil, and stapes. Vibrations from the eardrum are transmitted to vibrations within bone. Vibrations of the ossicles end at the oval window, another tissue-covered opening that vibrates. Finally, these tissue vibrations are transferred to the cochlear fluid that fills the cochlea, where they are waves within the cochlear fluid. Sound waves in air have been transduced into fluid waves in cochlear fluid (**FIGURE 5.18**). How does this cause an action potential?

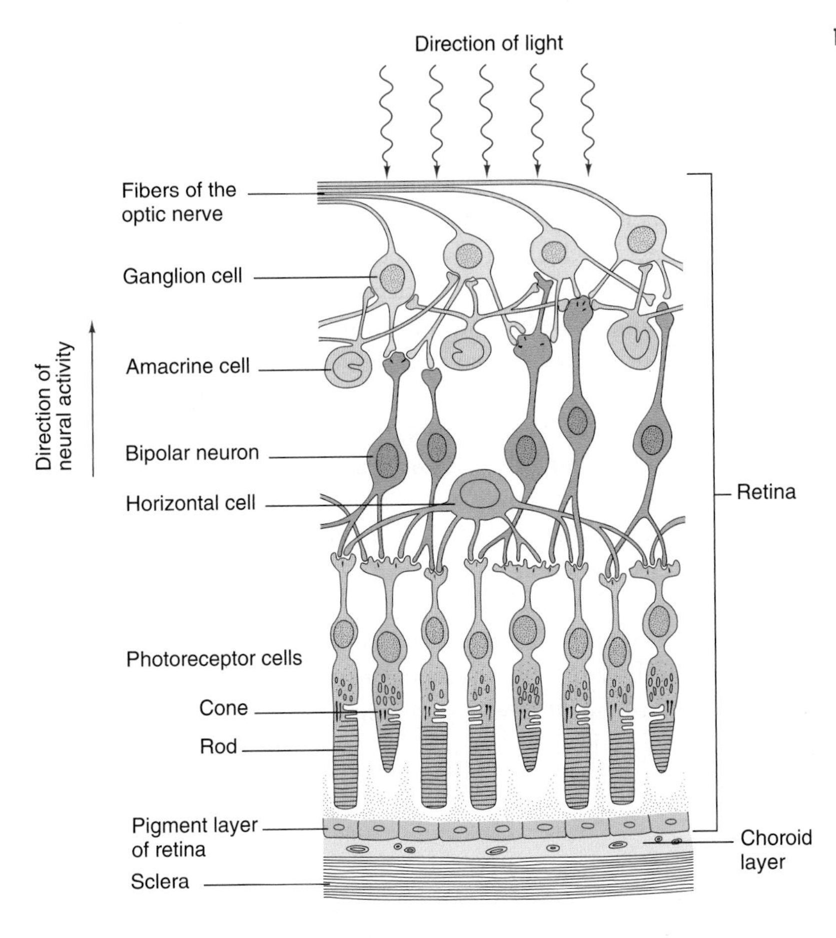

FIGURE 5.16 Cells of the retina.

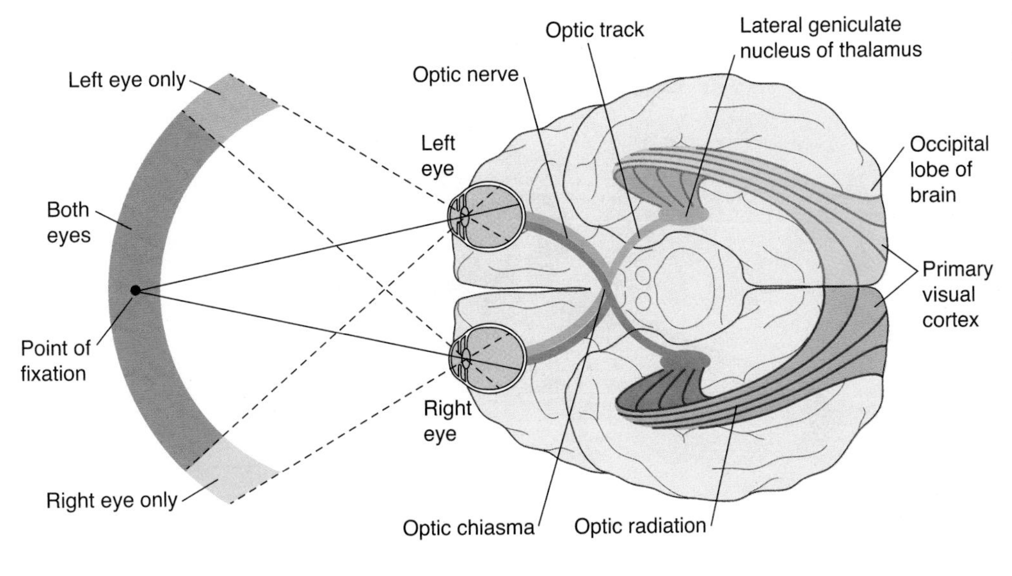

FIGURE 5.17 Pathway of the optic nerve to the visual cortex.

Within the cochlear duct, the organ of Corti lies embedded in the basilar membrane. The organ of Corti consists of the basilar membrane, some supporting cells, but most importantly, hair cells. Hair cells are positioned in such a way that the stereocilia projecting from the surface of the hair cells are embedded in the tectorial membrane. When a wave passes through the cochlear fluid, it moves the basilar membrane so that the stereocilia of the hair cells are bent by the tectorial membrane. Bending of the stereocilia in one direction causes depolarization by closing K^+ ion channels—a depolarization mechanism similar to the one we see in pancreatic β cells. Bending of stereocilia in the opposite direction causes hyperpolarization by opening K^+ and allowing positive charges to leave the hair cell. Depolarization of the hair cell membrane opens voltage-gated Ca^{2+} channels and causes neurotransmitter release (**FIGURE 5.19**). Hyperpolarization inhibits neurotransmitter release. Neurotransmitters begin the action potential in nervelets of the cochlear branch of the vestibulocochlear nerve, which travels to the cochlear nucleus of the medulla and finally to the auditory cortex. Unlike the pain or touch pathways, there is no organized pattern of neuronal crossing, so sounds we hear are a mixture of impulses from both ears. Like the sensory cortex, the auditory cortex is organized, in this case, by tone. The adjacent auditory association cortex helps us make sense of the raw sounds.

Hair cells are the actual receptors of the auditory system, and your hearing is dependent upon the health of these cells. Hair cells are not generally replaced, so if they are damaged by excessively loud sounds, for example, you simply lose those tones. The basilar membrane is not uniformly compliant along its length. At the first part of the cochlea, the membrane is stiffest, and only high-pitched sounds can vibrate this portion of the basilar membrane. At the apex, or end of the cochlea, the basilar membrane has more mobility, it is more flexible or compliant, and can respond to low-frequency vibrations. Thus, the sounds we hear are detected by hair cells at particular locations within the cochlea. Typically, hair cells at the base of the cochlea, where high pitches are heard, are the first to fail in older age (Figure 5.18). Thus, hearing loss may not be a diminution of all sound, but an inability to hear some sounds.

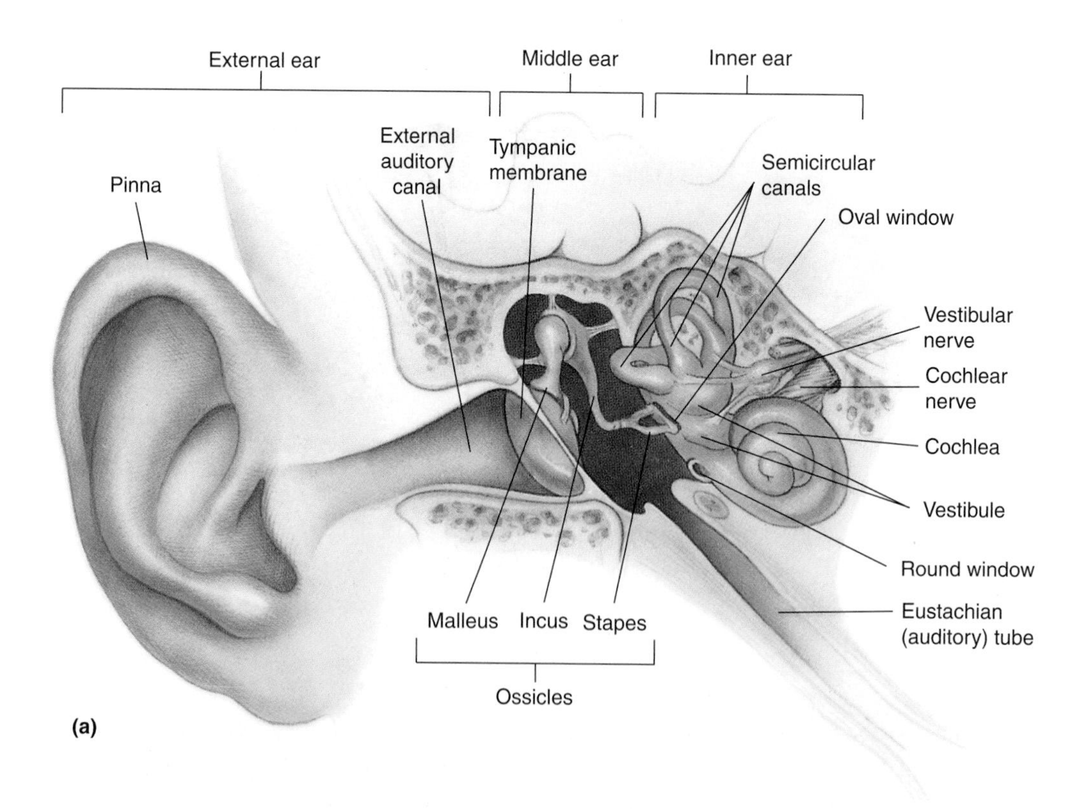

FIGURE 5.18 The anatomy of the ear.

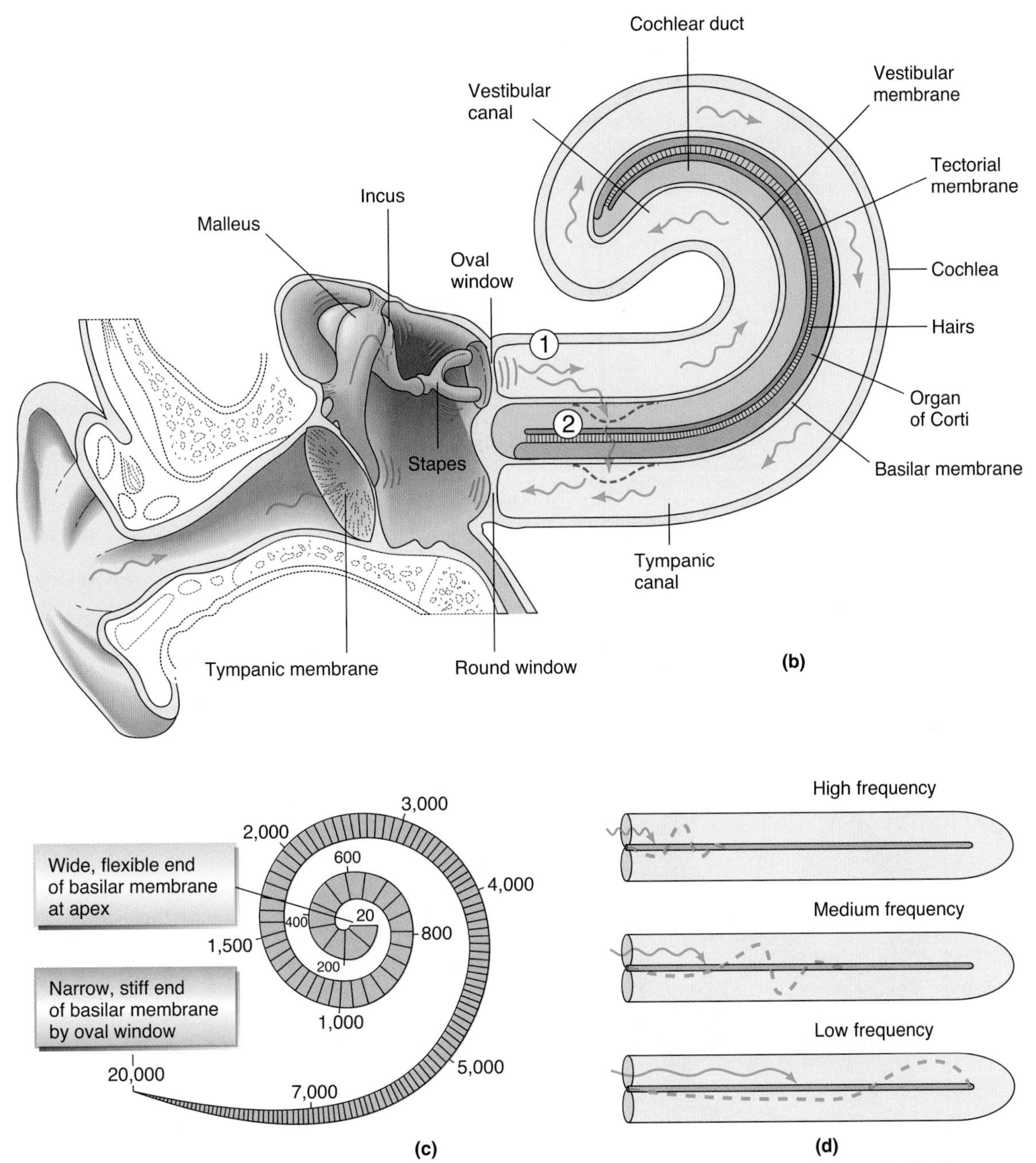

The numbers indicate the frequencies with which different regions of the basilar membrane maximally vibrate.

FIGURE 5.18 *(continued)*

The Vestibular System (Balance)

The vestibular system that controls our sense of balance lies very close to the cochlea. Like the cochlea, it consists of fluid-filled canals and chambers. This sensory organ responds to our change in position and helps us recognize where we are in space. Each of the semicircular canals is in a different plane, so that the X, Y, and Z planes are all represented. Movement of the head will cause the fluid to flow, and in a mechanism similar to that of the ear, it will cause hair cells within the ampulla to move their stereocilia and

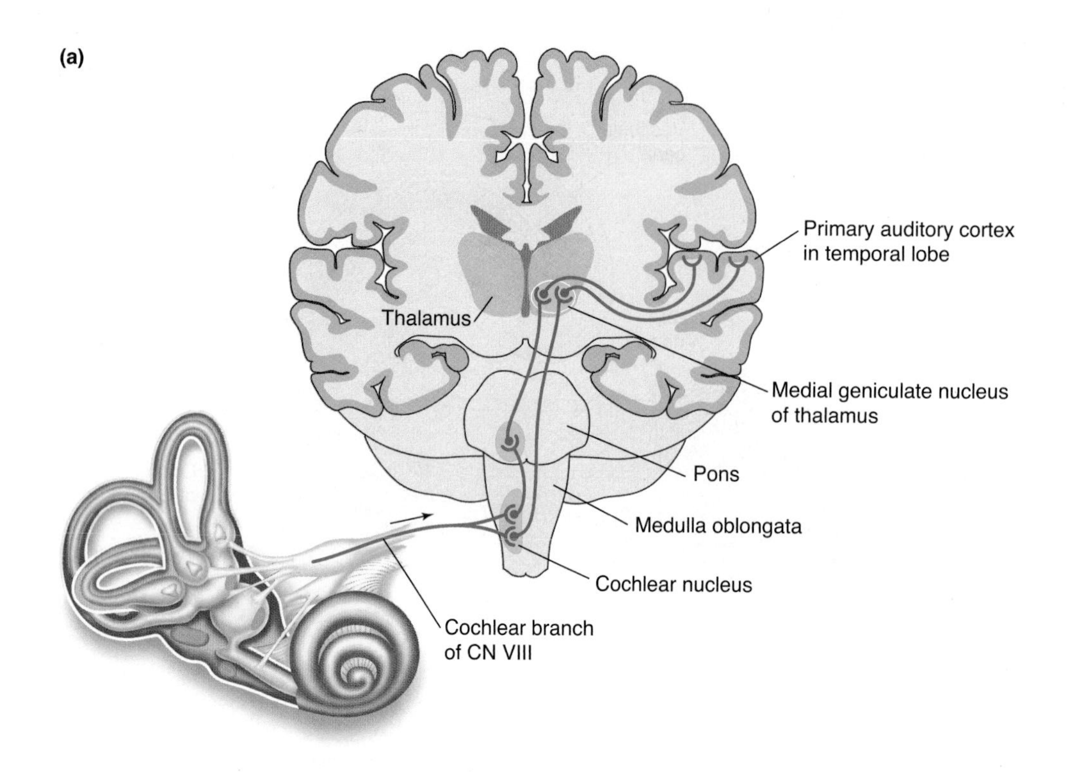

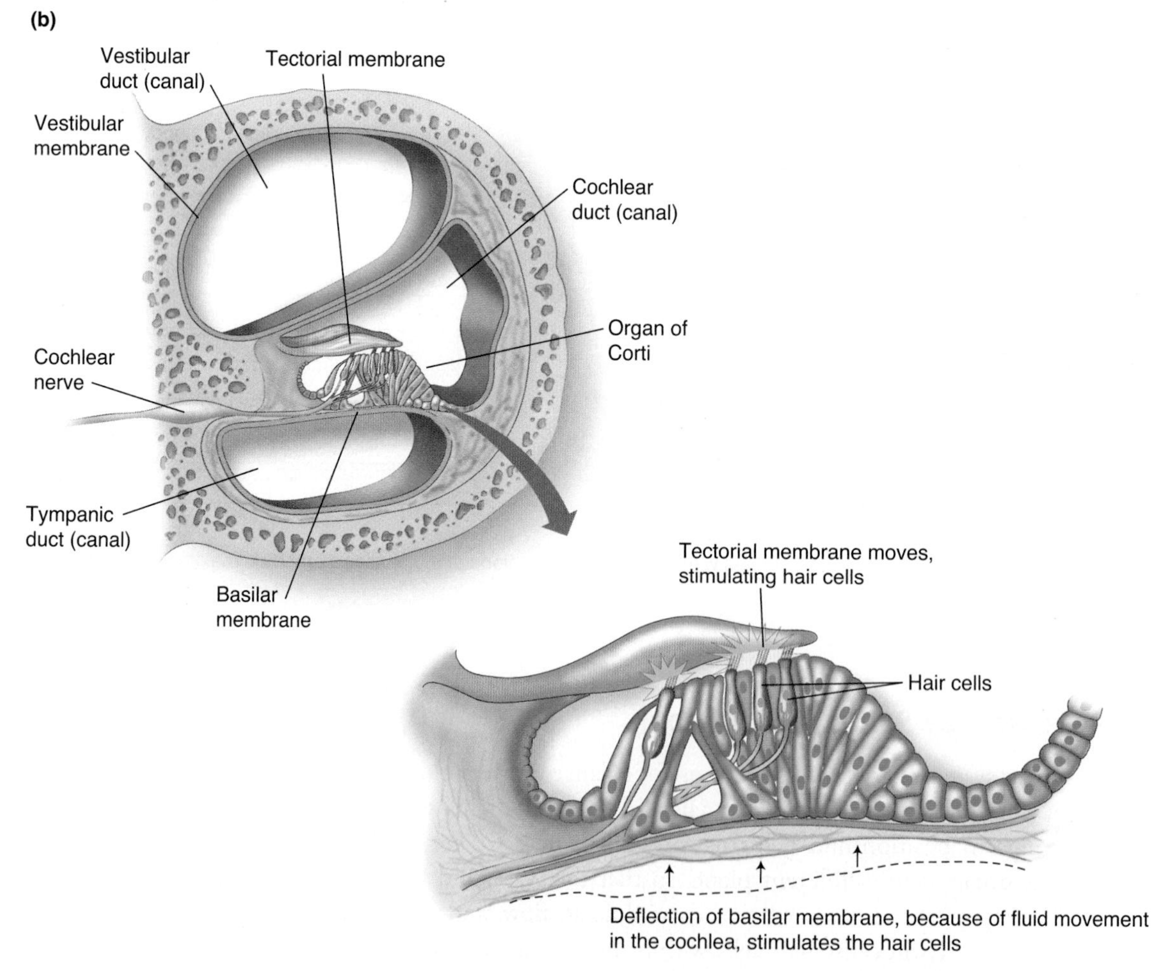

FIGURE 5.19 The mechanism of sound transduction. (a) Hair cells depolarize when they are bent by the tectorial membrane. This causes a depolarization that travels along the cochlear nerve to the auditory cortex. (b) Nerve pathway from the cochlea to the auditory cortex.

(a)

(b)

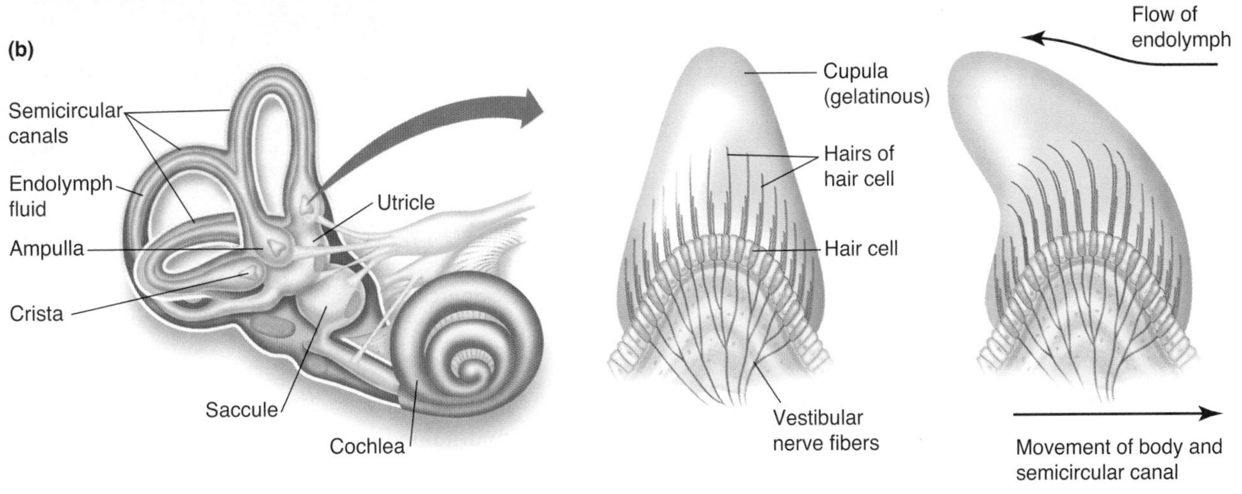

FIGURE 5.20 (a) The vestibular system helps us maintain balance as we move through space. (© Jiang Dao Hua/ShutterStock, Inc.) (b) The ampullae react to change in position by the movement of hair cells.

cause a depolarization or hyperpolarization. Once again, depolarization causes neurotransmitter release and generation of an action potential within the vestibular nerve. When you are still, this system is inactive, but as you walk along the beach, turn your head toward the sea, or glance around at a soaring seabird, the vestibular system will react to the motion (**FIGURE 5.20**).

There is another form of motion: linear acceleration or deceleration. When you are in a car and step on the gas, you can sense the change in velocity because of the otolith organs located within the utricle and saccule. Calcium carbonate crystals sitting on a bed of gel are the mechanism for triggering the hair cell receptors in this system. As the car accelerates, the crystals slide backwards on the gel bed, stimulating hair cells, which depolarize and release neurotransmitter. Nervelets carry the action potential along the vestibular branch of the vestibulocochlear nerve to the vestibular nucleus in the medulla. Receptors in the utricle and saccule react to any change in head position that is due to gravity, so these receptors will also be stimulated as you look down at the sand to watch the patterns made by incoming waves. Movement of the otolith crystals is all it takes to stimulate this

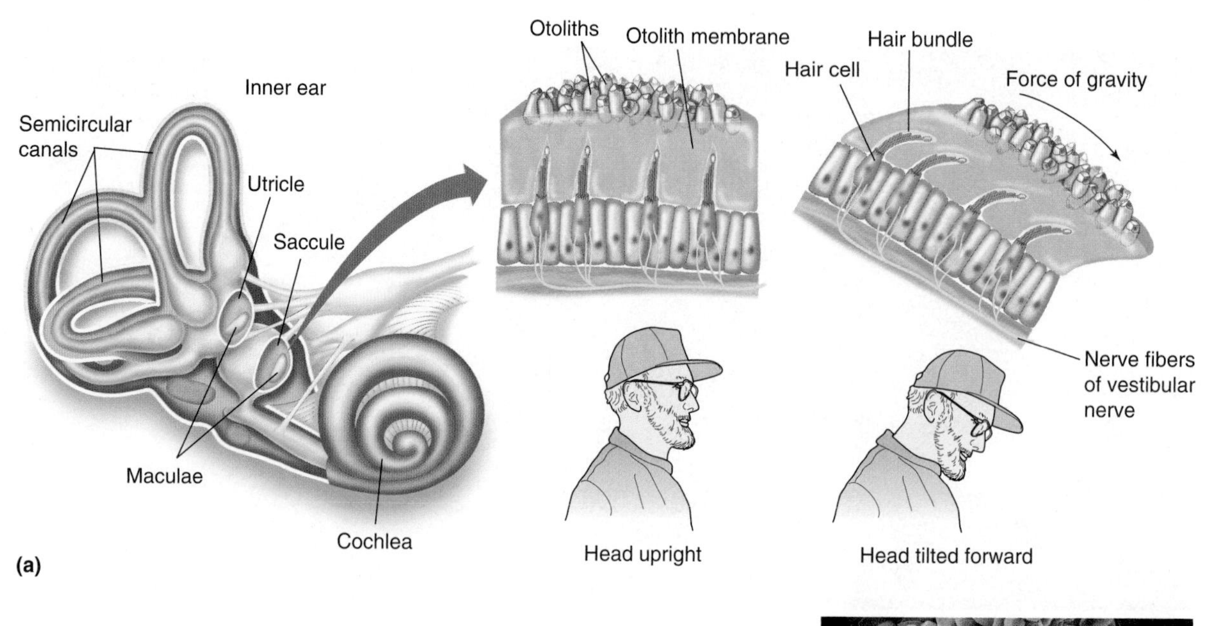

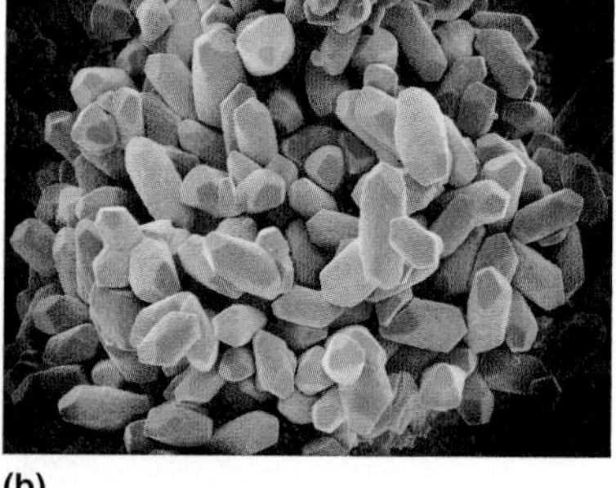

FIGURE 5.21 (a) The maculae of the utricle and saccule react to gravity, identifying changes in body position and acceleration. (b) Otoliths are crystals of calcium carbonate. (© Susumu Nishinaga/Science Source.)

sensory system (**FIGURE 5.21**). Why is this sense important? The movement of your head will affect your balance, which is generally maintained by postural muscles in the rest of the body. Motion sensed by both types of sensory cells of the vestibular system send their action potentials to the vestibular nucleus, which in turn stimulates areas of the brain and spinal cord that control muscle contraction. Input from the vestibular system allows us to maintain our balance (**FIGURE 5.22**). Notice that unlike any other sensory system pathway, this one ends at the vestibular nucleus and does not go to the cerebral cortex.

What Neural Pathways Allow Us to Respond to the Outside World?

We have a host of senses that allow us to gather information about the outside world. Now, how do we respond to that information? Often, our response to a stimulus is motion, so now we will explore how muscles are stimulated to contract, and by which pathways that response occurs.

Reflexes

Motion is mediated by several levels within the central nervous system—and the simplest of these motion responses is the reflex. The simplest of these is a reflex that requires only one synapse, or the monosynaptic reflex. You've undoubtedly experienced this reflex during a physical exam. The patellar reflex is stimulated by tapping the patellar tendon at the

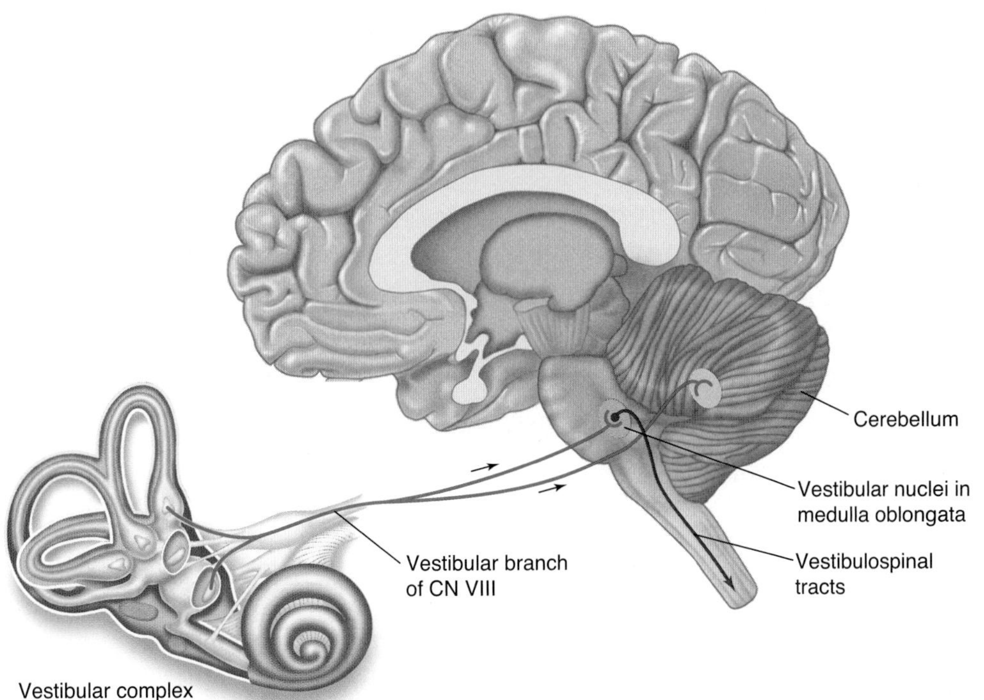

FIGURE 5.22 The vestibulospinal tract receives input from the vestibular apparatus. This tract coordinates your muscle tension in response to changes in body position.

knee. This causes the stretch of a muscle spindle in the muscles of the thigh. The sensory, afferent stimulus travels to the spinal cord of the appropriate section, enters through the dorsal horn, and synapses with the motor neuron in the ventral root of the spinal cord. This alpha motor neuron contracts the muscle of the thigh, causing the knee to extend (**FIGURE 5.23**). All of this occurs without any command from the brain. It is entirely a spinal cord reflex.

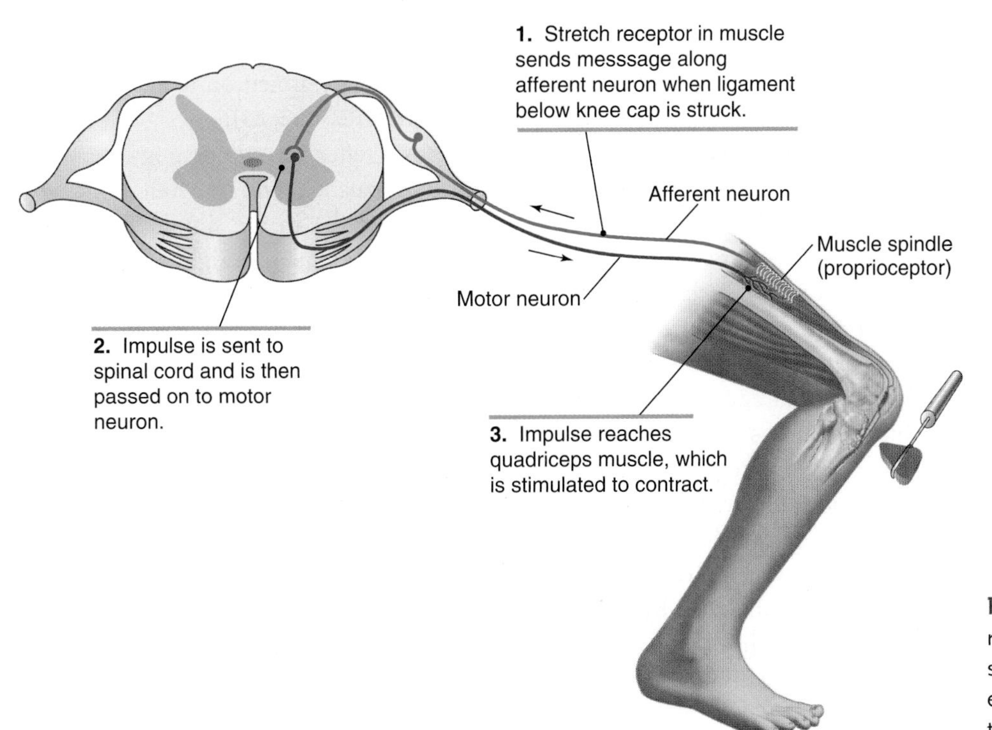

FIGURE 5.23 Even the simple monosynaptic pathway has multiple synaptic bulbs off the sensory nerve endings that will coordinate the input to several motor neurons.

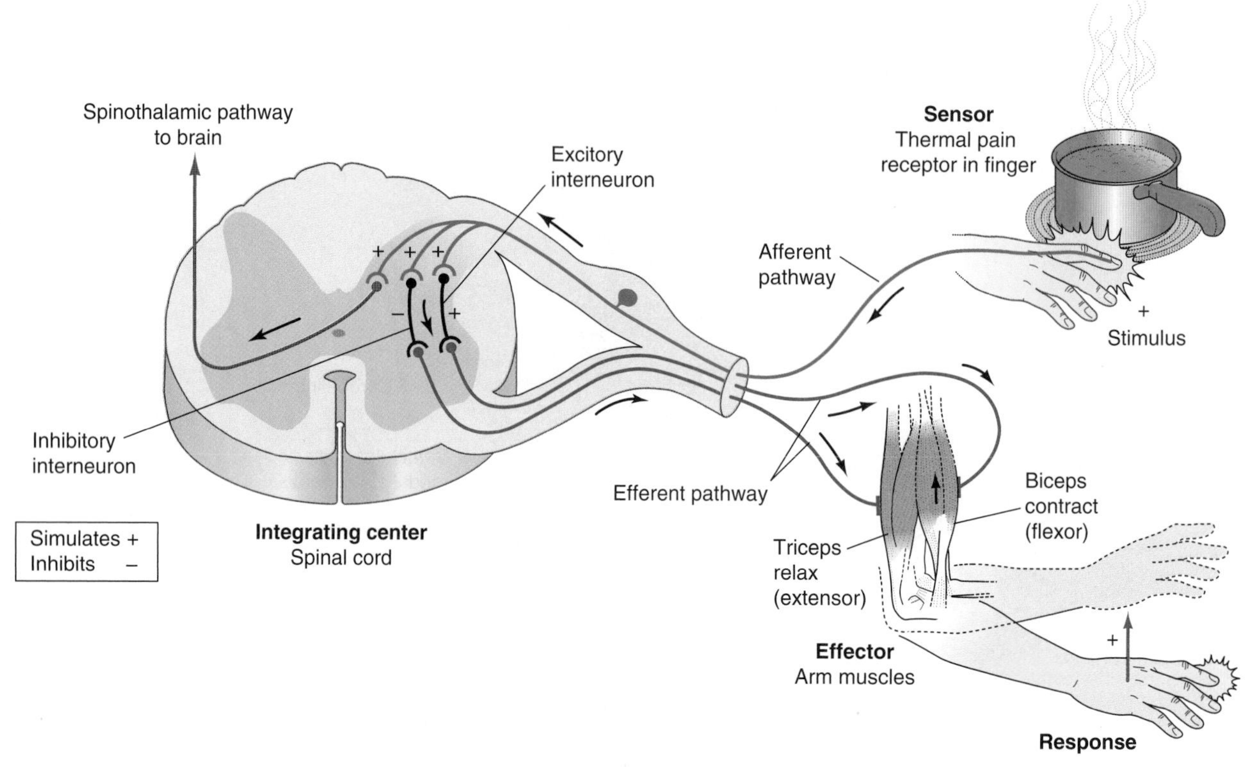

FIGURE 5.24 A simple withdrawal reflex occurs within the spinal cord, with collateral communication to the brain.

You have also probably experienced a simple withdrawal reflex. Imagine you are reaching into the microwave to retrieve the cup of cocoa you just made, but before you grip the cup handle, you withdraw in pain because of the heat of the handle. Like a monosynaptic reflex, this reflex arc has a sensory pathway that goes from the fingers to the spinal cord, via the dorsal root, where it synapses in the dorsal horn of the spinal cord with an interneuron. The interneuron functions to split the signal, sending an action potential to the motor neuron for withdrawal of the arm, and also sending a collateral neuron to the spinothalamic pathway, so you can be consciously aware of pain (**FIGURE 5.24**).

The last reflex we will consider is the crossed extensor reflex. In our opening case, when you stepped on a sharp object and withdrew your foot, balancing on one foot alone, you experienced a crossed extensor reflex. The pathway is similar to the simple withdrawal reflex, but it must also involve the opposite leg, so that you remain upright. The pathway goes as follows: (1) pain receptors in the foot stimulate an afferent neuron that ascends to the spinal cord and synapses with an interneuron within the dorsal horn of the cord; (2) the interneuron synapses on the α-motor neuron in the ventral horn and begins an action potential that ends in the flexor muscle of the thigh, and you withdraw your leg; (3) inhibitory neurons are also stimulated, allowing the extensor muscles of the thigh to relax; (4) the interneuron also crosses over to the opposite side of the cord and causes the inverse actions of the flexor and extensor muscles on the opposite side. This series of action potentials causes the extensor muscles to be activated on the unaffected side and the flexor muscles to relax (**FIGURE 5.25**). Thus, you are standing on one foot. Again, all of this motion is commanded by the spinal cord, without input from the cerebral cortex.

There is a related but separate spinal mechanism that allows us to walk without substantial input from the brain. Certain centers of the brain, parts of the basal ganglia and

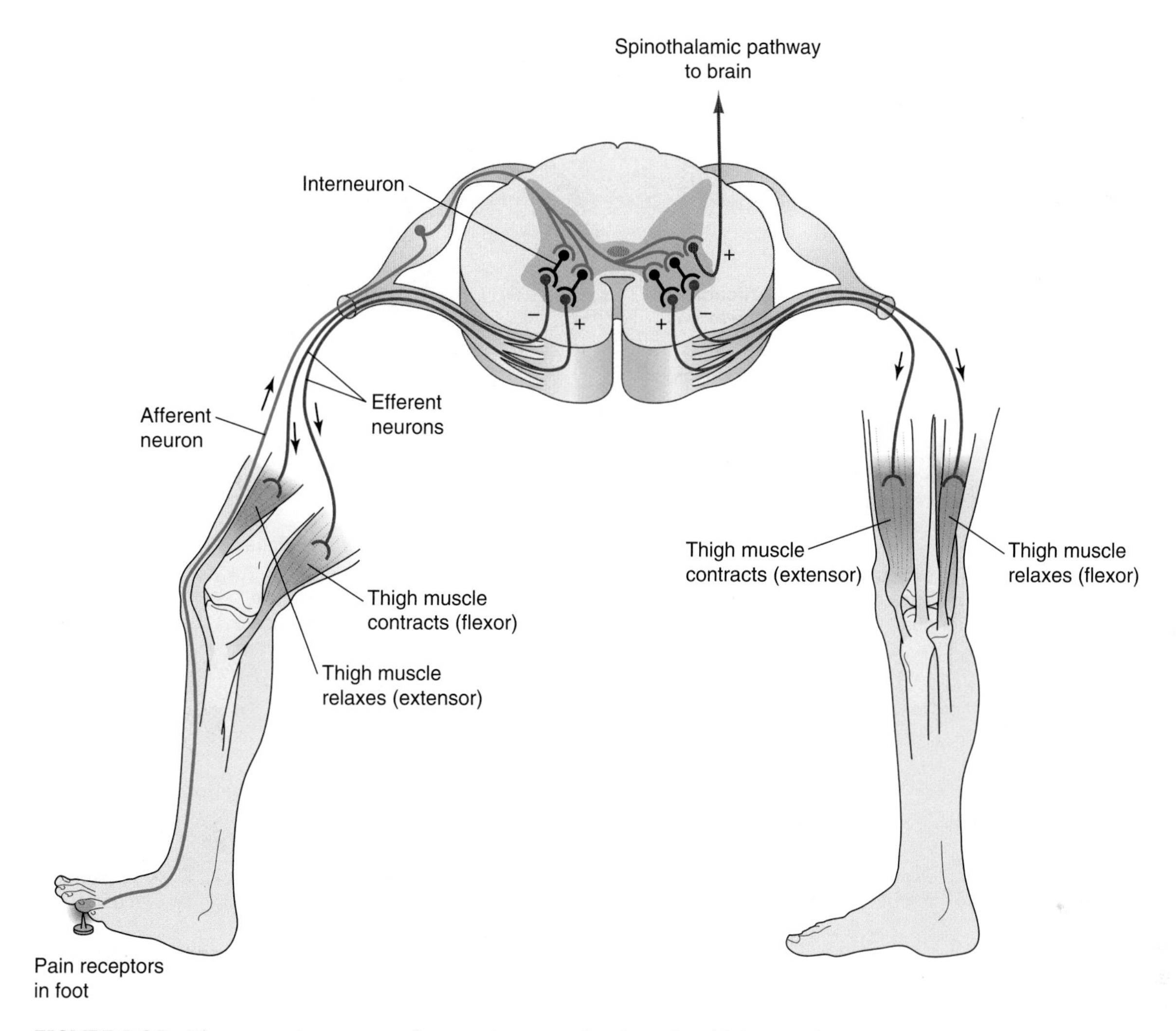

FIGURE 5.25 The crossed extensor reflex requires coordination of multiple muscle groups.

subthalamus, along with the motor cortex of the brain are important for initiation of voluntary movement. For example, when you decide to walk across the campus, the action potentials for that begin in the brain. However, once initiated, the movement of walking is spinally mediated. Clusters of neuronal circuits within the spinal cord, called pattern generators, create an oscillating circuit that alternately move our legs in a typical walking motion. While the brain begins the pattern, the spinal cord continues it without further input from the brain.

Voluntary Motion: Descending Pathways Command Motion

During your walk down the beach, your walking motion was coordinated and maintained by the pattern generators, but through which nerve pathways? Voluntary motion, beginning in the motor cortex, is mediated by action potentials in the corticospinal pathway, also sometimes called the pyramidal pathway. This is a simple two-neuron pathway. The upper motor neuron begins in the motor cortex located in the frontal lobe and ends in the ventral horn of the spinal cord. The lower motor neuron we are already familiar with as the α-motor neuron, which begins in the ventral horn of the spinal cord and ends at the neuromuscular junction on skeletal muscle (**FIGURE 5.26**). Your nonpatterned actions,

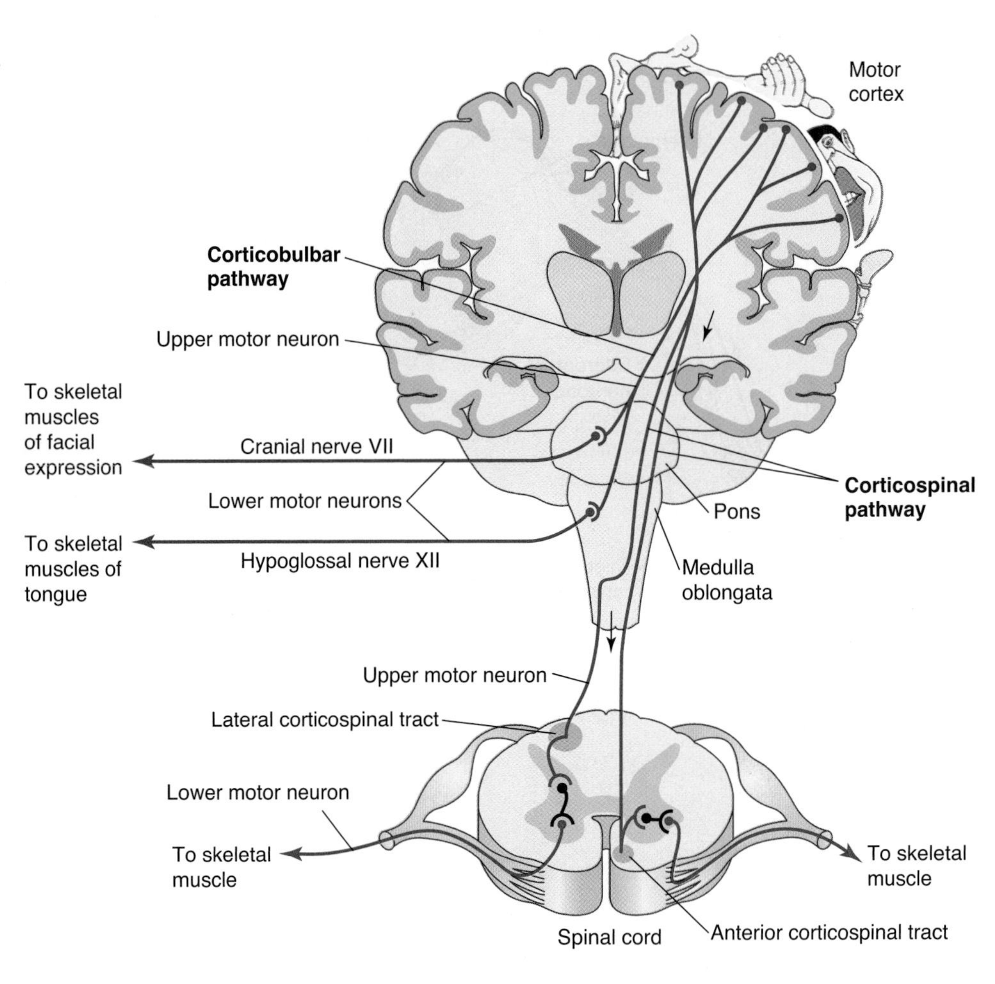

FIGURE 5.26 The corticospinal tract commands our voluntary motion. Other motor tracts beginning in the brainstem also mediate muscle action, but on an unconscious level.

such as hand gestures or stooping to pick up a stone from the beach and throwing it into the sea, are all voluntary skeletal muscle actions initiated by the motor cortex and accomplished by the corticospinal tract.

Above the spinal cord, the muscle contractions are mediated by a branch of the corticospinal tract called the corticobulbar tract. The corticobulbar tract ends at the cranial nerve nuclei that innervate the face and head. Thus, when you wrinkle your nose at the smell of a dead seabird on the beach, the motion is caused by cranial nerve VII, the facial nerve functioning as the lower motor neuron to the corticobulbar tract.

At the spinal cord, the upper motor neuron synapses with an interneuron, which in turn synapses with the lower motor neuron. What function is served by the interneuron? The corticospinal tract is not the only motor pathway we possess. We have several motor pathways that originate within nuclei of the midbrain. These motor pathways, called the lateral or medial pathways, carry motor signals to the lower motor neuron that are not consciously within our control. For example, the vestibular system, our balance organs, send input to the vestibular nucleus, where the sensory neurons synapse with motor neurons. Motor neurons of the vestibular nucleus travel to the spinal cord along the vestibulospinal tract and synapse with the interneuron and therefore the lower motor neuron (**FIGURE 5.27**). So, while you are balancing on one foot, the vestibular organs are sensing your changes of position and help to adjust your position via the motor input to the skeletal muscle. The lateral and medial motor pathways also go by the name of the extrapyramidal tracts, because they are outside of the pyramidal tract.

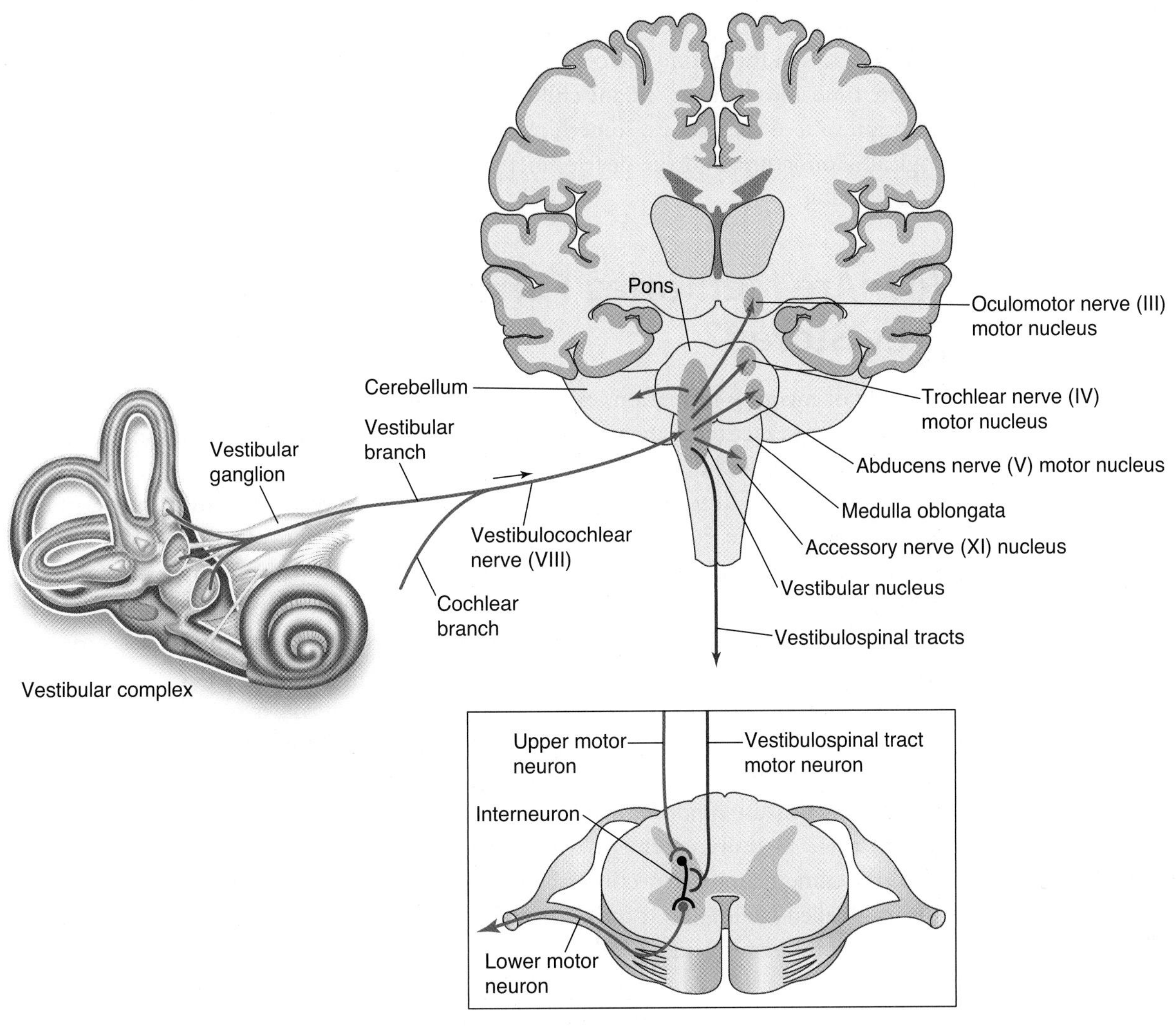

FIGURE 5.27 The interneuron between the upper motor neuron and the lower motor neuron sums the signals from multiple motor pathways.

Because the lower motor neuron receives input from many different upper motor neurons, the loss of one of the upper motor neurons does not result in complete paralysis. If the upper motor neuron of the corticospinal tract is severed, there will be no voluntary motion as initiated by the motor cortex, but there will be reflex action. Severing the lower motor neuron, however, results in complete paralysis because there is no input to skeletal muscle.

How Does the Cerebellum Influence Motion?

At the beginning of this chapter we mentioned that the cerebellum coordinated motion. What does this mean? Imagine that you are playing basketball. You run to the end of the court and stop under the basket to throw up a shot. How did you know when to decelerate your running so that you didn't run into the stands? How did you know how forcefully to throw the basketball so that it would reach the top of the basket? The calculations of time and distance are not functions we consciously perform; they are made by the cerebellum. The cerebellum does not directly send action potentials to the muscle but rather "informs" the motor cortex, the basal ganglia, and the thalamus of body position and motions required. The output from the cerebellum to the cerebrum is one of the largest nerve tracts in the brain. The cerebellum spends much of our childhood "learning" to calculate time and distance. Infant children are notorious for missing their mouth when they begin to feed themselves, something that rarely happens as adults. We are born neurologically immature, and the development of the cerebellum is evident when we watch the very young.

How Are Memories Formed, and Where Are They Stored?

The study of memory formation, essential to learning, is an ongoing one. While no one understands the exact mechanism of how we learn or remember, there are some pathways and events that are well described.

There are several different types of memory. There is very short-term memory, such as remembering a phone number just long enough to enter it into the phone. This type of memory is thought to exist within the sensory cortex of its modality. In the case of a phone number, it would be auditory (if someone is telling you the number) or visual (if you are seeing the number). In either case, very short-term memory is fleeting and lasts minutes at best.

Short-term memory lasts hours to days. As a student, you may utilize this form of memory for an exam! Memories that last longer than a few minutes all must be processed by the hippocampus. Even short-term memory requires hippocampal input. We know this primarily because hippocampal injury precludes even short-term memory formation. How are these memories formed? Memory formation requires repetitive action potentials of the same neural connections. This type of continuous action potentials between neurons is called facilitation if they last milliseconds, and potentiation if they last hours. A prolonged increase in intracellular Ca^{2+} in the presynaptic neuron, caused by multiple action potentials, is thought to be crucial to the memory formation process. Our practice of learning by repetition is repeated in the brain by repeated action potentials causing elevated Ca^{2+} levels (**FIGURE 5.28**).

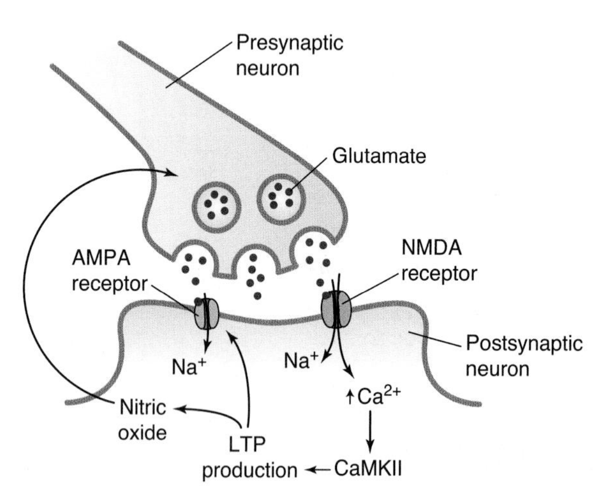

FIGURE 5.28 At the level of the neuron, learning occurs as glutamate binds to two different receptors, initiating signaling cascades as well as an action potential. A positive feedback loop causes the release of more glutamate, which potentiates the increase in Ca^{2+} in the postsynaptic neuron for a long period of time.

Long-term memory, which lasts for years or a lifetime (you remember your name for your entire life), is not stored in the hippocampus but in the cortex. Long-term memories are stored throughout the cortex in a nonrandom manner, and we retrieve memories from the cortex in an integrated way through myelinated nerve communication pathways: between hemispheres and within hemispheres. How memories are transferred from the hippocampus to the cortex is still unclear. We do know that synapses of long-term memories are unique. Neurons that encode a memory have a smaller synaptic space and are sometimes joined by gap junctions. In short, memories physically change the anatomy of the neurons at the synaptic space. Because long-term memories are stored in the cerebral cortex, damage to the hippocampus does not affect long-term memory, only the ability to create new memories.

How Is the Brain Protected from Injury?

Brain tissue is soft and is protected by the cranium. Metabolically, the brain uses a variety of neurotransmitters, including some amino acids like glutamate, so it must be protected from the contents of blood. How is this protection afforded? Both physical and metabolic protection is provided by the CSF produced by the cells of the choroid plexus. Let's look at the anatomy of the brain, cranium, and fluid production and how it serves to protect the brain.

The brain is covered with three layers of membrane. The first layer is called the pia; it is a layer of connective tissue, connected to neuronal glial cells. This membrane is so tightly adhered to the brain that it cannot be removed without damage to the brain itself. The next layer is the arachnoid membrane, in which the blood vessels that serve the brain are embedded. The arachnoid membrane also encloses the CSF. Finally, there is the dura, a tough connective tissue layer that lies closest to the bone of the cranium. The dura is itself comprised of two layers, with a venous sinus circulating between them (**FIGURE 5.29**).

The CSF is a unique fluid formed as a filtrate of blood, by the ependymal cells of the choroid plexus. The choroid plexus contains a cluster of leaky capillaries, located within the ventricles of the brain, which are surrounded by ependymal cells. The ependymal cells absorb water and ions from the exudate of the blood and release a dissimilar fluid into the space between the pia and arachnoid membranes. This fluid is closely controlled; it contains no proteins, few amino acids, and ion concentrations that differ from extracellular fluid. This fluid bathes the brain and allows the brain to "float" in a CSF sea. This physically protects the brain from physical impact. CSF is produced continually by the slight hydrostatic pressure provided by the capillaries of the choroid plexus. CSF removal occurs through arachnoid granulations, which release fluid into the dural sinuses, returning it to general venous circulation (Figure 5.29). Metabolic waste products from neurons also end up in CSF and are removed from the brain via CSF drainage.

Outside of the ventricles, blood vessels in the brain are surrounded by endothelial cells joined by tight junctions and completely covered by astrocyte feet (**FIGURE 5.30**). This is the blood-brain barrier that exists throughout the brain, except for a few isolated areas, such as the area of the hypothalamus that senses blood osmolarity. (We see this mechanism in the endocrine system as well, where an increase in blood osmolarity triggers the release of antidiuretic hormone from the posterior pituitary.) This is possible only if the receptor cells have access to the blood supply. However, the selected areas of the brain that are exposed to blood are few and small in area. Most of the neurons are protected by the blood-brain barrier.

If the feet of astrocytes cover brain capillaries, how do neurons get access to glucose and amino acids for making neurotransmitters? This anatomical arrangement makes it clear how important astrocytes are to neuronal function. Nutrient flow to neurons comes through astrocytes, not directly from capillaries. In this way, the neurons are protected from ions, toxins, or metabolites within the general circulation. Astrocytes are an important mediator of neuronal metabolism, because astrocytes metabolize glucose into lactic acid before exporting it to neurons. The brain is glucose-dependent, but in the brain, it is astrocytes that take up glucose first, processing it for the neighboring neurons.

Similarly, astrocytes can absorb glutamate, convert it to glutamine, and export it to neurons. Neurons will reconvert it to glutamate and use that as a neurotransmitter. Neurons cannot make glutamate themselves. Clearly, astrocyte function is essential to neuronal health.

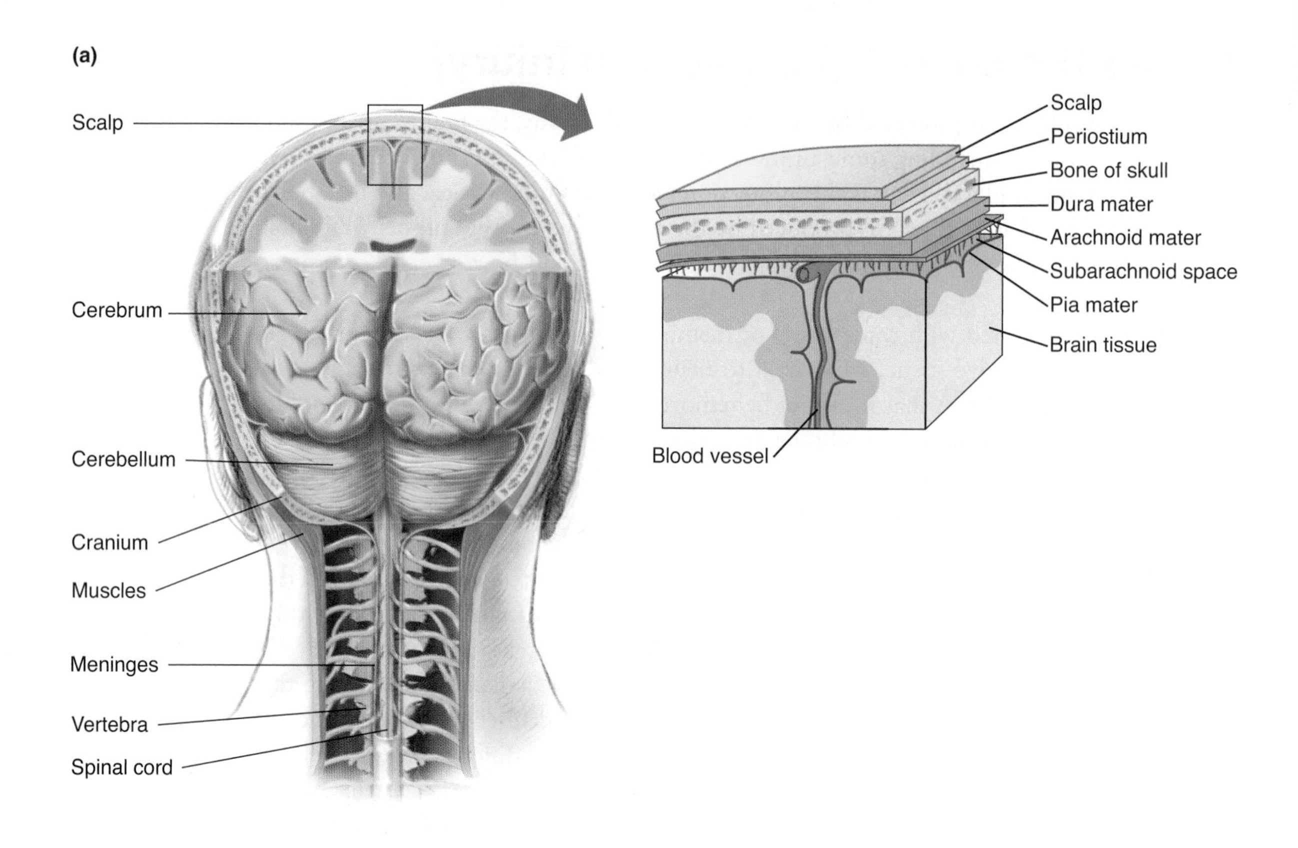

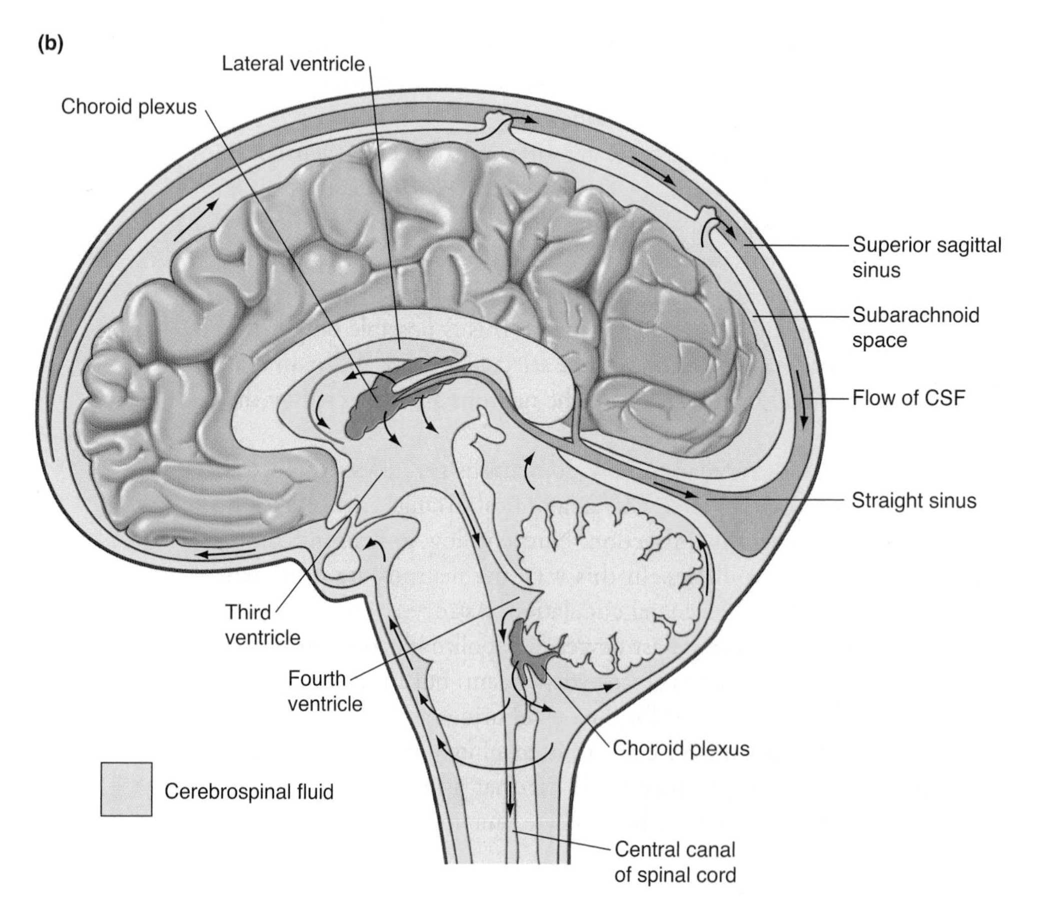

FIGURE 5.29 The arrangement of the dural layers (a) and the circulation of cerebrospinal fluid (b).

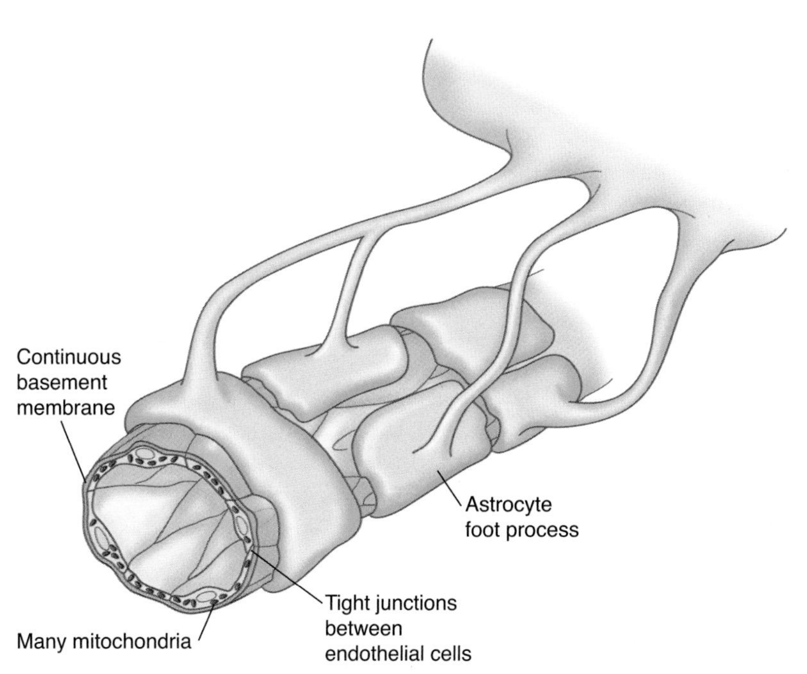

FIGURE 5.30 Astrocytic feet cover the blood vessels of the brain.

The blood-brain barrier is quite effective. In health, the blood-brain barrier insulates the neurotransmitter-sensitive neurons from extraneous chemicals. However, in disease (particularly in brain disease), delivering drugs to the brain is generally difficult because of this barrier.

Summary

The somatic nervous system is spectacularly complex, and we have looked at only the basic mechanisms that allow this system to function. However, you now know enough to understand how this system integrates with all the other organ systems.

Key Concepts

Sensory nerve pathways
Motor nerve pathways
Labeled line
Somatic skin sensory receptors
Dorsal column pathways
Anterolateral pathways
Spinocerebellar tracts
Corticospinal tract
Corticobulbar tract

Key Terms

Saltatory conduction
Blood-brain barrier
Cerebrum
Temporal lobe

Occipital lobe
Frontal lobe
Cerebellum
Graded potential
Spatial summation
Temporal summation
Acetylcholine
Norephinephrine
Serotonin
Organ of corti
Vestibular system
Choroid plexus
Arachnoid granulations
Cerebrospinal fluid (CSF)
Astrocytes

Application: Pharmacology

1. You are going to the dentist for a filling, and you are given Novocaine to deaden the pain. Novocaine blocks the voltage-gated Na^+ channel. Can you explain why that would prevent you from feeling pain?
2. Most drugs are unable to affect the brain because they cannot cross the blood-brain barrier. Describe this barrier and hypothesize why drugs have so much trouble crossing it.
3. Among its other effects, alcohol alters the functioning of the cerebellum. If a person were intoxicated, what symptoms would you associate with cerebellar dysfunction?

Clinical Case Study

Type 2 Diabetes Mellitus

BACKGROUND

One of the most common complications of diabetes mellitus is diabetic neuropathy. In this condition, nerves become damaged and fail to transmit a signal from sensory receptors in the skin to the brain. Damage is generally in the first-order neuron of the sensory pathway and may involve some demyelination. The exact mechanism of neuronal deterioration is unknown; however, it is well documented that diabetics lose sensation in the periphery, particularly in the lower limbs. Thus, diabetics tend to injure themselves without realizing it. This results in injuries, generally small injuries, going unnoticed. Untended injuries, which might heal spontaneously in a healthy individual, are much more likely to become infected in a person with diabetes, leading to ulceration and, if left untreated, gangrene.

Why are the injuries more likely to become infected? If you consider the immune system events of a cut finger, you will recognize that the response includes vasodilation and movement of cells and fluid out of the vasculature into the tissue. Diabetics with elevated circulating glucose will lose some glucose into the tissue, which can promote bacterial or fungal growth. This contributes to proliferation of the organisms, perhaps allowing them to grow faster than neutrophils and monocytes can eliminate them. Gangrenous wounds are the primary cause of amputations in diabetic patients. This is a clear example of how loss of a simple pain signal can have disastrous results.

THE CASE

Your aunt, recently diagnosed with Type 2 diabetes that is not yet well controlled, has been referred to a podiatrist for her foot care. To her tremendous dismay, this specialist has told your aunt that she can no longer trim her own toenails. Shocked by what she sees as an invasion of her personal cosmetic care, she immediately comes to you to complain. "Why can't I give myself a pedicure?" wails your aunt. Because you have become a physiology student, she is convinced that you know everything, and you try hard not to disappoint her.

THE QUESTIONS

1. What are the hazards of your aunt cutting her toenails?
2. What sensory pathways might be affected?
3. Why isn't she restricted from cutting her fingernails?

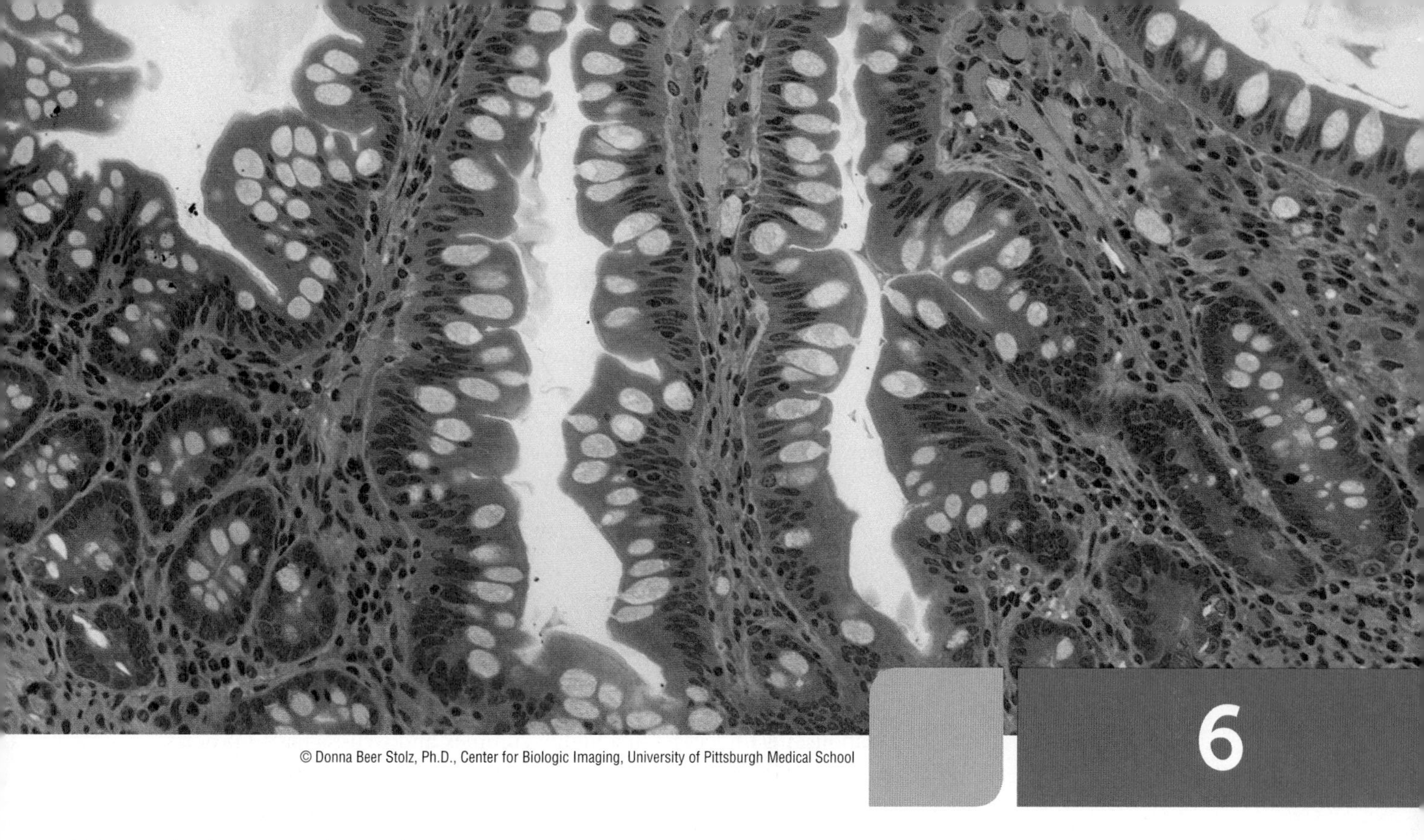
© Donna Beer Stolz, Ph.D., Center for Biologic Imaging, University of Pittsburgh Medical School

6

The Digestive System

Case 1

It is Thanksgiving, and you have been anxiously awaiting dinner, which smells wonderful. In fact, it *is* wonderful; you are enjoying turkey, mashed potatoes, candied yams, peas and onions, and cranberry sauce. You are just reaching for seconds when Uncle Ed begins to argue politics with your father. The discussion gets increasingly heated; you become angry, both about the issues and the fact that yet another family dinner is being ruined. The food sticks in your mouth and feels like lead in your stomach. Describe the digestive process, taking into account the stages of digestion, the hormones of digestion, and the processing of each of these food types. What is happening to cause your mouth to go dry and to cause your stomachache? Give hormonal and cellular mechanisms.

Introduction

Digestion seems like a familiar system because we all eat several times each day and have done so for our entire lives. Physiologically, it is a marvel of neural and hormonal regulation, all of which occurs without our conscious input. While we are casually enjoying food, its catabolism is underway in a highly complex and well-regulated process.

We must ingest food to live, and the digestive system has the daunting task of reducing turkey muscle and all the plants you consumed in the course of this dinner into their component parts: amino acids, monosaccharides, and fatty acids. Once in circulation, these building blocks can be used to make human skeletal muscle, epithelial cells, or any tissue, as well as adenosine triphosphate (ATP). As you might imagine, the path from turkey leg to human bicep is a long and complex one, yet by the time we finish, it is a path you will be able to describe!

The Beginning: What Happens in Your Mouth?

All digestion begins in the mouth, where we masticate food, physically breaking it down with our teeth and mixing it with saliva. Both parts of this process are essential. Our teeth reduce the size of food to be digested, increasing the surface area required for enzymatic catabolism. Saliva moistens the food, which will allow it to travel through the digestive system more easily. Saliva also contains enzymes: salivary amylase starts the hydrolysis of sugars, and salivary lipase begins the breakdown of fats.

Saliva is produced by three separate salivary glands within the mouth, but all are under control of the autonomic nervous system (**FIGURE 6.1**). The parasympathetic nervous system has the most profound effect on the salivary glands, stimulating the production of saliva, even before food enters our mouth. Acetylcholine, activating muscarinic receptors, causes an increase in inositol triphosphate (IP_3) and intracellular calcium concentration, which begins the signaling cascade that stimulates salivary production. The salivary glands make saliva as a filtrate of blood. Initially, saliva is isotonic to blood, but as the fluid moves through the duct, Na^+, K^+, Cl^-, and HCO_3^- are removed, making the final saliva that is released into the mouth hypotonic. The smell or sight of food can cause your mouth to "water," preparing the oral cavity for the reception of your first bite.

Transport to the Stomach

Swallowing is not something we must learn to do; it is controlled by a swallowing center in the medulla, which neurally controls the contraction of muscle and the movement of the epiglottis as it folds down and covers the tracheal opening, preventing food from entering

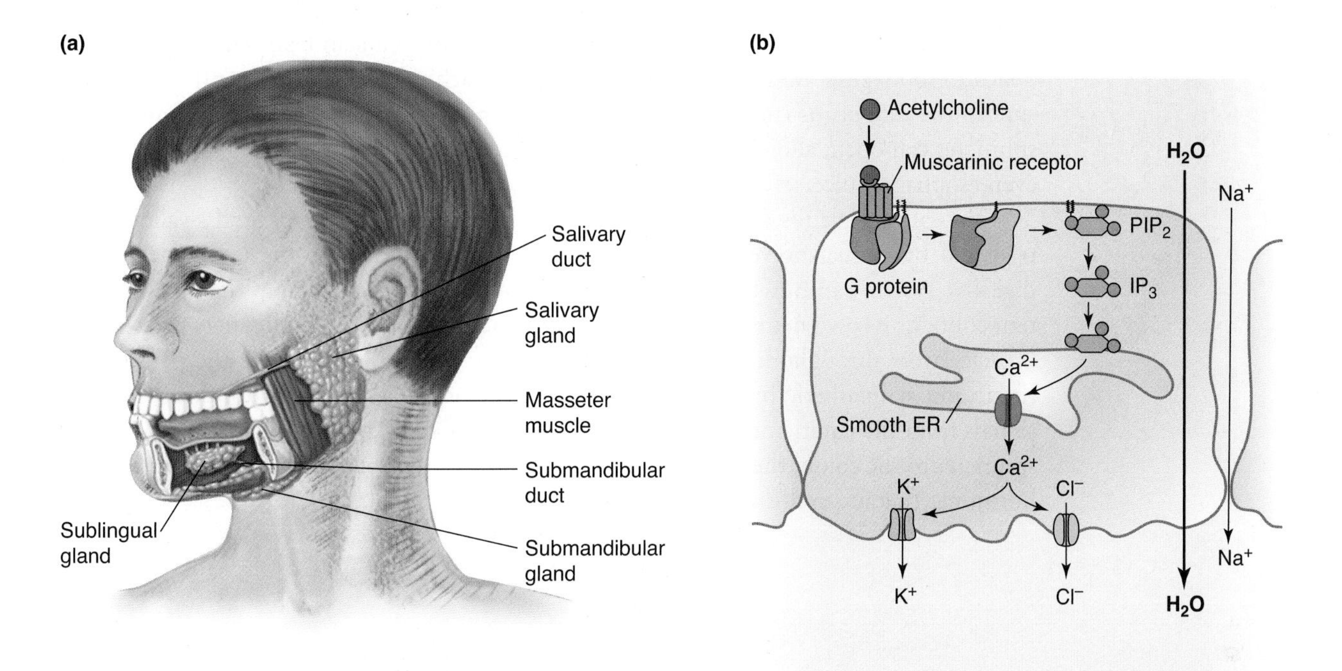

FIGURE 6.1 (a) The three salivary glands. (b) Initially, saliva produced is isotonic to blood (shown); but later within the duct, ions are reabsorbed (not shown) making saliva hypotonic to blood.

the lungs. Your mouthful of food, now reduced to a masticated, lubricated chyme, enters the esophagus and travels down to the stomach. The upper part of the esophagus contains skeletal muscle, which we have some conscious control over, but the lower half is all smooth muscle.

At the junction between the esophagus and the stomach lies the lower esophageal sphincter. This sphincter opens in response to increased pressure in the esophagus, relative to the stomach on the other side. So, as a bolus of food moves down the esophagus, the sphincter opens, allowing food to move into the stomach (**FIGURE 6.2**). The sphincter has

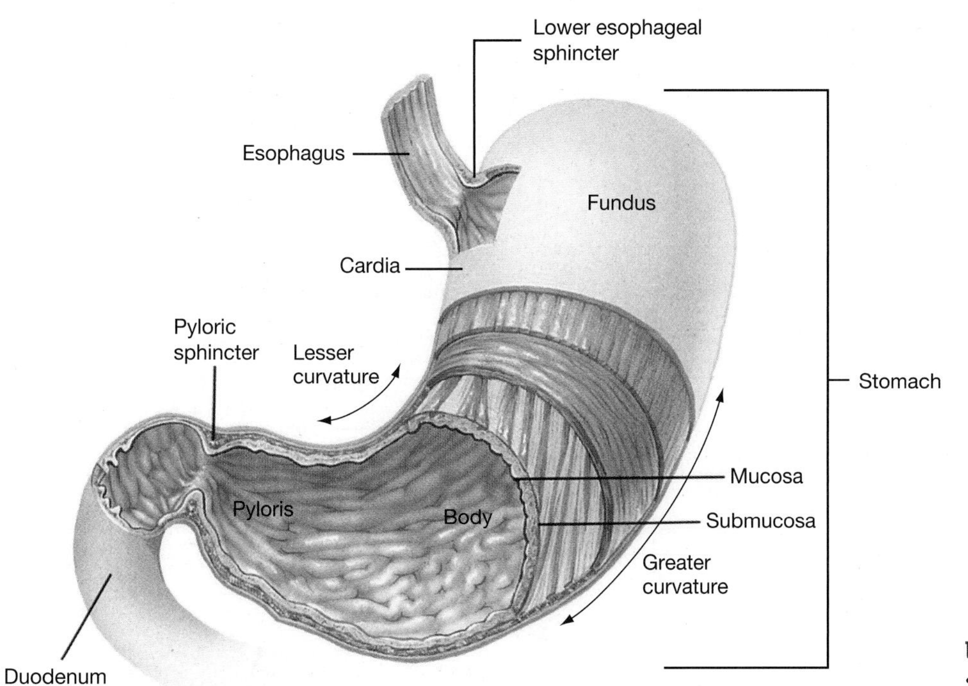

FIGURE 6.2 The lower esophageal sphincter separates the esophagus from the stomach.

an important "gate-keeper" function, as it prevents stomach acid from refluxing into the delicate esophagus. Should the pressure on the stomach side become greater than the pressure in the esophagus (maybe you really ate too much at that Thanksgiving dinner), the sphincter can open, allowing acidic chyme to enter the esophagus, causing "heartburn" or even esophageal ulcers.

The salivary glands are under autonomic nervous system control, and swallowing is triggered by cranial nerves originating in the medulla. However, the remainder of the digestive system can function independent of any central neural input. The digestive system contains its own neural network, the enteric nervous system, that lies between smooth muscle layers throughout the entire digestive tract (**FIGURE 6.3**). While the parasympathetic and sympathetic branches of the autonomic nervous system may influence neural signals from the enteric nervous system, they are not required. From the lower esophageal sphincter to the colon, the process will be controlled by the enteric nervous system and the hormones that are exclusive to the digestive system.

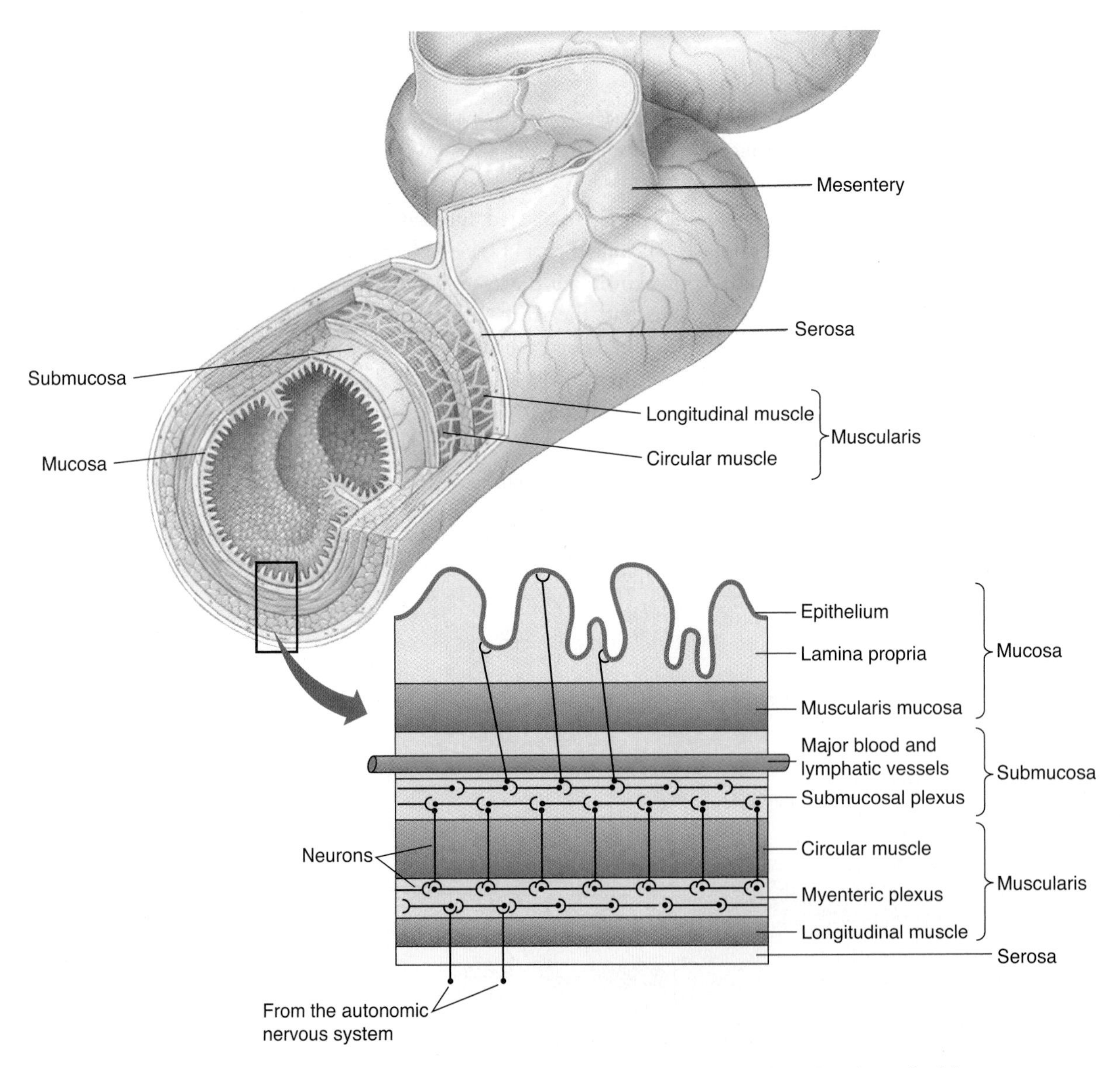

FIGURE 6.3 The enteric nervous system stimulates circular and longitudinal muscle within the wall of the intestine coordinating motility.

Smooth Muscle Contraction in the Digestive Tract

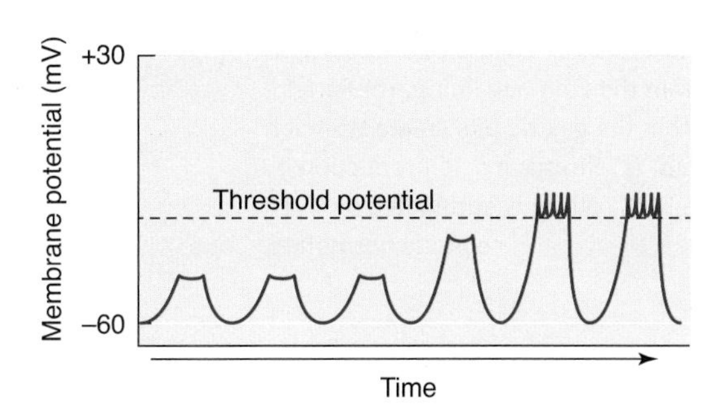

FIGURE 6.4 Smooth muscle contractions within the intestine are generated by pacemaker cells and are slow to reach threshold.

Smooth muscle contraction in the digestive tract takes a unique form. The enteric nervous system contains pacemaker cells that rhythmically open Ca^{2+} channels and allow muscular depolarization. However, these slow waves, as they are called, do not reach threshold and do not propagate as action potentials. They are more akin to long, graded potentials (**FIGURE 6.4**). However, because of the nature of smooth muscle, an increase in intracellular Ca^{2+} will cause some actin–myosin interaction, and some muscle contraction occurs even during these sub-threshold slow waves. The slow waves bring the membrane potential of the smooth muscle closer to threshold, so action potentials occur on top of the slow waves, causing fast waves and normal smooth muscle contraction. This combination of rhythmic slow waves and periodic fast waves, both of which cause smooth muscle contraction, is the mechanism for moving food through the digestive tube. Parasympathetic stimulation from the vagus or pelvic nerves will increase the frequency of contractions, while sympathetic stimulation will reduce the frequency of contractions. So, when your uncle begins the argument and your sympathetic stimulation increases, movement of food within the stomach and intestine slows. Remember, however, that in the complete absence of central autonomic function, the enteric nervous system will continue to move food from mouth to anus.

The Cephalic Phase: Preparation for Food Reaching the Stomach

We know our mouth waters when we are hungry and smell food, but perhaps you were unaware that your stomach is also preparing for your upcoming meal. Parasympathetic stimulation via the vagus nerve and acetylcholine receptors causes the activation of H^+-K^+ ATPase pumps within the cell membranes of parietal cells in the gastric glands. These cells secrete H^+ ions in exchange for K^+ ions, decreasing the pH of stomach fluids and preparing the stomach to initiate its role in digestion (**FIGURE 6.5**). Simultaneously, vagal stimulation will cause the release of the hormone gastrin from G cells located within the stomach. Gastrin also stimulates H^+ secretion by parietal cells. Acetylcholine released by parasympathetic neurons will stimulate motility, i.e., smooth muscle contraction of the three layers of smooth muscle within the stomach, which is potentiated by gastrin. Chief cells of the gastric glands secrete pepsinogen, a proenzyme, which is converted to pepsin at low pH (**FIGURE 6.6**). Low pH in the stomach also stimulates pepsinogen release from chief cells. Pepsin breaks down protein, which is the primary digestion that takes place within the stomach. All this begins when you see or smell food!

Gastric Phase: What Happens When Food Arrives in the Stomach?

Once swallowed, turkey from your Thanksgiving dinner will be reduced to peptides by pepsin. Peptides will stimulate G cells to release gastrin, so stomach motility and acid production continues. Pepsinogen release also increases, stimulated by peptides within the stomach. You have undoubtedly noticed that many of the mechanisms described so far are positive feedback loops, with multiple stimuli all augmenting the release of gastrin,

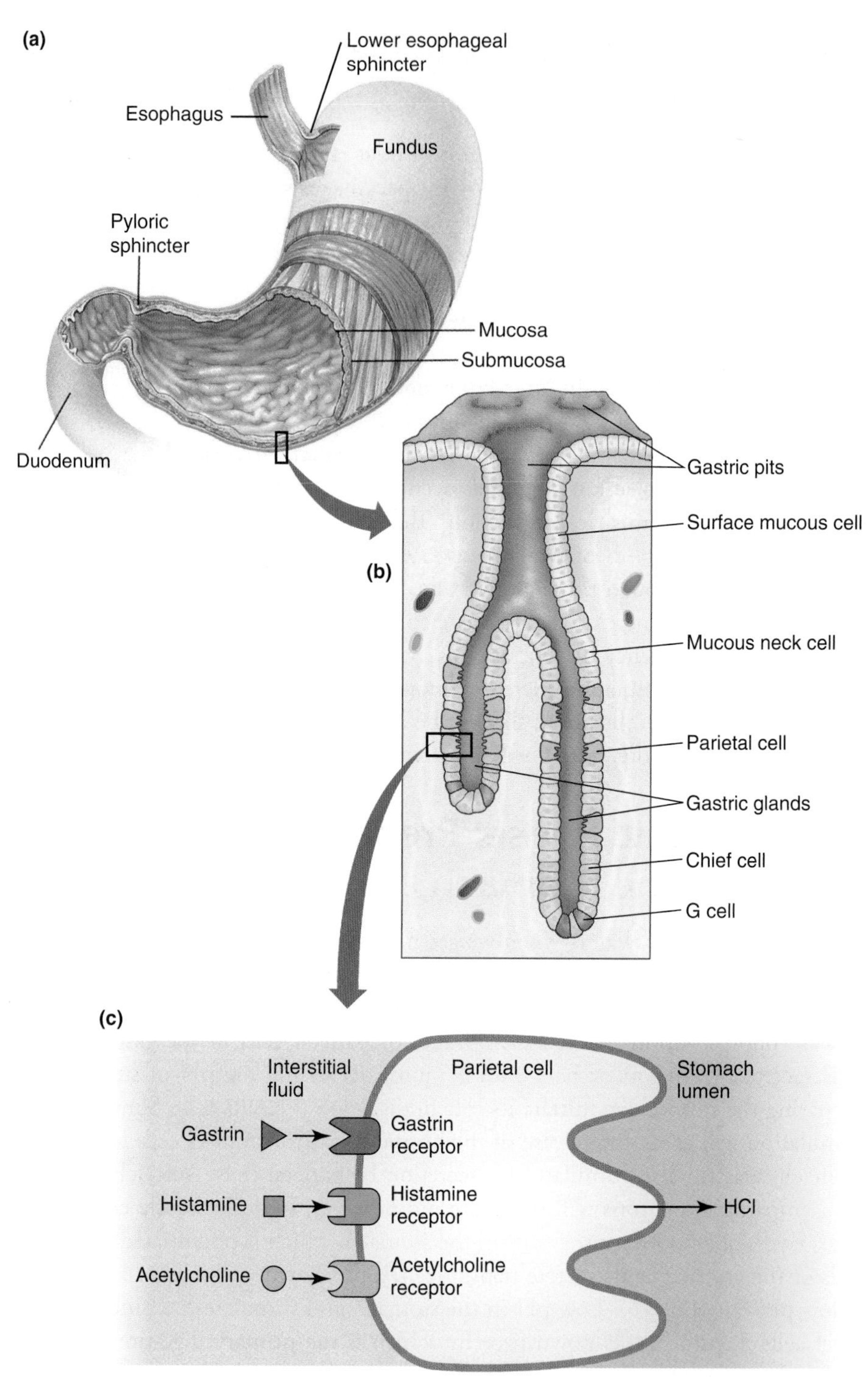

FIGURE 6.5 (a) Gastric pits are located within the stomach lining. (b) Parietal cells within the gastric pits create stomach acid. (c) Stomach acid production by the H^+-K^+ pump is stimulated by three separate hormones.

pepsinogen, and H^+ and increasing gastric motility. These positive feedback mechanisms change the stomach from its quiescent state into a powerful, active digestive organ.

The stomach muscle is itself made of protein and could potentially be digested by pepsin as well, but is protected by a thick layer of mucus. Specialized cells in the stomach produce mucus for just this purpose. However, this also means that amino acids are not absorbed within the stomach because the epithelial layer is too well insulated to allow

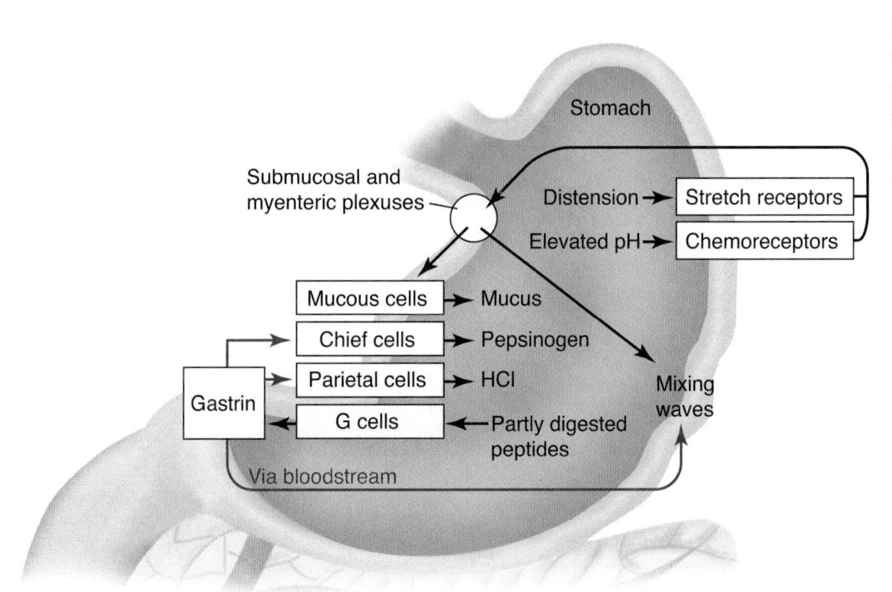

FIGURE 6.6 Gastric glands contain parietal cells that produce HCl, chief cells that produce pepsinogen, and G cells that release the hormone gastrin. Stretch receptors, chemoreceptors, and gastrin stimulate mixing waves of contraction.

transport. Actually, very little absorption occurs in the stomach. Only water and alcohol undergo significant transport at this point in digestion.

Chyme does not stay in the stomach but is gradually released into the duodenum. This presents several problems. Chyme is acidic, yet it is released into the first portion of the small intestine, the duodenum, which, as a transport epithelium, is not as well protected by mucus as the stomach. How is the duodenum protected from this acidic chyme? Anatomically, the duodenum is a busy place (**FIGURE 6.7**). It contains ducts from the pancreas as well as the gallbladder, both of which will contribute fluids to the chyme. An increase in peptides or H^+ will cause I cells in the duodenum to release cholecystokinin (CCK), a hormone that targets the exocrine pancreas. CCK stimulates exocrine glands of the pancreas to release a bicarbonate-rich fluid into the duodenum. Mixed with the acidity of chyme, this will neutralize the pH, so the chyme acquires a neutral pH. Also in response to CCK, the pancreas will release pancreatic lipase and amylase along these same ducts to begin the digestion of fats and sugars. These exocrine functions of the pancreas are quite distinct from the endocrine functions normally associated with the pancreas. Exocrine secretions are released into ducts that end in a different tissue but are never released into the blood supply. So, the bicarbonate solution and enzymes are targeted to the duodenal area. CCK also inhibits gastric emptying, so the release of acidic chyme from the stomach to the duodenum is slowed—the duodenum isn't hit with all that acid at once. Finally, CCK will cause contraction of the gallbladder and relaxation of the bile duct, so that bile flows into the duodenum. This is essential to fat digestion, as we will describe in more detail later on.

Secretin is another hormone, produced by S cells of the duodenum, which duplicates many of the functions of CCK. It also stimulates bicarbonate release from the pancreas, slows gastric emptying, and reduces gastric motility. In addition, it inhibits H^+ ion secretion in the stomach.

The last duodenal hormone we will discuss is glucose-dependent insulinotrophic peptide (GIP). GIP reduces gastric H^+ ion secretion, but more importantly, it targets

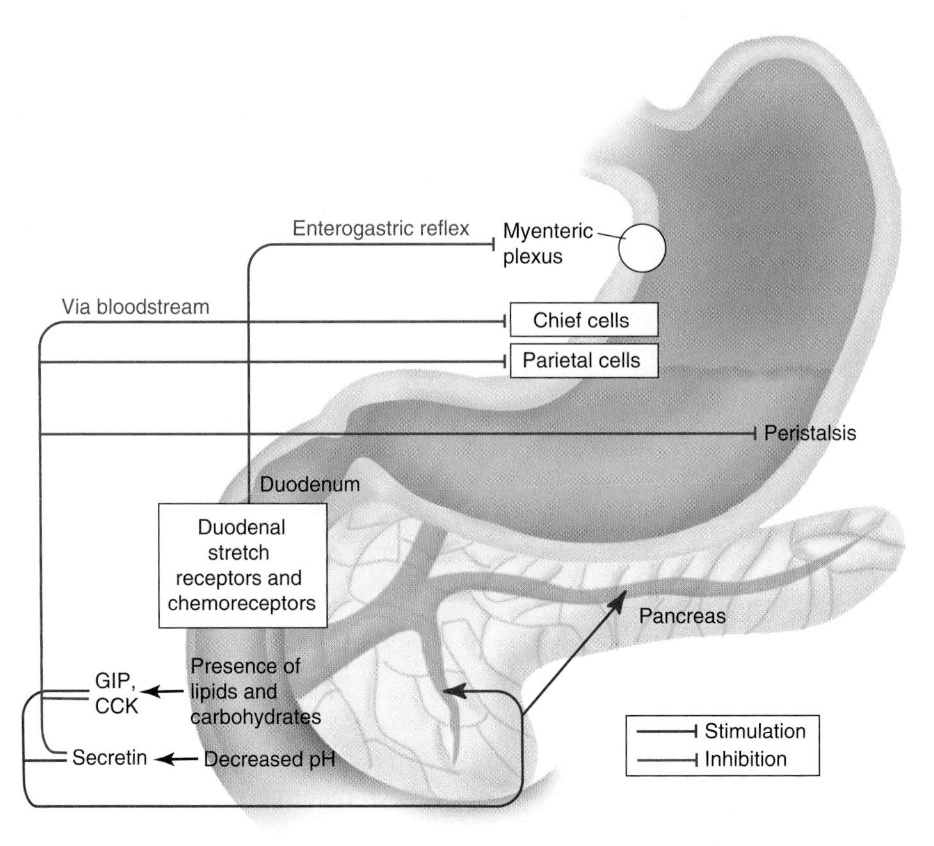

FIGURE 6.7 Hormones, lipids, carbohydrates, and pH coordinate the release of pancreatic secretions, of bile into the duodenum, and of chyme from the stomach.

endocrine cells of the pancreas and causes insulin release into the bloodstream. GIP is released from cells in the duodenum in response to amino acids, glucose, or fatty acids. So, even before glucose is absorbed by the intestinal tract, blood insulin concentrations are already increasing, preparing cells of the body for glucose, amino acid, and fatty acid transport.

Bile Emulsifies Fats and Facilitates Hydrolysis

Fats are not water soluble, and yet, all the enzymatic digestion within the intestinal tract occurs within an aqueous environment. In order to be digested, fats must be emulsified, i.e., broken down into smaller entities so that they are molecularly available for enzymatic attack. Bile is produced by the liver as a metabolite of cholesterol. Bile acids are amphipathic molecules, able to bind to hydrophobic portions of fatty acids on one end and hydrophilic molecules, like water, on the other. This makes bile acids ideal for breaking large droplets of fat into smaller micelles. The other component of bile, bile salts, is excellent at solubilizing phospholipids. Once fats have been emulsified into smaller particles, they can be enzymatically broken down by pancreatic lipase. You can think of bile as a detergent. If you were to drop a teaspoon of dish detergent into a dishpan covered in a slick of grease, it would break the cohesive grease slick into thousands of smaller fat droplets. Bile does essentially the same thing within the duodenum.

By the time your meal reaches the duodenum, all nutrient substrates are being hydrolyzed. Proteins were broken down in the stomach, carbohydrates are being digested by pancreatic amylase, and fats are being emulsified by bile salts and digested by pancreatic lipase. Gastric emptying is regulated by an interplay of the hormones gastrin, CCK, and secretin.

Intestinal Phase: Absorption and Motility in the Small Intestine

The perfectly browned and juicy turkey, sweet red cranberry sauce, and creamy mashed potatoes you sat down to eat have now been grossly broken down by mastication and partially digested by enzymes in the saliva, stomach secretions, and duodenal secretions. While this chyme is not as attractive as the original meal, it is much closer to providing you with nutrition. The essential steps in nutrient absorption occur in the intestine, primarily the small intestine. A look at the anatomy of the small intestine will make understanding absorption much easier (**FIGURE 6.8**). While the exterior of the intestine is smooth, the interior of the tube is invaginated with thousands of villi, which greatly increase the total surface area of the small intestine. Each villus is in turn covered with microvilli, so that the total absorptive epithelial surface of each villus is maximized. The epithelial surface of the microvilli is covered with enzymes, which break down proteins and sugars, and transporters for moving nutrients into the blood supply. Within each villus lies a lymphatic capillary, called a lacteal, surrounded by a capillary bed arising from arterioles and going to venules. Cells at the base of the villi secrete a bicarbonate solution into the lumen of the intestine, further neutralizing the pH of chyme and making it a thinner, more watery solution. This anatomy is optimal for absorbing nutrients from the watery chyme.

Carbohydrates are hydrolyzed to the disaccharides treholose, lactose, or sucrose. In turn, these are hydrolyzed to monosaccharides: glucose, galactose, and fructose. Glucose and galactose are transported by a Na^+-glucose cotransporter into the cytosol of the epithelial cell. As you may know, this transporter uses the steep driving gradient of sodium, which is high in the extracellular space and low in the intracellular space, to move glucose up a concentration gradient. Because the Na^+ concentration gradient must be maintained

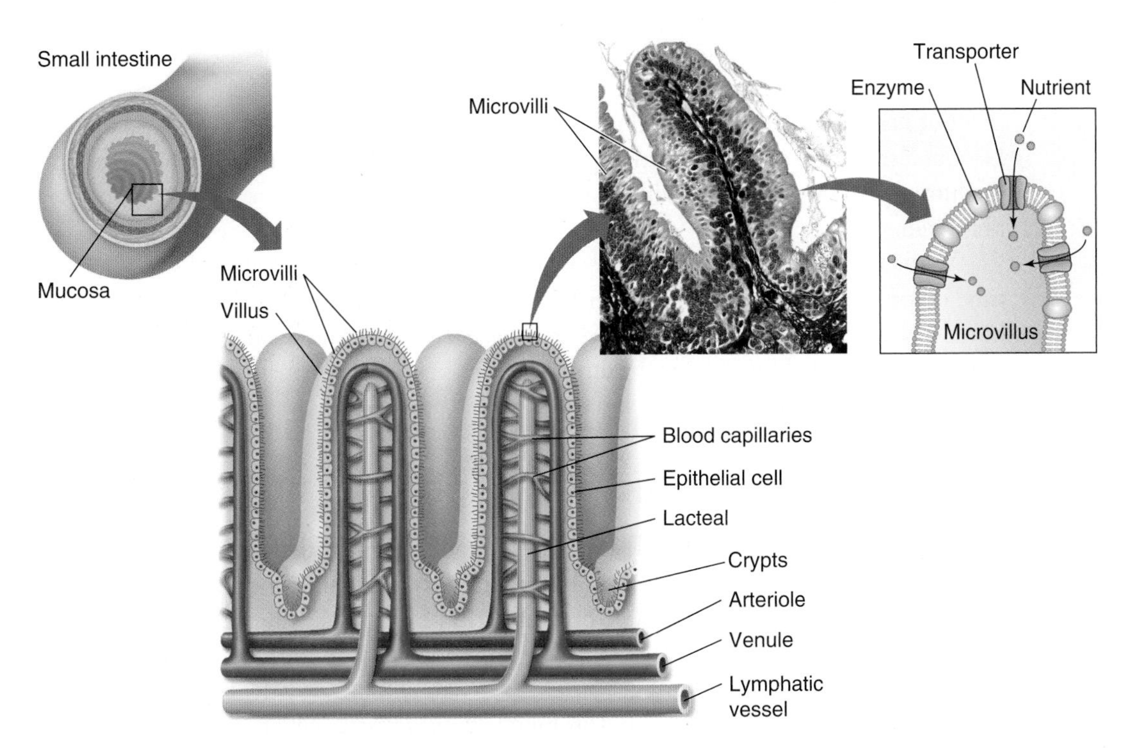

FIGURE 6.8 The anatomical structure of the intestinal wall, including the lacteal. Enzymes and transporters for sugars and amino acids are embedded in the membrane of the microvilli. (Photo © Steve Gschmeissner /Science Source).

by the Na^+/K^+ ATPase, and ATP is required for this pump, this form of transport is often called secondary active transport. Fructose is carried by a separate transporter. Once inside the cytosol, glucose is transported to the blood vessels by facilitated transport.

Glucose, along with galactose and fructose, can also transit from the lumen of the intestine to the vasculature within the villi via another route. Glucose dissolved in water can be carried by solvent drag through the intercellular spaces between epithelial cells. This route is osmotically driven and accounts for a significant amount of glucose uptake by the small intestine.

Protein digestion began in the stomach, leaving peptides to be digested within the small intestine. The luminal epithelium of the microvilli contains peptidases, which hydrolyze peptides into their component amino acids. Amino acids are transported by Na^+-amino acid cotransporters into the epithelial cytosol and by facilitated transport to the blood space. Unlike glucose, amino acids are poorly transported across intercellular space.

Lipids were emulsified by bile salts in the duodenum, increasing the surface area for hydrolysis by pancreatic lipase. Fatty acids can transit the plasma membrane of the enterocyte. Inside the enterocyte, fatty acids are combined with cholesterol, phospholipids, and apoproteins to make a chylomicron. Chylomicrons traverse the basal membrane of the enterocyte and enter the lymphatic system through the lacteal (**FIGURE 6.9**). The lymphatic system drains into venous circulation, so chylomicrons are ultimately released to the venous system. Capillaries within skeletal muscle and adipose tissue have the enzyme lipoprotein lipase on their luminal surface that will cleave fatty acids from the chylomicrons, absorbing them into muscle or adipose tissue for metabolic use or storage.

Water and Ions Are Absorbed in the Small Intestine

Water that we drink makes up only a small portion of the water in the small intestine. Salivary juice, gastric secretions, intestinal secretions, and pancreatic and gallbladder secretions all contribute to production of a very watery chyme in the small intestine. Most of this water is reabsorbed into the blood supply in the small intestine. Recall that glucose and amino acids are transported using a Na^+-glucose or Na^+-amino acid transporter. Movement of Na^+ out of the lumen and into the intestinal epithelial cells not only reclaims the Na^+ but also sets up an osmotic driving gradient, which will also cause water to move from the lumen into the epithelial cells and then across the basal membrane into the blood supply (Figure 6.8).

While epithelial cells of the villi reabsorb ions and water, epithelial cells of the intestinal crypts can secrete ions and water back into the lumen. Secretion is generally under hormonal control but can also be triggered by bacterial toxins, such as cholera, causing a watery diarrhea. However, in health, 75% of the water in small intestinal chyme is absorbed before it enters the large intestine.

How Is Food Moved Through the Intestine?

If chyme didn't move through the intestine, the efficiency of the absorptive epithelium would soon be lost. While the autonomic nervous system can influence intestinal motility, the enteric nervous system is capable of contraction on its own. Intestinal contraction of smooth muscle is a very coordinated event and has several important patterns. Smooth muscle contraction in the intestine occurs in circular smooth muscle, which changes the diameter of the intestinal lumen and in longitudinal muscle, which changes the muscle length. Pacemaker cells of the enteric nervous system connect to smooth muscle through gap junctional proteins, setting up an intrinsic rhythm. Each section of the digestive tract has a characteristic number of slow waves per minute.

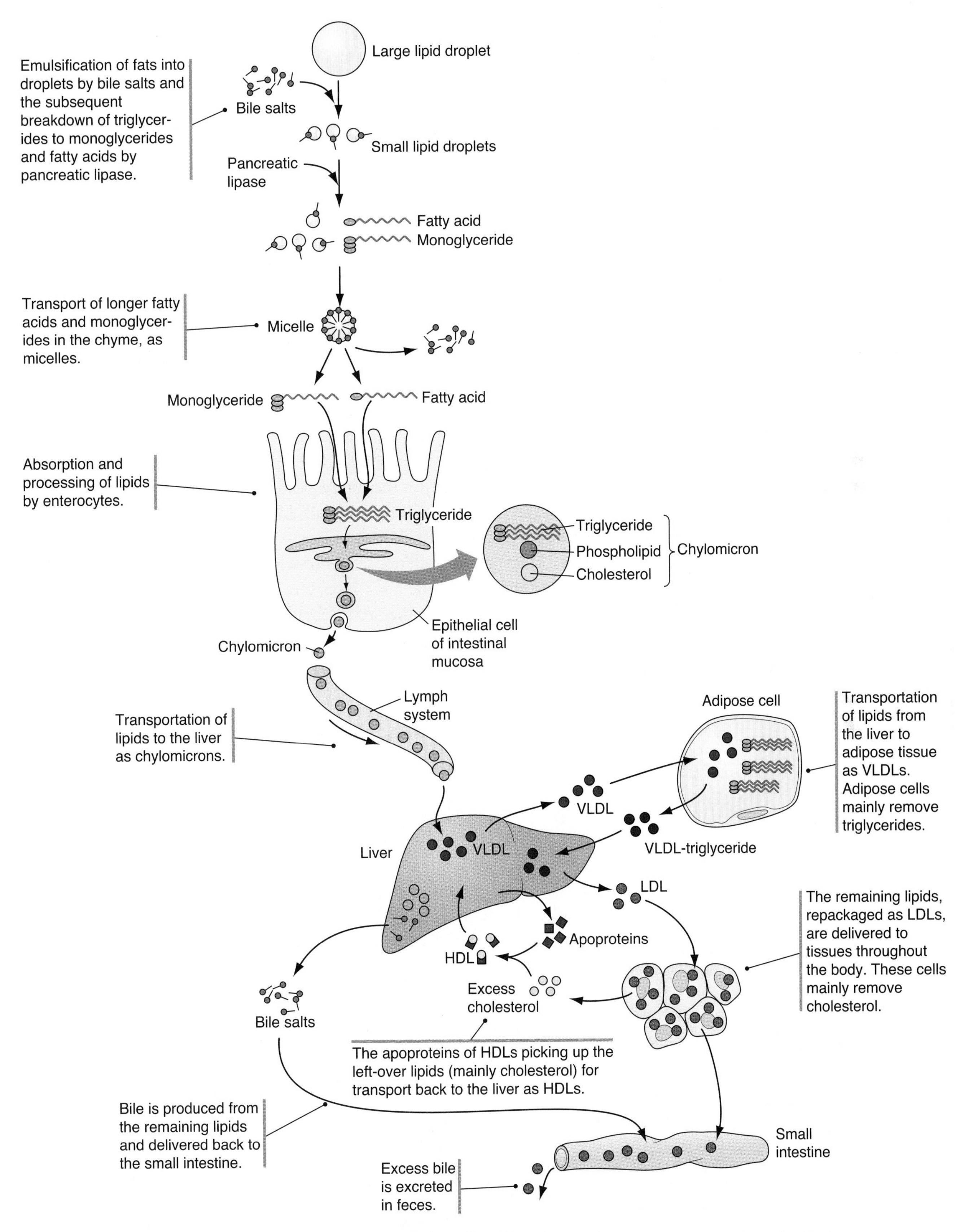

FIGURE 6.9 Chylomicrons are formed within the epithelial cells of the villus and then enter the lacteal. HDL = high-density lipids; LDL = low-density lipids; VLDL = very-low-density lipids.

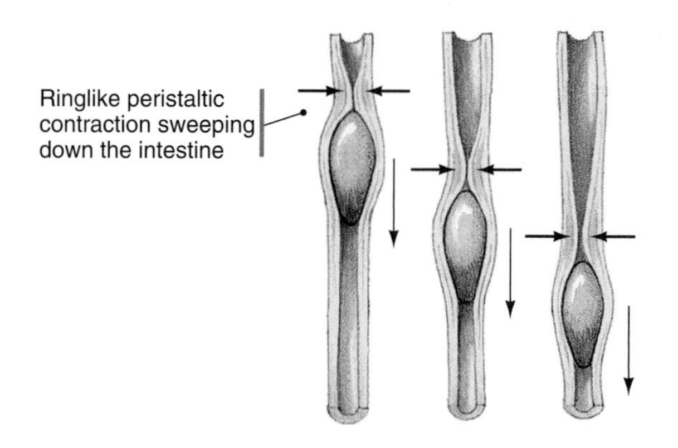

FIGURE 6.10 Segmentation mixes chyme, and peristalsis moves it through the intestine.

The intestine adds to this basic rhythm with two distinct patterns of muscle contraction: segmentation and peristalsis. Segmentation occurs when circular muscle in small sections of small intestine contract, while muscle in neighboring sections relaxes. This causes the chyme to squirt in both directions, analogous to squeezing a tube of toothpaste in the center. Segmentation promotes mixing of the chyme so that all parts of the bolus are exposed to the luminal epithelium. Without segmentation, much of our food would move through the intestinal tract without enzymatic digestion.

Peristalsis moves chyme "forward" in the intestine, toward the large intestine (**FIGURE 6.10**). In peristalsis, a section of circular muscle contracts while the longitudinal muscle in that same section relaxes. At the same time, in the next distal section, circular muscle relaxes and longitudinal muscle contracts, making the lumen of this section larger. Chyme then moves down a pressure gradient, from the section where circular muscle is contracted to where the lumen is larger. This coordinated activity of intestinal smooth muscle moves the chyme bolus through the digestive tract. This process is so vital that if the enteric nervous system fails in a section of the intestine, chyme will not move through that section but will form a block where chyme accumulates, stretching that portion of the intestine. This will not self-resolve but requires surgical intervention.

By the time chyme transits into the large intestine, most of the nutrients, ions, and water have been absorbed from the chyme. Your Thanksgiving dinner has been chewed, broken down to its component parts, and absorbed, ready to be used for energy or to become the building blocks of new human proteins, fats, or glycogen.

The Large Intestine: An Emerging Role for an Ancient Organ

Traditionally, the large intestine has been viewed as a site of water absorption and was known to harbor a variety of bacterial species that eventually come to compose the bulk of our feces. The large intestine contains a large number of goblet cells that secrete a thick mucus layer in the large intestine, thus protecting the intestinal epithelium from being invaded by these resident bacteria. It has long been appreciated that bacteria contribute some vitamins and digest some complex carbohydrates that we do not possess the enzymes to hydrolyze. For example, beans contain some carbohydrates our bodies cannot process, but bacteria can. Bacterial hydrolysis of these carbohydrates produces a by-product of methane, which we expel as gas. Bacteria within the large intestine are symbiotic nutrient processors, and increasingly they are seen as an important component of our nutritional balance and our immune system.

Intestinal bacteria are initially acquired from our mothers during childbirth. The types of bacteria change throughout our lifetime and may change with diet or stress. Gut bacteria help us to digest and recycle bile acid; they produce vitamin B_{12}, biotin, folate, thiamine, riboflavin, pyridoxine, and vitamin K; and they produce short-chain fatty acids that we can readily absorb, as well as phospholipids and polyamines. In short, bacteria in the large intestine provide us with additional nutrients that we cannot glean from the food.

While it is a new field of investigation, intestinal bacteria are now thought to contribute significantly to metabolic regulation and immune system function, although the specific actions of each bacterial species is not well described. In the process of digesting the remaining foodstuffs in the large intestine, small molecules are released that serve as signaling molecules for both metabolic regulation (leptin signaling, for example), or immune system activation. A variety of diseases are now associated with altered gut microbial populations.

Motility in the Large Intestine and Defecation

The large intestine is anatomically distinguished by haustra, obvious segments of the large intestine formed by longitudinal smooth muscle bands. Like the small intestine, segmentation is an important component of motility in the large intestine, mixing the digested chyme. A unique form of motility exists in the large intestine—mass movement. In healthy people, this occurs once or twice a day. Segmental contractions cease and the enteric nervous system depolarizes, causing contractions that move chyme toward the rectum, the terminal portion of the large intestine.

As the rectum fills, stretch receptors in the rectal walls are activated and relay this signal to the brain. Now you are aware of the need to defecate. Parasympathetic stimulation from the sacral portion of the spinal cord activates the pelvic splanchnic nerves that cause contraction of the musculature of the rectum and relaxation of the internal sphincter.

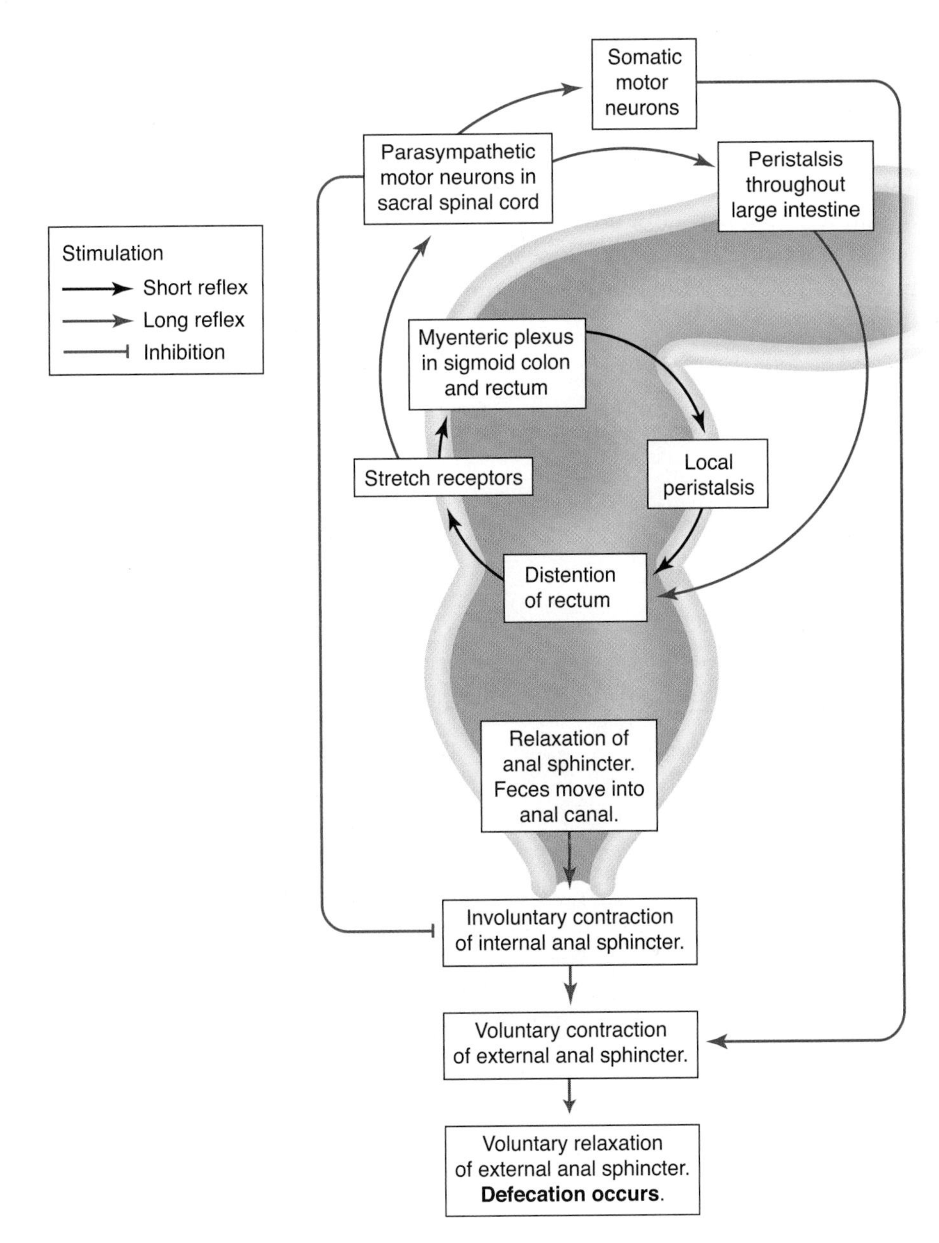

FIGURE 6.11 Defecation requires the coordination of the enteric nervous system, the parasympathetic nervous system, and voluntary motor pathways of the somatic nervous system.

Pudendal and levator ani nerves, motor neurons that control the external sphincter and the levator ani muscles, are activated voluntarily by the motor cortex to prevent defecation until it is convenient. When these muscles relax, defecation occurs (**FIGURE 6.11**).

How Are Nutrients Distributed?

To this point, we have followed your Thanksgiving meal on only part of its journey. We've moved nutrients from the mouth through the stomach to the lumen of the gut and into the blood. Where do they go next?

At this point, all of the nutrients go into a specialized circulation, the hepatic portal circulation. The stomach, small intestine, and large intestine are all served by this vasculature, which begins with a capillary network at the gut and ends in the hepatic portal vein in the liver, where it immediately distributes into another capillary bed. Thus, all nutrient-rich blood from the stomach and intestines goes to the liver, where nutrients can be used, stored, or distributed.

The anatomical arrangement of the liver lobules is an important component of its function. Each liver lobule is organized with a triad of vessels: hepatic portal vein, hepatic artery, and bile duct. The hepatic portal vein anastomoses into sinusoids that flow toward the central vein, past hepatocytes. The hepatocytes are never more than one cell away from a sinusoid, so there is a very short diffusion distance for nutrient absorption. Hepatocytes nearest the triad are primarily responsible for clearing hepatic portal blood of nutrients. They are also the most highly oxygenated, being close to the origin of the hepatic artery. As blood travels through the sinusoids, it flows over Kupffer cells, which are resident macrophages attached to the sinusoidal walls. These macrophages clear damaged cells and bacteria from the blood. Hepatocytes nearest the central vein are specialized for detoxification,

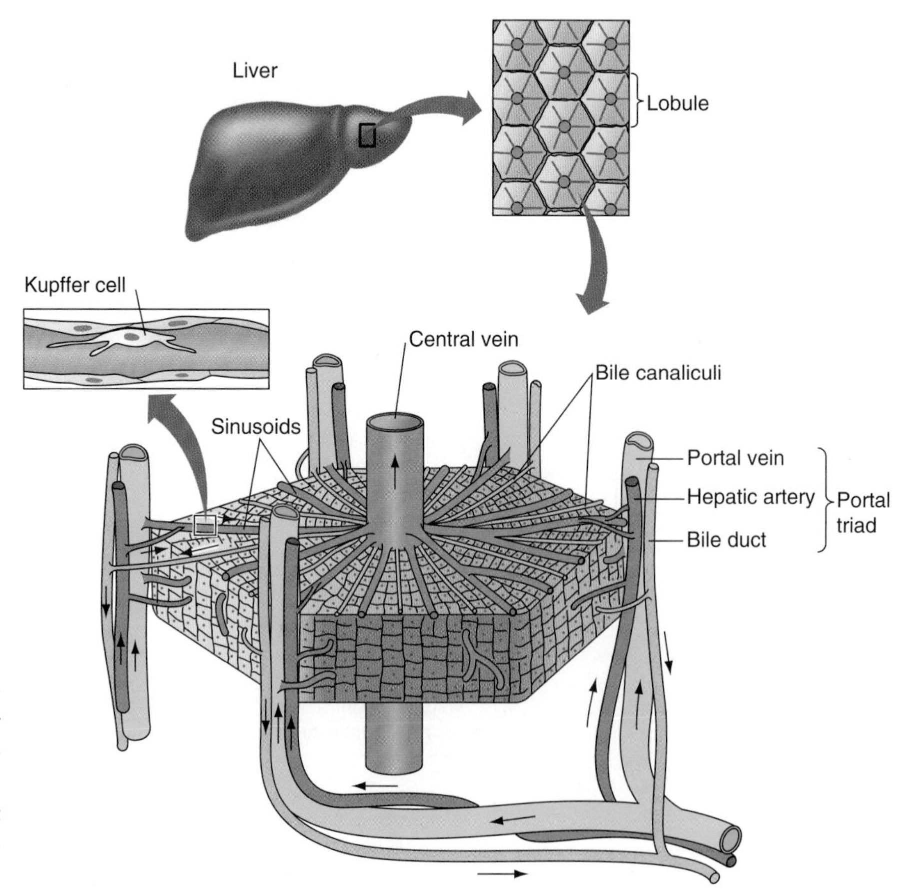

FIGURE 6.12 Blood from the hepatic portal vein flows across the liver lobule toward the central vein. Hepatocytes closest to the outside of the lobule are specialized for nutrient absorption, while those closest to the central vein are specialized for detoxification. Bile produced by the hepatocytes flows out the bile ducts. Arterial blood, nourishing the hepatocytes, flows toward the central vein.

targeting drugs, alcohol, and toxins for elimination. These hepatocytes receive less oxygen than those at the periphery of the lobule and are more at risk for hypoxia. Finally, hepatic portal blood empties into the hepatic vein, which will fuse with the inferior vena cava for return of blood to the heart (**FIGURE 6.12**).

Summary

Now, finally, about 24 hours after it began, your Thanksgiving dinner is complete, and you are gaining the benefits of all the nutrients you ingested during the feast. The argument at the table that activated the sympathetic nervous system may have slowed your digestion temporarily but did no long-term damage. You are now well supplied with protein in muscle, fatty acids in adipose tissue, and glycogen stored in muscle and liver. What was once turkey meat and sweet potatoes are now being used for human structure or energetics.

Key Concepts

Smooth muscle contraction in the digestive system
Mechanism of stomach acid production
Hormonal control of digestion
Role of bile
Absorption of nutrients within the intestine
Hepatic portal circulation

Key Terms

Salivary glands
Gastrin
Gastric glands
Parietal cells
Chief cells
Pepsinogen
Cephalic phase
Gastric phase
Cholecystokinin
Secretin
Glucose-dependent insulinotrophic peptide
Intestinal phase
Nutrient transporters
Chylomicron formation
Segmentation
Peristalsis
Defecation
Hepatocytes

Application: Pharmacology

1. The primary cause of stomach ulcers is the bacterium *Helicobacter pylori*. These bacteria release inflammatory molecules that stimulate gastrin production. This not only increases gastrin production but also proliferation of parietal cells. In addition to a course of antibiotics, people suffering from stomach ulcers are

often prescribed H^+/K^+ ATPase inhibitors. What is the effect of *H. pylori* on stomach acidity, and what would happen under the influence of the H^+/K^+ ATPase inhibitors?

2. Over-the-counter antidiarrheal medications work by reducing smooth muscle contraction in the gut. Why would this reduce diarrhea?
3. Sometimes the lower esophageal sphincter, which lies between the esophagus and the stomach, doesn't stay closed, which allows chyme from the stomach to reflux back into the esophagus. This can damage the delicate epithelial layer of the esophagus. What do you imagine would be an effective treatment for such a condition?

Digestion Clinical Case Study

Type 2 Diabetes Mellitus.

BACKGROUND

The digestion of fats is less straightforward than that of proteins or carbohydrates. While a small amount of fat digestion begins in the mouth with salivary lipase, most fat digestion takes place in the small intestine. Medium- and short-chain fatty acids can be absorbed by tissues directly, but most of the fats we consume contain long-chain fatty acids, with 12 or more carbons in their fatty acid tails. Long-chain fatty acids require the emulsifying action of bile and the enzymatic action of lipase to be absorbed. Fatty acids are taken up by enterocytes lining the wall of the small intestine. Within the cytosol of the enterocyte, the long-chain fatty acids are converted to triglycerides, combined with cholesterol, lysophospholipids, monoglycerides, and apoproteins to create chylomicrons (see Figure 6.9). The newly formed chylomicron is released from the enterocyte into the extracellular space, where it enters the lymphatic system. Chylomicrons, along with the lymphatic fluid, are then released into the venous circulation at the thoracic duct.

Once in the systemic circulation, chylomicrons travel through capillaries. Capillaries within adipose tissue and skeletal muscle have the enzyme lipoprotein lipase on the endothelial surface of the capillaries, which is capable of cleaving triglycerides into fatty acids. Fatty acids are taken up by adipose tissue for reconversion into triglycerides for energy storage. Muscle tissue also takes up fatty acids for use in ATP production. The glycerol backbone of the triglyceride is also taken up by adipose and muscle tissue for energy use via glycolysis.

The chylomicron now has a very different composition, being triglyceride poor and cholesterol rich. These chylomicron remnants are taken up by liver in an low-density lipoprotein (LDL) receptor mediated endocytosis. The endocytotic vesicles fuse with lysosomes, and the remaining molecular species are converted to free fatty acids. These fatty acids may have one of several fates: They can undergo β-oxidation for ATP production or be converted to acetyl coenzyme A or to acetoacetate, a keto acid. Acetoacetate can be converted to β-hydroxybutyrate and acetone. Together, acetoacetate and β-hydroxybutyrate are known as ketone bodies. When fatty acids are in lavish supply, they may be reconverted to triglycerides and stored within the liver.

Cholesterol is a vital component of our cellular membranes and is the parent molecule of all the steroid hormones. It is so essential to life that the liver synthesizes cholesterol. If we have an abundant supply of dietary cholesterol, this inhibits liver synthetic pathways of cholesterol. Cholesterol, being poorly soluble in water, is packaged by the liver into a protein-cholesterol-triglyceride particle known as very low-density lipoprotein, or VLDL. It is low in density because it is very high in triglycerides and cholesterol. VLDLs released from the liver enter the circulation, where once again, lipoprotein lipase in the capillaries of adipose tissue and muscle can cleave the triglycerides into free fatty acids and glycerol for use by those tissues. In the process, VLDLs are converted to intermediate-density lipoproteins (IDL) and LDL particles. LDLs are high in cholesterol and can be taken up by many cells, delivering cholesterol needed for membrane or hormone synthesis. The remaining, unused LDL particles are taken up by the liver and can either be used to create bile acids or be excreted in bile, which is produced by the liver.

High-density lipoproteins (HDL) are high in protein and low in cholesterol. HDL particles remove unesterified cholesterol from plasma membranes and transfer it to VLDL, IDL, or LDL molecules for transport back to the liver. Thus, HDL scavenges excess cholesterol from the tissues and arranges for its return to the liver for redistribution.

How does this system become deranged in diabetes mellitus? To begin with, it's important to recognize that lipoprotein lipase, located on the endothelial membranes of capillaries within adipose and muscle tissue, is normally insulin sensitive. In type 2 diabetes mellitus (T2DM), this enzyme becomes resistant to insulin's actions and fails to break down triglycerides into fatty acids for storage, or hydrolysis. Thus, triglycerides stay in circulation. In addition, hormone-sensitive lipase (HSL) is also insulin-regulated. This enzyme within adipose tissue breaks down stored triglycerides, releasing fatty acids into circulation. Insulin normally inhibits this enzyme, but in T2DM, the inhibition is partially released and fatty acids in circulation, already abundant, are increased. This stimulates the liver to create even more VLDLs. Peripheral tissues store relatively fewer triglycerides, and the liver stores proportionately more. The net result is high circulating triglycerides and non-alcoholic fatty liver disease.

THE CASE

Your aunt, recently diagnosed with type 2 diabetes, has regretfully stopped eating high-fat meals like burgers and fries, but after six weeks, she still has been unable to significantly lower her blood triglyceride levels. Her physician strongly recommends that she lose weight, which has been shown to improve insulin sensitivity in all tissues. If she cannot or will not lose weight, he has written her a prescription for metformin, a drug that will improve peripheral insulin sensitivity. She leaves the doctor's office faced with two unpleasant choices: dieting or more drugs. As she often has in the past, she comes to you for advice.

THE QUESTIONS

1. Can you explain to your aunt why insulin sensitivity is important?
2. What might it do for her glucose levels?
3. What might an increased insulin sensitivity do for her triglyceride levels? Why? Explain the mechanism.
4. Would a diet of complex carbohydrates help lower her triglyceride levels? Explain why or why not.

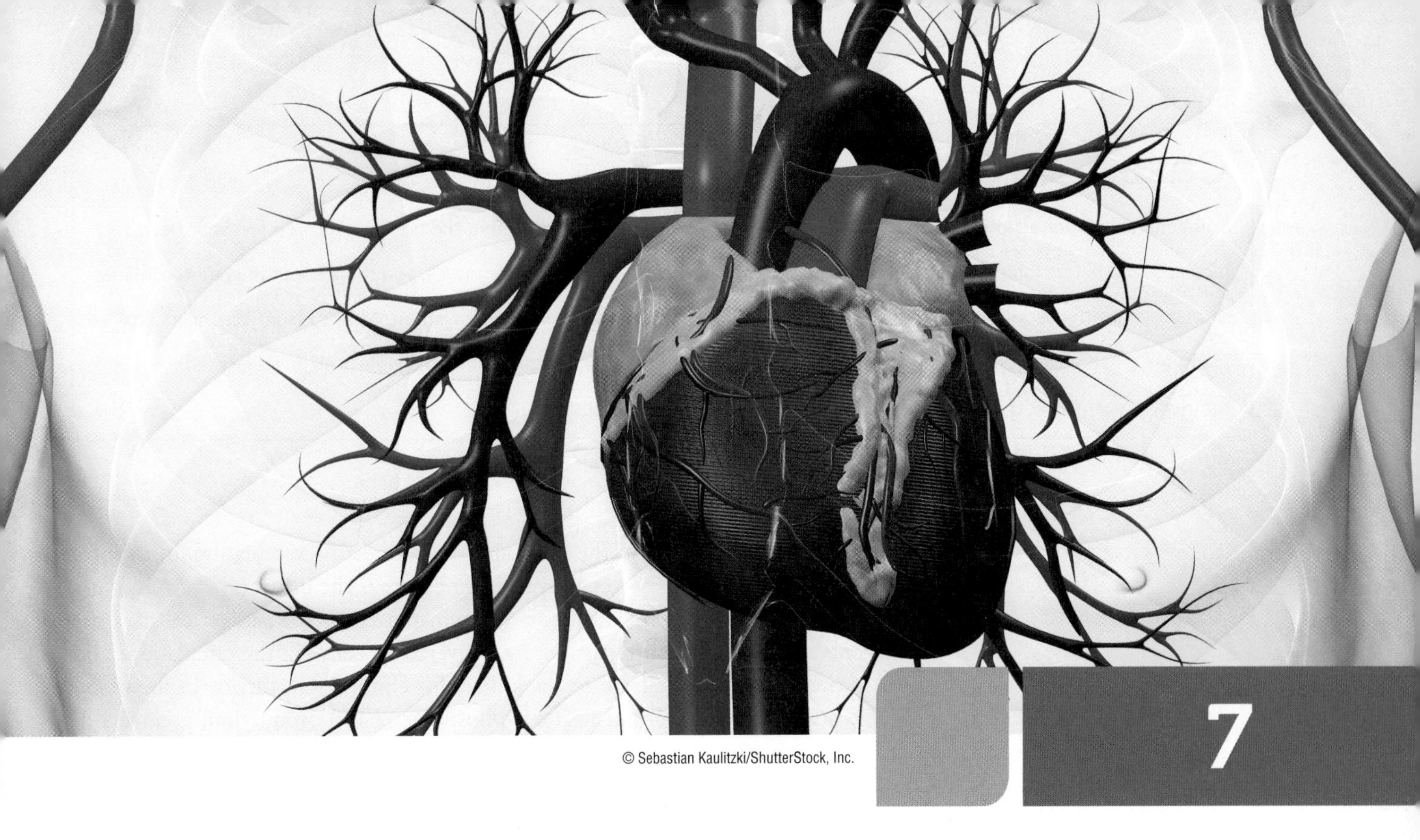

© Sebastian Kaulitzki/ShutterStock, Inc.

7

Cardiovascular Physiology

Case 1

You generally go for a 3-mile run in the morning, and today is no exception. Within the first minute, you feel your heart rate increase, and before long you can feel the pounding of your heart. As your run continues, you think about the blood coursing through your arteries and veins and wonder at the workings of the cardiovascular system. What caused your heart rate to increase? What makes your heart beat harder? What makes you flush by the end of the run? How is all this regulated, so that it returns to "normal" at the end of your run?

Introduction

The cardiovascular system consists of the heart and the connecting vasculature, from aorta to arterioles to capillaries to veins to vena cavae. It functions as the distributor of molecules to the billions of cells in the body. Hormones are transported to their target cells via the blood. Nutrients absorbed during digestion are delivered to cells via the circulation, and some of the waste products of cells go to the kidney for elimination, carried in the blood supply. Oxygen (O_2), essential for adenosine triphosphate (ATP) production, is carried in the blood, as is the gaseous metabolic waste product, carbon dioxide (CO_2). Transportation of all of these molecules is dependent upon the constant movement of blood within the circulatory system. This movement is achieved through the actions of a pump, our heart, and a series of non-rigid, living pipes, the vasculature.

The Anatomy of the Cardiovascular System: The Heart

The heart is a muscular organ, approximately the size of your fist, comprising four chambers: left and right atrium and left and right ventricles. It is composed of striated muscle, similar to skeletal muscle, but instead of contracting against fixed attachments to bone, as skeletal muscle does, it contracts against an incompressible fluid, blood. So the heart contracts against a hydrostatic skeleton. The walls of the atria are thin, reflecting the low pressure exerted by blood returning to the heart. The atria connect to their respective ventricles by atrioventricular (A-V) valves, which open whenever the pressure in the atrium exceeds that of the ventricle. These valves are in turn connected to the ventricular wall by papillary muscles and chordae tendineae, lengths of connective tissue that gave rise to the name "heartstrings" (**FIGURE 7.1**). These muscles contract whenever the ventricles contract, keeping the A-V valves closed during ventricular contraction. This prevents blood from returning to the atria.

The right ventricle is a thin-walled chamber, as it only needs to exert enough pressure to force blood from the ventricle to the nearby lung. The conducting vessels of the lung quickly divide into a vast capillary bed, so there is little resistance to the flow of blood. The left ventricle, however, must generate enough force to pump blood to the entire systemic circulation. To achieve this, the left ventricle is a thick-walled chamber that contracts in a spiral fashion (remember that the heart resembles a cone in shape) with sufficient force to open the semilunar valve of the aorta and pump blood throughout the body. Like skeletal muscle, cardiac muscle can hypertrophy after repeated bouts of heavy use, increasing the thickness of the ventricular wall and improving its ability to generate forceful contractions. This means your morning exercise increases your heart strength as well as your leg strength!

The heart serves as the pump of the cardiovascular system, providing the pressure head to move fluid, blood. As mentioned earlier, the heart is four-chambered, but it is best thought

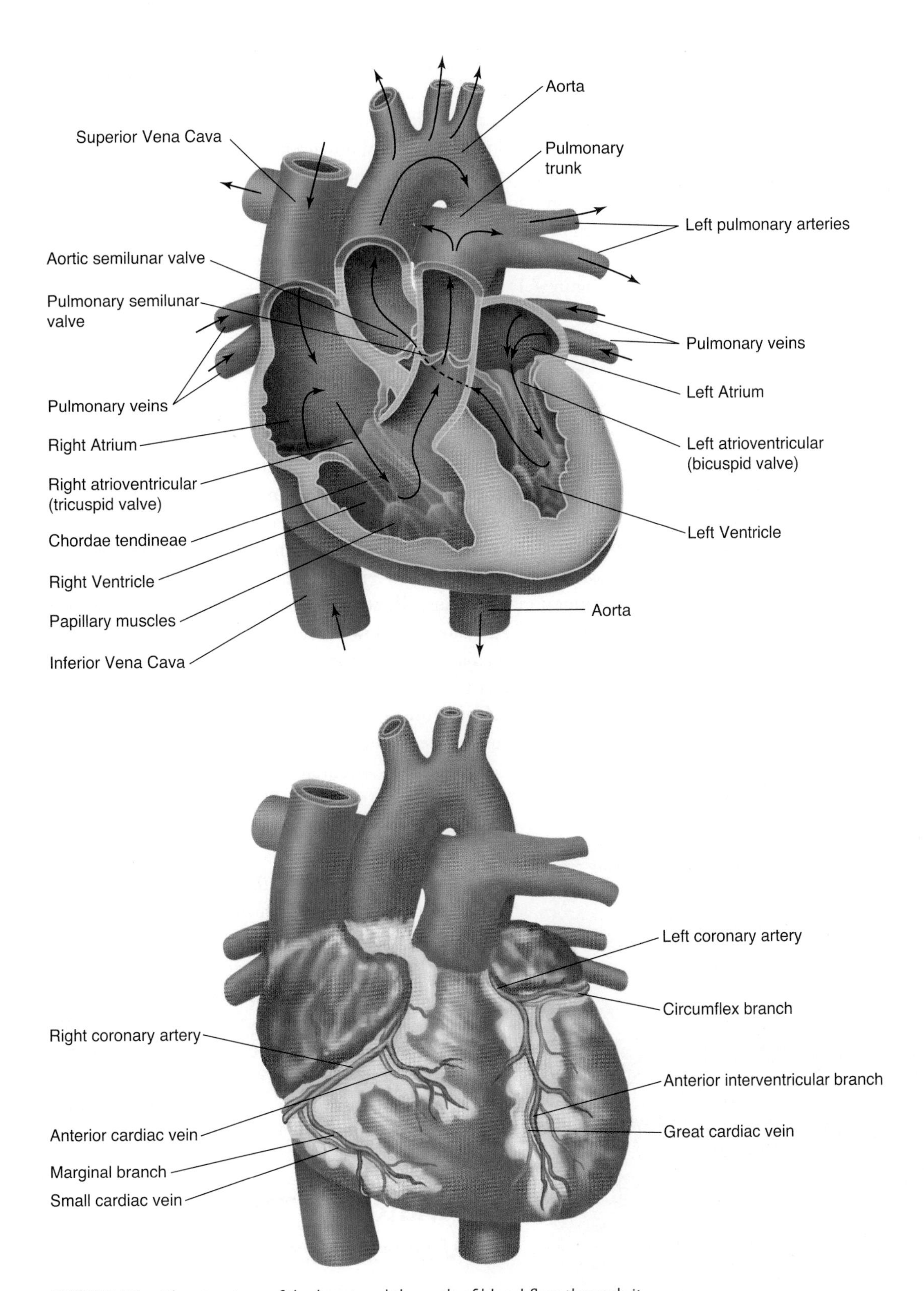

FIGURE 7.1 The structure of the heart and the path of blood flow through it.

of as two separate pumps functioning in parallel, with each atrium filling simultaneously, draining into the ventricles simultaneously and the ventricles contracting simultaneously. The right atrium receives blood from the superior and inferior vena cavae. Blood moves passively into the right ventricle following a simple pressure gradient. Contraction of the right ventricle pumps blood into the pulmonary artery and then into the nearby pulmonary circulation of the lung. There, blood is intimately exposed to air in the alveoli, releasing CO_2 and binding O_2. Pulmonary veins collect this oxygenated blood and return it to the left atrium through the pulmonary veins. Again, the blood moves passively, down a pressure gradient, through the A-V valve, into the left ventricle. Upon contracture, the left ventricle pumps blood through the semilunar valve into the aorta, for its transit through the vast systemic circulation. Blood will ultimately reach each cell of the body, delivering nutrients, hormones, and O_2 while picking up CO_2, metabolic waste products, and metabolites before being returned to the right atrium via the vena cava (Figure 7.1).

Cardiac Muscle: Cellular Level of Organization

At the tissue level, cardiac muscle resembles skeletal muscle, in that it contains regular arrays of thick filaments (myosin) and thin filaments (actin and associated regulatory proteins) arranged within sarcomeres, bounded by Z disks. During contraction, cardiac muscle works on the same sliding filament mechanism described in skeletal muscle. However, there are important differences. Cardiac muscle fibers contain only one nucleus, instead of the multiple nuclei of skeletal muscle. Cardiac muscle fibers branch instead of being consistently linear. Most importantly, cardiac muscle fibers are connected to one another by intercalated disks, which contain gap junctional proteins. These proteins are most abundant on the longitudinal axis of the muscle cells and allow very low resistance conduction between two cardiac muscle cells. It is analogous to adjacent hotel rooms joined by a connecting door—transit between the two is extremely fast, becoming essentially a single room. Gap junctional proteins allow the heart muscle to function as a syncytium, that is, as though heart muscle cells were one large muscle fiber, instead of thousands of individual muscle cells (**FIGURE 7.2**). Simultaneous contraction is essential if coordinated force is to be generated to pump blood out of the heart on each beat. The heart is the first organ to function in a human embryo, and it continues throughout our lives. To fuel this continual muscle work, heart cells are very rich in mitochondria, which provide the ATP required for contraction. Glycogen stores in human cardiac muscle are minimal, capable of fueling contraction for only a minute or two. The primary substrate for cardiac metabolism is fatty acids. Fatty acids do not undergo glycolysis, so there is no anaerobic component to their catabolism. This means that cardiac muscle is dependent upon O_2 for muscle contraction.

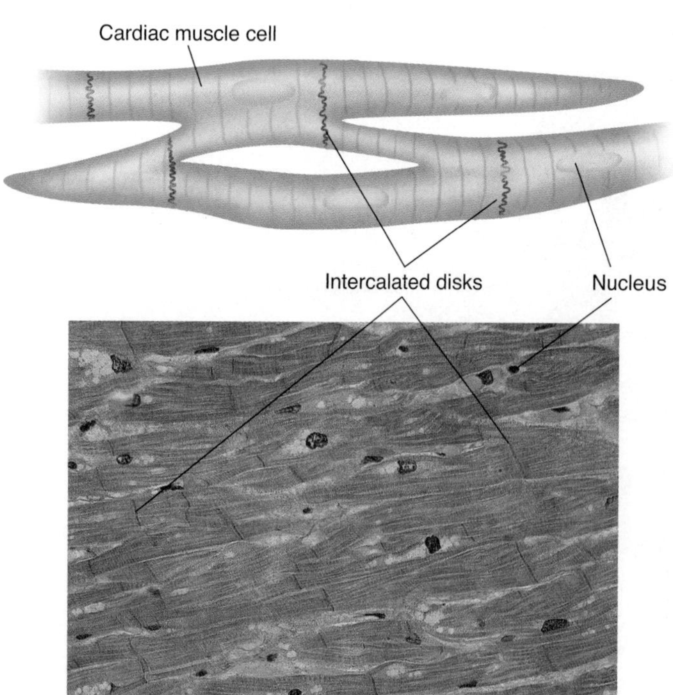

FIGURE 7.2 Gap junctions allow the heart to depolarize as a syncytium and mitochondria fuel its activity. (Photo © Donna Beer Stolz, Ph.D., Center for Biologic Imaging, University of Pittsburgh Medical School.)

Anatomy: The Vasculature

When blood leaves the left ventricle, it enters the aorta, a large, muscular vessel with a rich supply of elastic tissue. Imagine the contents of the left ventricle suddenly and forcefully entering a much smaller diameter vessel—the aorta. If the aorta were rigid, it would create an enormous resistance to flow. However, it is elastic, which allows the aorta to expand as it fills with blood and spring back to its resting length as blood continues down the vascular tree. The elastic recoil of the aorta contributes energy to continuous flow of blood through the remainder of the blood vessels. From the aorta, the blood flows to conducting arteries, which are also fairly thick walled, and then to the arterioles. With each level of branching, a vessel divides into multiple vessels, expanding the total cross-sectional area of the arterial vessels. The walls of these vessels contain not only connective tissue but also smooth muscle, which continually alters the diameter of the vessel (**FIGURE 7.3**). You may know that the autonomic nervous system (ANS) has control over arteriole diameter; it is here that the ANS's control is exerted, constricting arterioles and reducing flow or dilating arterioles and increasing flow. The innermost layer of the blood vessels is a layer of endothelial cells that are in intimate contact with blood. Long thought to be simply a quiescent lining of the vessel, we now realize that these endothelial cells are also hormonal cells and active modulators of arteriole diameter. When the hydrostatic pressure of flow within a vessel increases, the endothelial cells experience shear stress, stress parallel to the surface of the vessel exerted by blood itself. Shear stress causes endothelial cells to release nitric oxide (NO), a gaseous signaling molecule that relaxes smooth muscle and increases vessel diameter.

Arterioles branch into thin-walled capillaries, which is where the exchange of ions, nutrients, and gases occurs. The capillaries are simply the endothelial layer surrounded only by a basement membrane. This is where our closed circulatory system comes closest to being open, and where fluid moves from capillary to interstitial space and back again. Capillaries must lie close enough to each cell in the body to perfuse that cell and provide it with the nutrition and waste removal that it needs. Remember, diffusion is a slow process, so capillary distance from a cell cannot be great.

How Does Fluid Exchange Occur within the Capillary Bed?

There are two primary forces responsible for fluid movement across the capillary: hydrostatic pressure, that is, the pressure generated by the heart, and osmotic pressure, the force generated by solutes within the blood. Hydrostatic pressure not only pushes blood through the vessels but also exerts force on the vessel walls (**FIGURE 7.4**). This pressure tends to move water out of the capillary into the interstitial space. The movement of water can occur across the plasma membrane of the endothelial cell, but it mostly travels through the perivascular spaces between endothelial cells. As water leaves the capillary bed, moving into the interstitial space, the hydrostatic pressure within the capillaries goes down, the driving gradient is reduced, and less water leaves the capillary bed.

The second force, osmotic pressure, is exerted by osmotically active particles within the blood. This can be ions and nutrients, but an important contributor to osmotic force in the blood is the protein albumin. Albumin, made in the liver, is a normal circulating protein within the blood; at the capillary, albumin is a powerful attractor of water. As blood flows through the capillary bed, hydrostatic pressure moves some water out at the arterial end of the capillary network, and albumin and other osmotically active particles draw water back into the blood at the venous end. Throughout the capillary bed, these two forces are in dynamic opposition, and together they maintain fluid balance.

There are two other forces that also contribute to vascular volume: tissue hydrostatic pressure and tissue osmotic pressure. The amount of water within the tissue will exert a hydrostatic pressure of its own, opposing the hydrostatic pressure generated by the heart, pushing blood against the capillary walls. This is one reason that more fluid leaves the capillaries at the arterial end of the capillary bed. The more fluid that leaves the vessels, the greater the tissue hydrostatic pressure will become. The second force, tissue osmotic

FIGURE 7.3 Arteries and veins differ in the amount of elastic tissue and smooth muscle each contains.

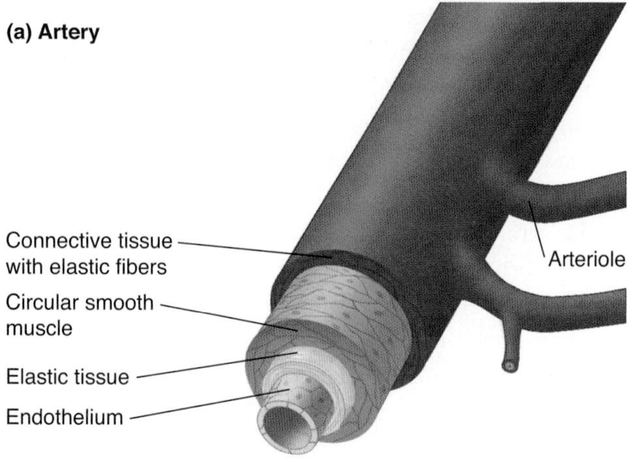

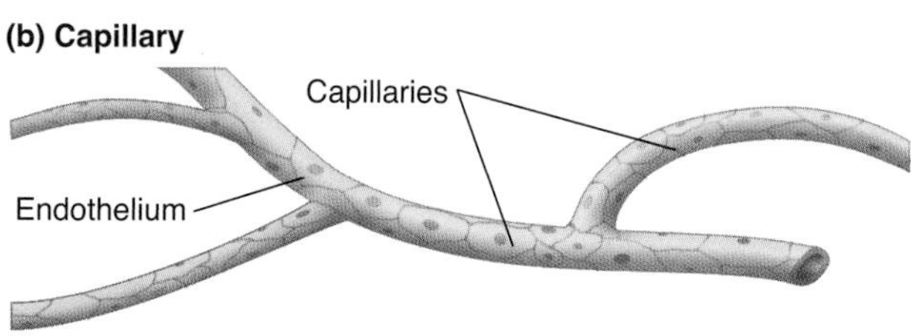

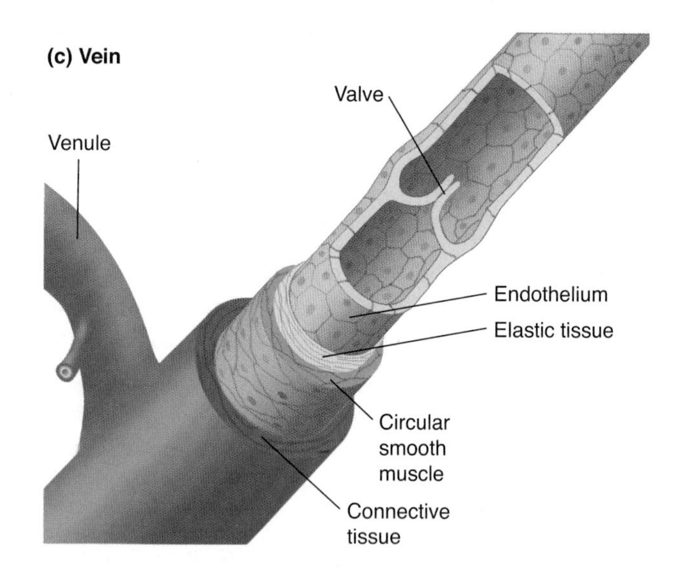

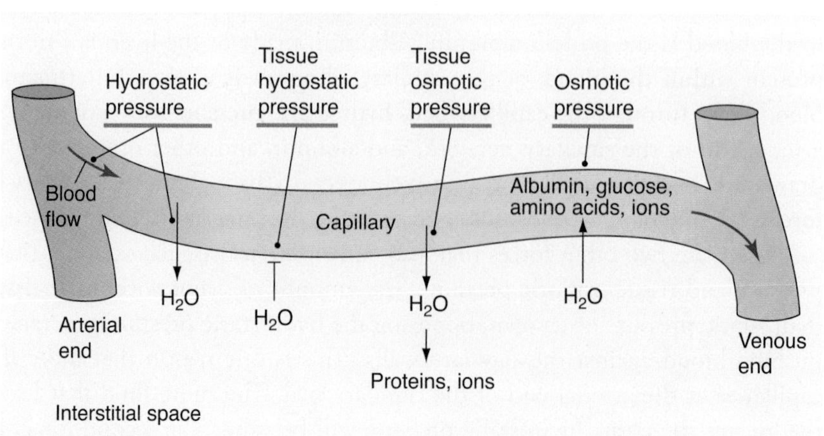

FIGURE 7.4 Movement of water into and out of capillaries is a result of four separate forces.

pressure, is exerted by proteins and ions within the interstitial space. This force tends to pull water from the vasculature into the surrounding tissue. This force is generally small, but should capillaries become leaky, allowing protein into the interstitial space, this force can become significant. This is one of the mechanisms behind tissue swelling following a bee sting, for example.

Role of the Lymphatic System in Fluid Balance

Actually, there is a small net loss of fluid from the capillaries to the interstitial fluid. This fluid moves, primarily by interstitial hydrostatic pressure, into the blind-ended lymphatic vessels. The lymphatic vessels have valves, much like the veins, and move fluid from these terminal sacs along a low-pressure, fluid-filled third vascular space back to the venous circulation (**FIGURE 7.5**). During the transit, the fluid will pass through lymph nodes, and the fluid will be monitored by the immune system. Independent of its role in the immune system, fluid recovery from tissues by the lymphatic system is essential. Malfunction of the lymphatic system—as in elephantiasis, where the lymphatic system is blocked by parasites—causes vascular fluid loss and tissue swelling.

Nutrient Exchange

Gases like O_2 and CO_2 can pass through a plasma membrane unimpeded, moving down a partial pressure gradient. Because cells use O_2 during ATP production, partial pressure of oxygen in arterial blood (PaO_2) will be higher than the tissue, and O_2 will move to the tissues from the capillary bed. Because CO_2 is produced as a cellular metabolic by-product of metabolism, CO_2 will move in the opposite direction, and, thus, the exchange of gases at the capillary bed is accomplished.

Nutrient exchange at the capillary beds can occur by diffusion. Glucose is dissolved in water in the vasculature and diffuses into the interstitial space and out again along with water. Small proteins or amino acids can also be dissolved in plasma; these also leave the capillaries by diffusion. In contrast to glucose, these nutrients rarely move back into the capillary. If they are not transported into the surrounding tissues, they are taken up by the lymphatic system and returned to the venous vasculature at the thoracic duct, near the heart (**FIGURE 7.6**).

How Does Blood Return to the Heart?

The blood leaving the capillaries goes to venules, then veins, and then great veins, and finally the inferior or superior vena cava. The venous vessels are parallel in pathway to the arteries but are dissimilar vessels. Unlike the arteries, the veins are poorly supplied with muscle but still contain elastic fibers. This makes the venous circulation a very compliant one, that is, there is a large increase in volume of the vessel for any given pressure. As a result, venous vessels serve to store blood and actually form our largest reservoir of blood. As you recall about the autonomic nervous system, sympathetic stimulation causes venoconstriction, contraction of vascular smooth muscle in the veins, which will increase the return of venous blood to the heart. However, even in the absence of sympathetic stimulation, blood is moved from the extremities to the heart by skeletal muscle, which squeezes the veins during contraction, pushing blood upward. The skeletal muscle pump plus one-way valves within the veins keep the blood moving back toward the heart (**FIGURE 7.7**). Finally, negative pressure within the thoracic cavity functions as a form of suction to bring blood back to the heart.

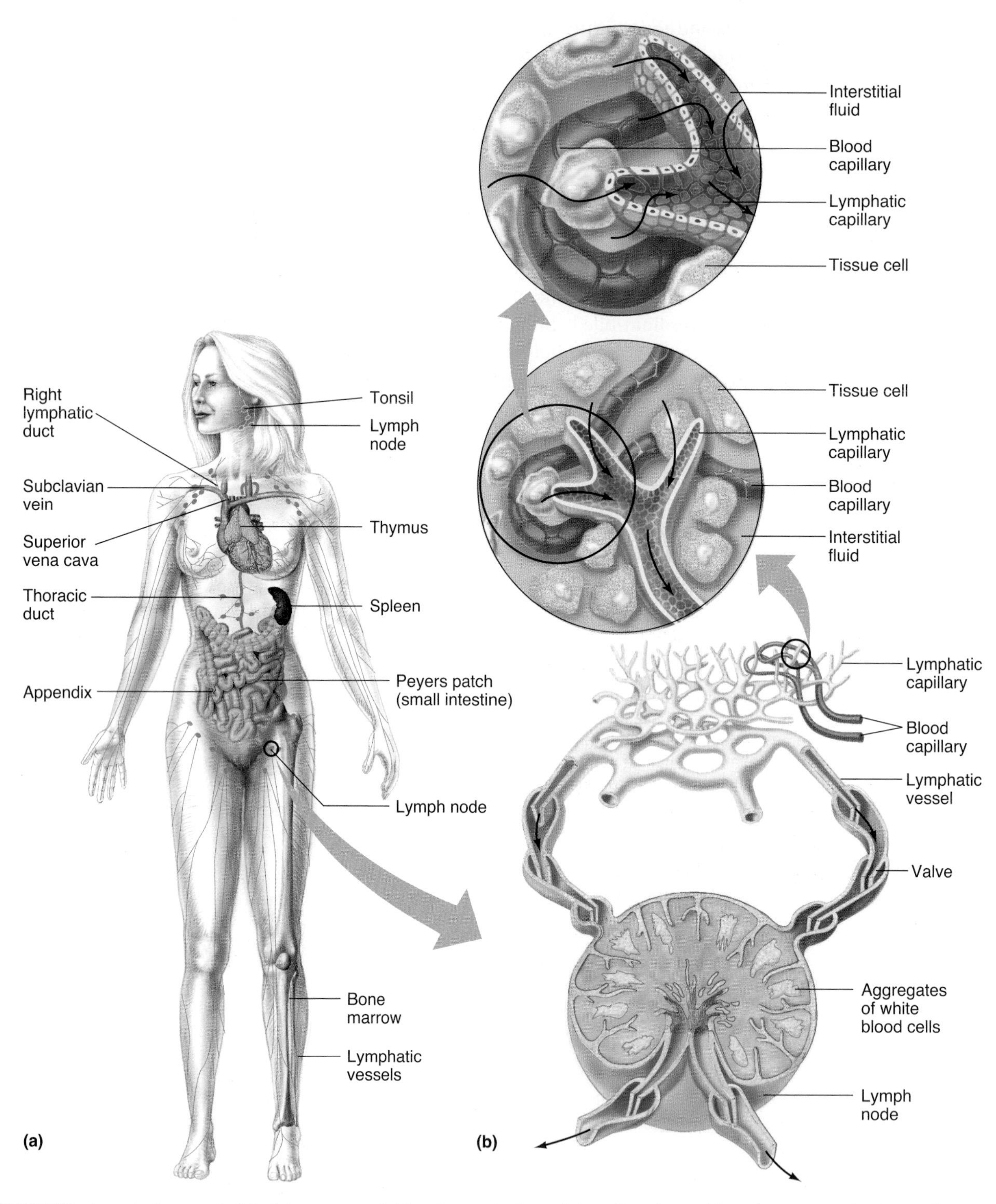

FIGURE 7.5 Lymphatic vessels (a) pick up excess fluid and move it through lymph nodes and into the venous circulation (b).

Blood: What Is Flowing Through These Vessels?

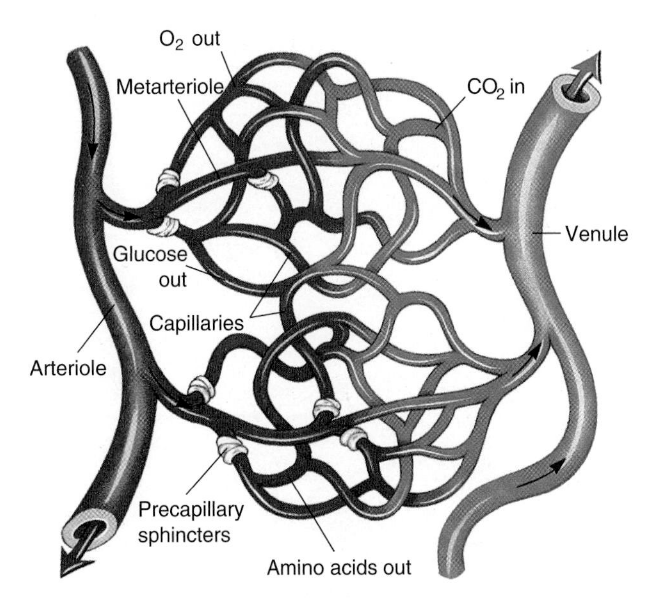

FIGURE 7.6 Gases, glucose, and amino acids all move from the capillaries into the surrounding tissue.

Blood is a complex fluid, containing red blood cells (erythrocytes), white blood cells (leukocytes), and plasma. Plasma is a solution in which hormones, proteins, ions, nutrients, waste products, and gases are all dissolved. In short, blood is both the supply line and the waste disposal system for the cells of your body. Selective transport at the cell membrane allows uptake of nutrients or binding of hormones to receptors, while waste products, with no available transport, stay in the blood plasma until they are either metabolized or excreted.

As we discussed above, the proteins, ions, and nutrients in blood plasma contribute to the osmotic forces that keep blood in vessels. So, in addition to serving as an avenue for communication and nutrient delivery, plasma functions to maintain itself within the blood vessels.

Erythrocytes are specialized for carrying O_2 and are essential if our cells are to be supplied with the quantity of O_2 necessary to sustain life. Erythrocytes are derived from stem cells within the bone marrow that also give rise to leukocytes: T and B lymphocytes, neutrophils, eosinophils, basophils, macrophages, megakaryocytes, and monocytes. (Although they are a component of blood, the roles of the various white blood cells will not be discussed at length here, as these functions primarily relate to the immune system.) Erythrocytes develop from a common stem cell but undergo a remarkable transformation during their maturation (**FIGURE 7.8**).

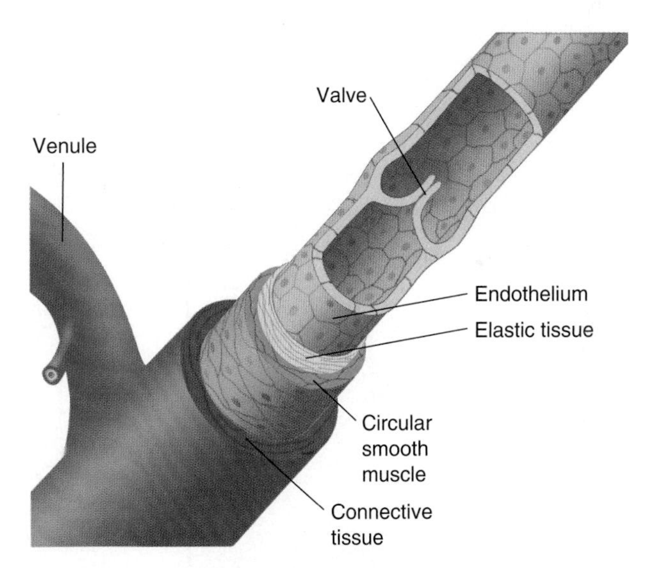

FIGURE 7.7 Veins have less smooth muscle than arteries (see Figure 7.3), making them more compliant. They also have one-way valves that move blood in one direction—toward the heart.

After their initial differentiation from the blood stem cell, erythrocytes contain all the normal cellular components: nucleus, mitochondria, endoplasmic reticulum (ER), ribosomes, and Golgi bodies. In addition to their traditional role in metabolism, mitochondria within developing erythrocytes transport iron from the cytoplasm into the matrix of the mitochondria and incorporate it into a porphyrin ring. The porphyrin ring is then exported out of the mitochondria into the red blood cell cytosol, where it is incorporated into a globin protein made in the cytoplasm. Four of these globin proteins unite to form one large, four-subunit protein known as hemoglobin. One red blood cell may contain as many as 300 molecules of hemoglobin formed in this way.

Once the hemoglobin molecules are created, however, the final stage of maturation for the red blood cell is exclusion of the nucleus, mitochondria, ER, and Golgi. For this reason, a mature red blood cell is small; its size allows it to move through the tiniest of capillaries, but it lacks the cellular machinery to create more hemoglobin or even to repair itself. As a result, erythrocytes have a limited life expectancy, approximately 120 days.

What can damage a red blood cell? Erythrocytes do not float quietly within the blood but are forcefully propelled through the vessels, where they are subjected to shear stress as

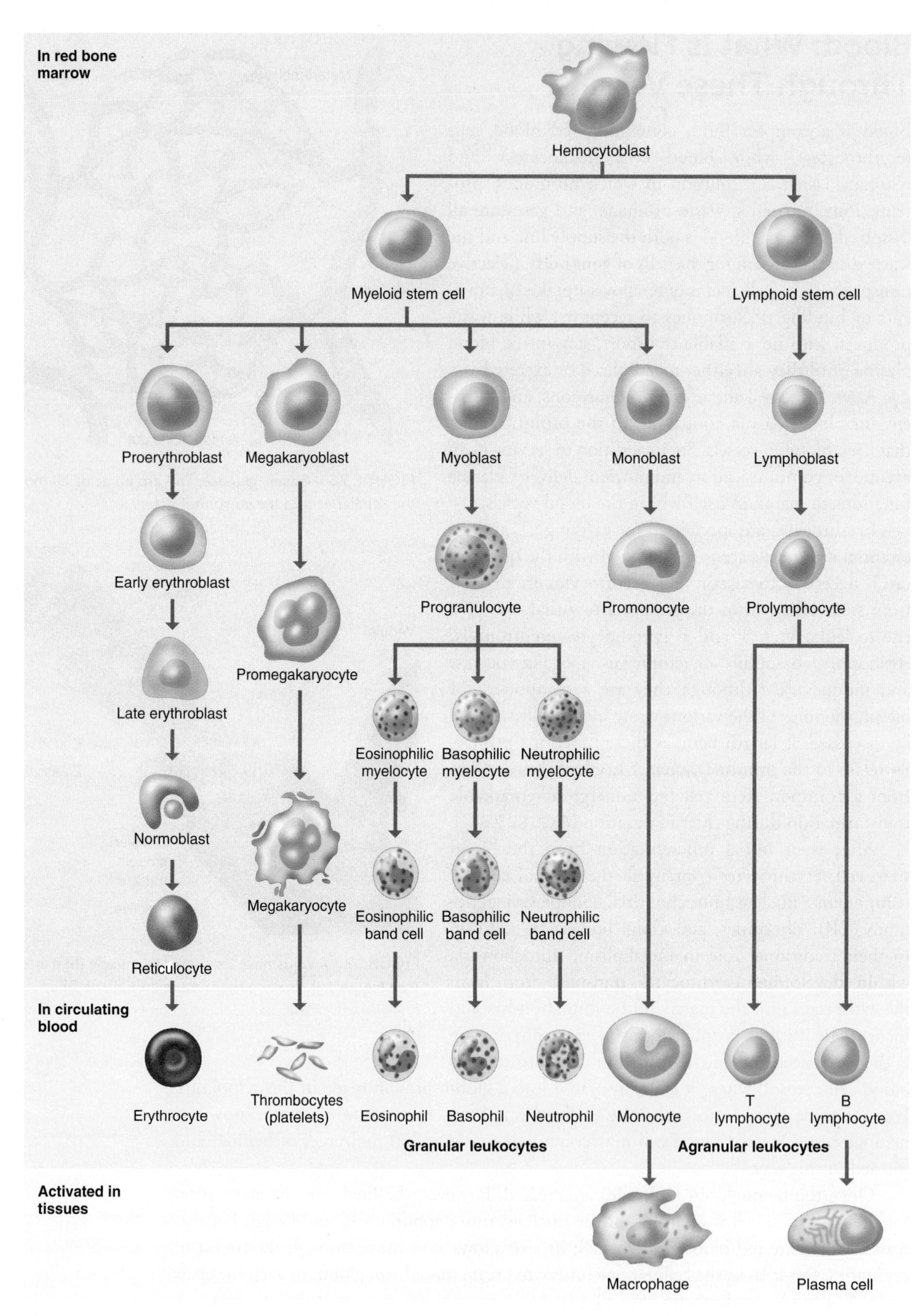

FIGURE 7.8 Red blood cells differentiate from a stem cell and mature into cells without organelles. They are specifically designed to carry hemoglobin. (Adapted from Shier, D. N., Butler, J. L., and Lewis, R. *Hole's Essentials of Human Anatomy & Physiology*, Tenth edition. McGraw-Hill Higher Education, 2009.)

they slide along vessel walls. Shear stress will eventually damage the red blood cell, causing its destruction.

Hemoglobin—A Protein with a Brain?

Hemoglobin within erythrocytes circulates in the cardiovascular system, where it binds O_2 quickly within the pulmonary capillaries and then delivers that O_2 by unbinding it just as quickly in the tissues. How can this protein bind O_2 in one location and unbind it in another? Hemoglobin is a perfect example of protein structure allowing protein function, so let's take a closer look at how hemoglobin works (**FIGURE 7.9**).

Hemoglobin consists of four protein chains, each with a porphyrin ring in its center. The porphyrin ring has a central Fe^{2+} iron, the only form that can bind O_2. The protein chains, with the Fe^{2+}-porphyrin ring centers, function to shield the Fe^{2+} from O_2, allowing O_2 close enough to have an affinity for the iron, but not close enough to irreversibly bind to Fe^{2+}, oxidizing it to Fe^{3+}. Each of the globins has two stable conformational states, tensed, which moves the porphyrin ring so that it has a poor attraction to O_2 and a relaxed state, where the protein changes shape enough to flatten the ring, which brings Fe^{2+} and O_2 into closer proximity, allowing O_2 binding. Movement of the globin proteins thus alters O_2 affinity for the caged Fe^{2+} ion.

What causes the structural transition between these two states? Hemoglobin has multiple allosteric binding sites, binding sites that are different from the site that binds O_2. These binding sites influence the shape of the protein and its transition from the tensed to the relaxed state. Hemoglobin can bind H^+ ions and CO_2, at separate sites, and each will cause the globin protein to move toward the tensed state, that is, the state where O_2 is released from the hemoglobin. Note that H^+ and CO_2 are in higher concentration at the tissues, so O_2 is released where it is needed, at the tissues. In the lungs, where H^+ and CO_2 are in lower concentration, these allosteric binding sites are emptied and hemoglobin enters a relaxed state, conducive to O_2 binding.

Mature erythrocytes have no mitochondria, so they rely exclusively on glycolysis for the production of ATP. If you recall the reactions of glycolysis, an intermediate step forms 1,3-bisphosphoglycerate. Erythrocytes contain the enzyme bisphosphoglycerate mutase,

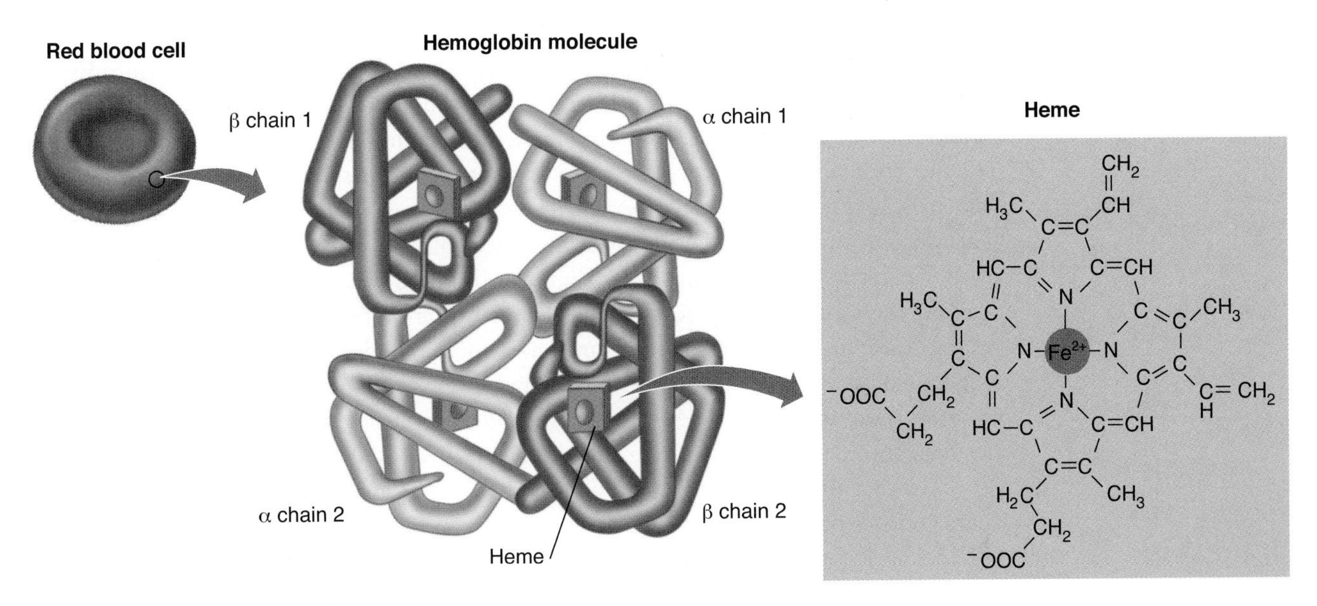

FIGURE 7.9 The structure of hemoglobin.

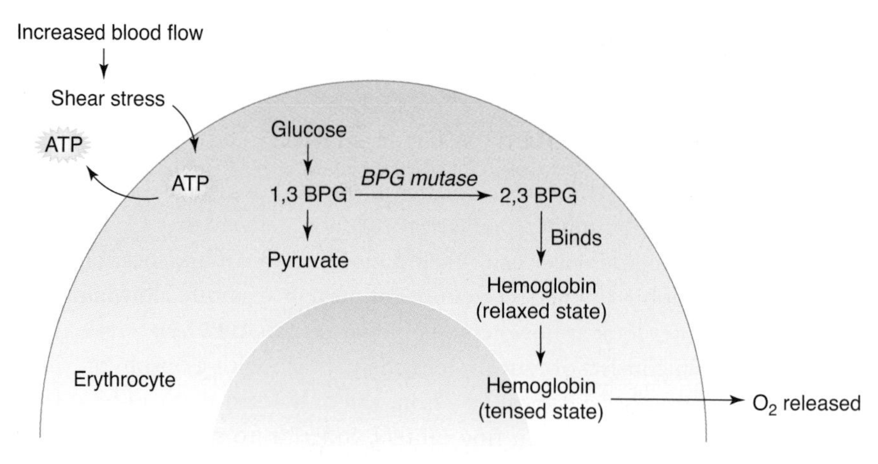

FIGURE 7.10 BPG is made from products of glycolysis. Once bound to hemoglobin, it reduces its affinity for O_2.

which converts 1,3-bisphosphoglycerate to 2,3-bisphosphoglycerate, or BPG. Hemoglobin within the erythrocytes possesses a binding site for BPG, which, when bound, moves hemoglobin toward the tensed state, decreasing its affinity for O_2.

When is BPG produced? Like most cells, erythrocytes have Na^+-K^+ ATPases and Ca^{2+} ATPases in their plasma membrane, both of which require ATP to function. In addition, erythrocytes release ATP in response to shear stress. Therefore, an increase in the rate of blood flow in the vessels will cause an increase in the rate of ATP release and production by erythrocytes. ATP production increases the number of glycolytic cycles, the potential for BPG synthesis, and binding to hemoglobin (**FIGURE 7.10**). Thus, increased blood flow increases O_2 release from hemoglobin, improving O_2 delivery to the tissues.

How Is Erythrocyte Synthesis Regulated?

Because of their limited lifespan, erythrocytes are continuously made. However, the rate of production is not static but is linked to blood oxygen levels. The sensor for PaO_2 lies within the kidney, in fibroblast-like cells of the interstitium of the renal cortex. The kidney receives 25% of the total output from the heart, so the cells of the kidney are in an excellent position to "sample" the composition of blood. If the fibroblast cells sense hypoxia (lowered O_2 levels), then they release the hormone erythropoietin into circulation. Erythropoietin targets stem cells within bone marrow and causes an increase in red blood cell differentiation. In this way, O_2 carrying capacity is maintained within the blood.

Isn't an increase in O_2 carrying capacity always good? As an active person, wouldn't it be beneficial to your athletic performance if you had more erythrocytes? Certainly it seems that way, but there is a downside to an increased number of red blood cells. The number of erythrocytes in blood contributes significantly to its viscosity, or stickiness. Increased viscosity increases resistance to flow. Therefore, greater blood viscosity requires more work by the heart to overcome this resistance and force blood through the vasculature. The heightened performance gained by increasing O_2 is lost via increased cardiac effort needed to move more viscous blood. So there is an ideal level of hematocrit, or red blood cell number, and that is actually the normal range of 40% to 50%.

Now that we understand the structure of the heart and vessels, the composition and characteristics of the fluid flowing through them, and some of the forces that limit flow, let's look at the pump, the heart, in more detail.

What Causes the Heart to Beat?

When we examine skeletal muscle, we see that the stimulus for contraction is the neuronal action potential from an α-motor neuron. While cardiac muscle also contracts in response to an electrical signal, it is an action potential that is spontaneously and continuously generated within the heart itself. The cellular mechanism that initiates the cardiac action potential lies in the sinoatrial node (SA node) and is elegant in its simplicity. We know that ion channel openings and closings cause action potentials. This action potential is no different in that respect. What is unique are the suite of ion channels required for the SA nodal current and their order of opening and closing.

Four voltage-gated ion channels are required for the pacemaker action potential, each with overlapping voltage ranges. The first is a Na^+ channel called hyperpolarization-activated cyclic nucleotide-gated channel (HCN), which yields a current called the "funny current" (I_f)—so named because it looked "funny" (unusual) to the researchers who identified it—which opens at approximately –60 mV. This is a different Na^+ channel than the voltage-gated channel we saw in neurons or muscle. As Na^+ ions enter the cells of the SA node, the membranes are depolarized and gradually the membrane potential reaches the voltage range of a Ca^{2+} channel, called I_{CaT}, for transient calcium current. This channel opens at about –50 mV. As calcium enters the cells, the membranes depolarize further, until the long-lasting calcium channel opens at –40 mV, generating I_{CaL}. In the SA node, the final depolarization and propagation of the action potential comes from calcium channels.

As the cell depolarizes toward 0 mV, the K^+ channels open, K^+ ions leave the cells, and repolarization begins. This is the difference between the SA nodal action potential and any other: as current flowing through open voltage-gated K^+ channels repolarizes the cells, and the membrane potential returns to –60mV, the funny channels open again, to begin another action potential. So, once begun, the SA nodal action potential is self-generating simply due to the overlapping voltage ranges of the ion channels that generate the action potential. The other consequence of this mechanism is that there is no resting membrane potential in the SA node (**FIGURE 7.11**).

This SA nodal action potential propagates in several directions: to the Bachman's bundle in the atria and to the internodal pathways, which lead to the A-V node. The action potential in the Bachman's bundle causes atrial depolarization and then contraction of atrial muscle. The action potential in the internodal pathways causes depolarization at the A-V node. The ion channel distribution in the A-V node is similar to that in the SA node; however, the intrinsic speed of depolarization in the A-V node is slower, so in health, the SA node sets the pace of the heartbeat. Should the SA node fail, the spontaneous depolarizations of the A-V node will take over, but at a slower rate.

From the A-V node, the action potential travels down the bundle branch fibers in the septum of the ventricles to the Purkinje fibers, which run up the exterior cardiac wall. Finally, the action potential moves between ventricular cells, until the entire ventricle is depolarized. However, this action potential is now generated by a completely different set of ion channels. The ventricular action potential is initiated by our more familiar voltage-gated Na^+ channels, similar to the ones that initiate neuronal or skeletal muscle action potentials. Once depolarized to positive voltages, two ion channels open: the L-type (long-lasting) Ca^{2+} channel and a voltage-gated K^+ channel. Ca^{2+} ions enter the cells, depolarizing them further, while K^+ ions simultaneously exit the cells, causing repolarization. These two currents electrically negate one another, so that the ventricular action potential has a plateau and depolarization is maintained for approximately 150 milliseconds, even though the Na^+ channels have inactivated. As the L-type Ca^{2+} channels inactivate, the K^+ repolarization predominates

FIGURE 7.11 (a) The action potential in the SA node is continuous and triggers action potentials in the atria and ventricle of the heart. (b) Four ion currents generate the continuous SA nodal action potential.

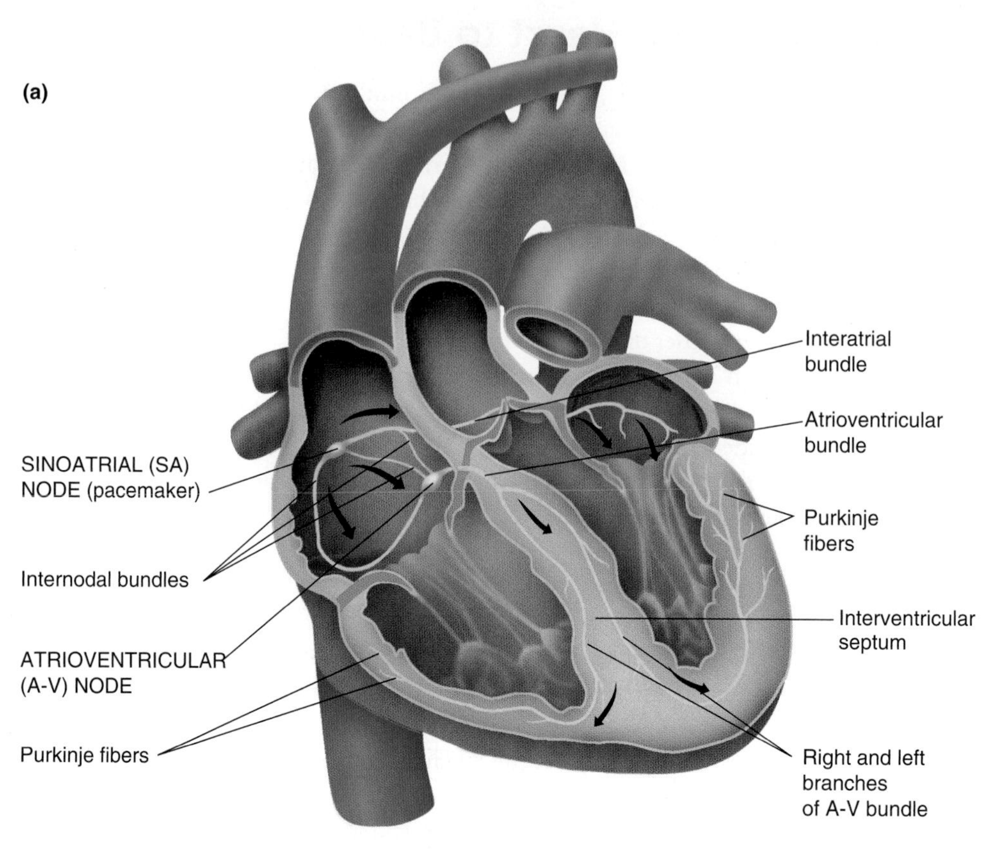

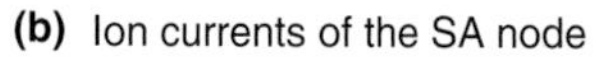

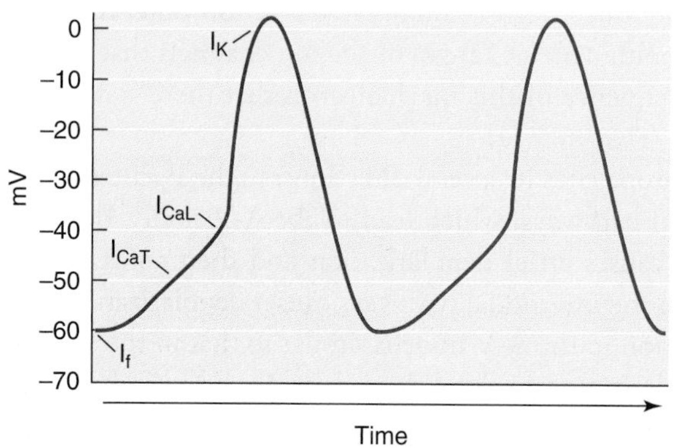

and the ventricular action potential returns to its resting membrane potential of –90 mV (**FIGURE 7.12**). Ventricular cells will remain at this membrane potential until another action potential is triggered by the SA node.

Inactivation of Na^+ channels, a normal behavior of that channel as we have already seen in the somatic nervous system, allows for a refractory period. In cardiac tissue, the absolute refractory period, when most of the Na^+ channels are inactivated, occurs during the plateau phase. The relative refractory period occurs during the K^+ repolarization phase of the action potential (Figure 7.12). Since an action potential can be initiated during the relative refractory period, any action potential that begins inappropriately, stimulating ventricular cells during this relative refractory period, may cause a lethal ventricular arrhythmia.

What Makes the Heart Rate Increase or Decrease?

Our heart rate is not always the same, as you noticed on your run. Exercise or any physical activity will cause heart rate to increase, while relaxation slows it. We know that the autonomic nervous system can influence heart rate, through the sympathetic and parasympathetic branches. Rate change is accomplished simply by modifying the conduction of the SA nodal channels.

As you begin your run and sympathetic stimulation begins, as it always does with exercise, remember that norepinephrine is released from sympathetic nerve fibers and has the heart as its target. One specific cardiac target is the SA node. Norepinephrine, binding to β1 at the SA node, will cause phosphorylation of ion channels, via protein kinase A. Phosphorylated ion channels increase heart rate in two ways: (1) by increasing the I_f current, causing it to reach the threshold for calcium channels faster, or (2) by increasing I_{Ca} currents, causing the cell to reach 0 mV faster. Steepening the slope of depolarization for all three currents results in a faster overall rate, that is, your heart beats faster (**FIGURE 7.13**). Circulating epinephrine released from the adrenal medulla will work the same way, so as your workout continues, you will experience heart rate acceleration from two separate sources: one neurotransmitter and one hormone.

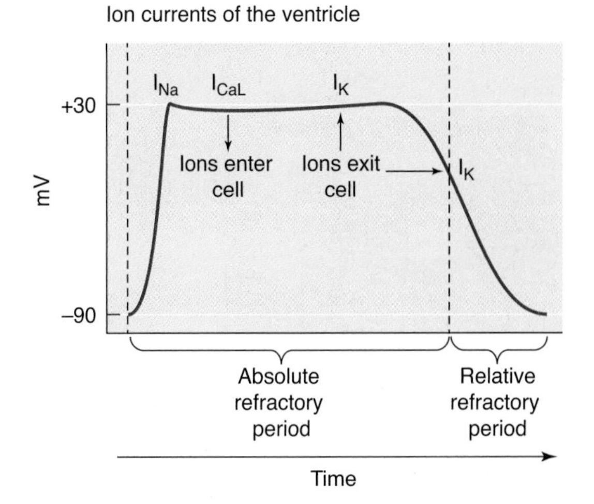

FIGURE 7.12 The action potential of the ventricle uses a different set of ion currents for depolarization. The refractory periods of the ventricular action potential can be important during dysrhythmias.

At the conclusion of your run, you feel your heartbeat slowly return to normal. As you recall, the heart is dually innervated by the sympathetic and parasympathetic autonomic pathways. While the sympathetic nerves increase heart rate, as described above, the vagus nerve, from the parasympathetic system, releases acetylcholine, which slows the heart. The primary mechanism for slowing the heart comes from activation of GIRK channels. Remember that acetylcholine binds to muscarinic acetylcholine receptors at the target tissues and that the muscarinic receptors are G-protein coupled receptors. At the SA node, acetylcholine binding causes a unique signaling mechanism. The G protein uncouples from the receptor and binds to a K^+ channel, the G protein-regulated inward rectifying K^+ channel, or GIRK. As a result, the K^+ channel opens and K^+ ions flow outward, hyperpolarizing the membrane potential, moving the SA node further from depolarization. Thus, it takes longer for I_f and I_{Ca} to depolarize the pacemaker cells, the slope of depolarization becomes shallower, and the heart rate slows.

Exercise provides a graphic example of heart rate change, but there are normal variations in heart rate even when you are at rest. This is due to the dual innervation of the heart by both the sympathetic and parasympathetic nervous systems. So, even at rest, the phosphorylation reactions will increase your heart rate for a few beats, while opening of GIRK channels will slow your heart rate for a few beats. We do not notice this variation, but it can easily be seen using an electrocardiogram, or ECG.

What Is an ECG and What Information Does It Provide?

An ECG is a noninvasive measure of the sum of all electrical activity of the heart. It is not showing us an action potential, but the total electrical change generated by all the action

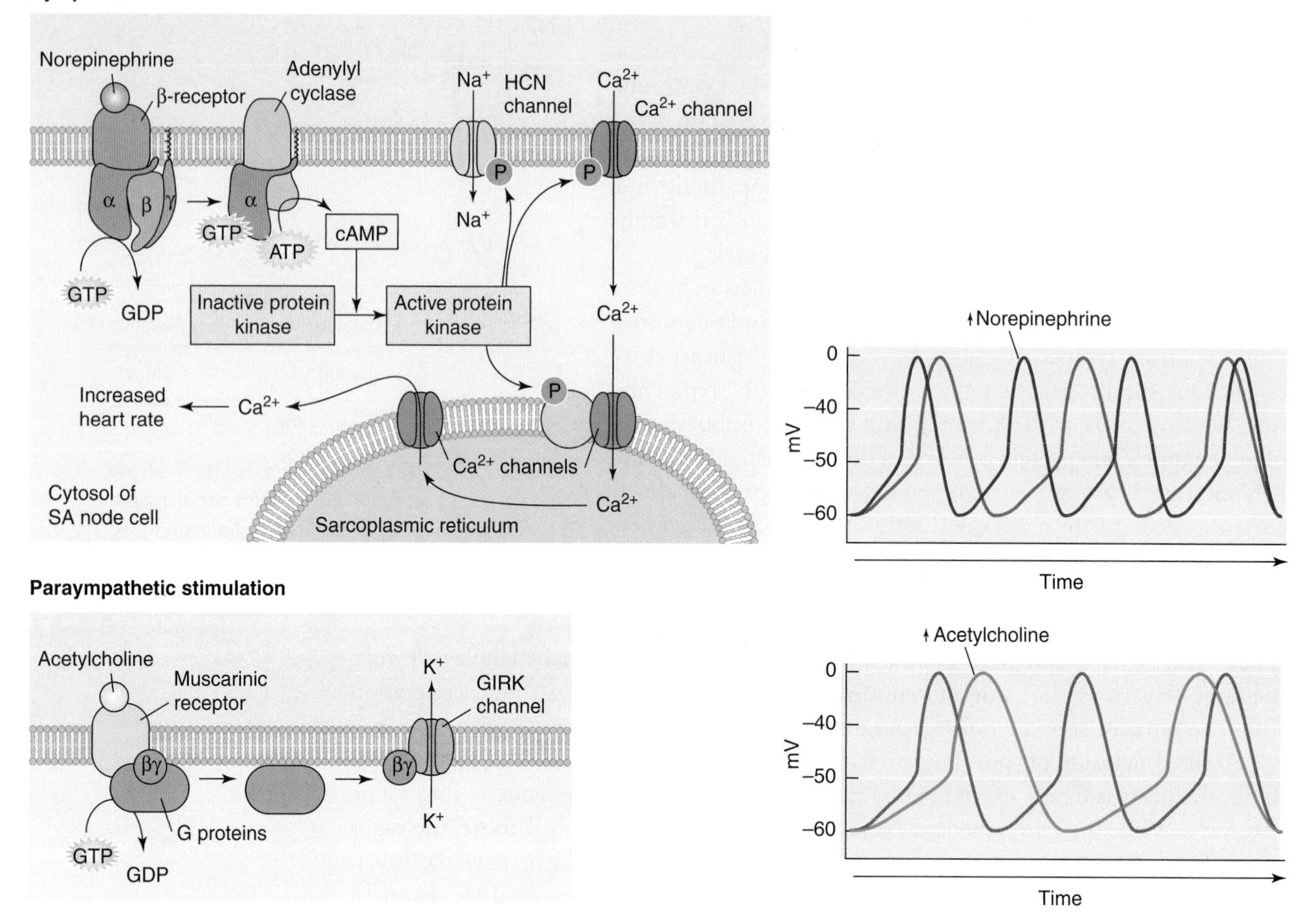

FIGURE 7.13 Sympathetic and parasympathetic nerve fibers modulate heart rate at the SA node, using adrenergic or muscarinic receptors to modify ion channel activity.

potentials that travel through the heart during a depolarization, from the SA node to the atria to the A-V node to bundle branch fibers to the ventricle and then the reverse trip of repolarization. An ECG tells us nothing about contractility or performance but can nonetheless provide an amazing amount of information. Let us first examine the tracing of lead II in a three-lead ECG, as seen in **FIGURE 7.14**.

As the pacemaker cells begin to depolarize, the ECG tracing starts up the P wave, which is completed as the electrical depolarization travels across the atria. The interval between the P and R waves is a straight line, as the impulse is slowed at the A-V node. The Q wave is the movement of the action potential through the bundle branch fibers, and the large R wave is the depolarization of the ventricle. The ST segment is relatively flat, because the entire ventricle is depolarized at this point. Finally, during the T wave, the ventricle repolarizes (Figure 7.14a). There is a resting membrane potential, before the entire complex begins again.

Each of the ECG leads provides an electrical "snapshot" of one area of the heart. Three-lead ECGs are now used only in teaching laboratories, however; clinical ECGs use 12 to 32 leads, giving the physician electrical views of the heart from every angle. If the heart has ischemic tissue or necrotic zones, these will show as electrical aberrations. But the most powerful use of ECG is in tracking cardiac rhythm and rhythm disturbances.

Consider an ECG recorded on you before and during your run. Prior to the run, your R-R interval, the duration between depolarizations of the ventricle, may have been one second apart. That would mean your heart rate was approximately 60 beats/minute. If you were to measure every R-R interval during a five-minute rest period, you would quickly notice that the R-R intervals are not always the same duration. They will actually alternate between runs of long intervals and runs of slow intervals. If you recall that heart rate is under control of both the sympathetic and parasympathetic nervous systems, you will realize that neither system is always dominant, but they trade off being "in charge," causing this oscillation of R-R interval duration. As you begin your run, the R-R intervals would get closer together as your heart rate increased, because sympathetic stimulation becomes more consistently dominant. Thus, R-R interval can be used as a measure of instantaneous heart rate and an indicator of autonomic function.

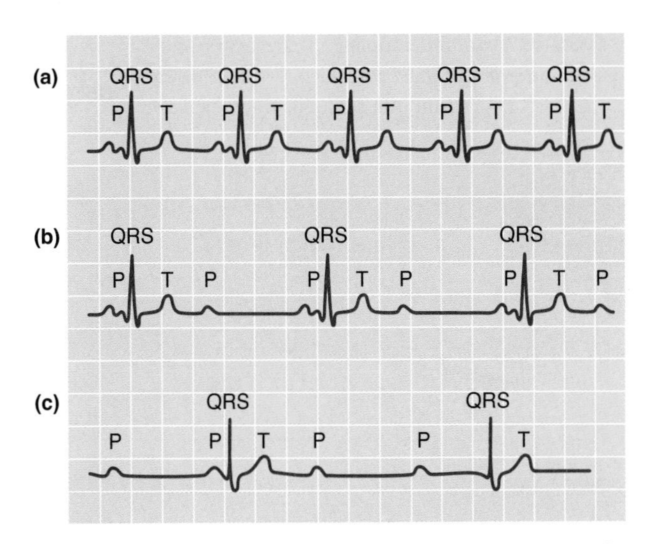

FIGURE 7.14 (a) An ECG measure the electrical activity of the heart. (b) During partial A-V block only some P waves are followed by QRS waves. (c) During complete block, P waves and QRS waves occur independently.

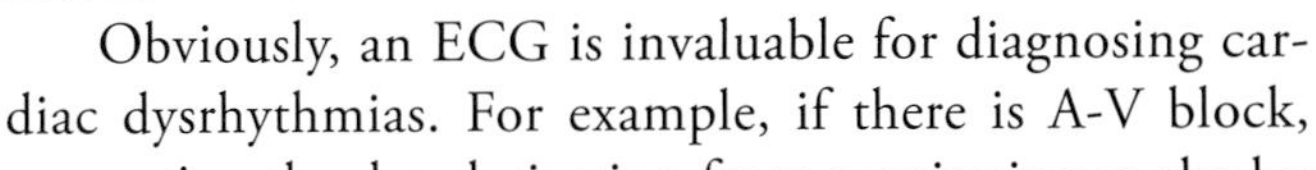

Obviously, an ECG is invaluable for diagnosing cardiac dysrhythmias. For example, if there is A-V block, preventing the depolarization from continuing to the bundle branch fibers, a P wave may occur singly, without being followed by a QRS complex. This may happen all the time in complete block (Figure 7.14c), or only sometimes, as is first-degree block (Figure 7.14b). Ventricular dysrhythmias can also be diagnosed and are the most serious cardiac dysrhythmias since they will compromise cardiac contraction and cardiac output.

An ECG is a measure of electrical activity, but in most cases, once the cardiac tissue is depolarized, the cardiac muscle will contract. What happens during cardiac contraction, and how does it differ from skeletal muscle contraction?

Cardiac Excitation-Contraction Coupling

How does the depolarization of cardiac myocytes cause cardiac contraction? Excitation-contraction (E-C) coupling in cardiac muscle is similar to E-C coupling in skeletal muscle, but it has some unique features. Let's follow the process in detail. The ventricular action potential begins with a voltage-gated Na^+ current, which provides the depolarization. Remember that voltage-gated Na^+ channels inactivate over time, so this current ceases after several milliseconds. Voltage-gated Ca^{2+} channels open during the sodium depolarization, and, just as in skeletal muscle, the Ca^{2+} channels are located within the T-tubules, which lie very close to the sarcoplasmic reticulum (SR). In skeletal muscle, a change in SR membrane voltage caused the opening of calcium-release channels in the SR. In cardiac muscle, it is Ca^{2+} ions flowing through L-type Ca^{2+} channels of the T-tubule that trigger SR calcium release channels to open. This trigger calcium, from the extracellular fluid, is absolutely required for cardiac E-C coupling. After intracellular $[Ca^{2+}]$ rise, the binding of Ca^{2+} to troponin C, movement of tropomyosin, revealing the actin binding site, and binding of myosin heads to cause the sliding filament mode of contraction continue as in skeletal muscle. Relaxation occurs as Ca^{2+} is resequestered in the SR by the SERCA pump, an ATPase (**FIGURE 7.15**). Just as in skeletal muscle, myosin requires ATP for unbinding from actin, and ATP is also required for pumping Ca^{2+} up a concentration gradient, back into the SR.

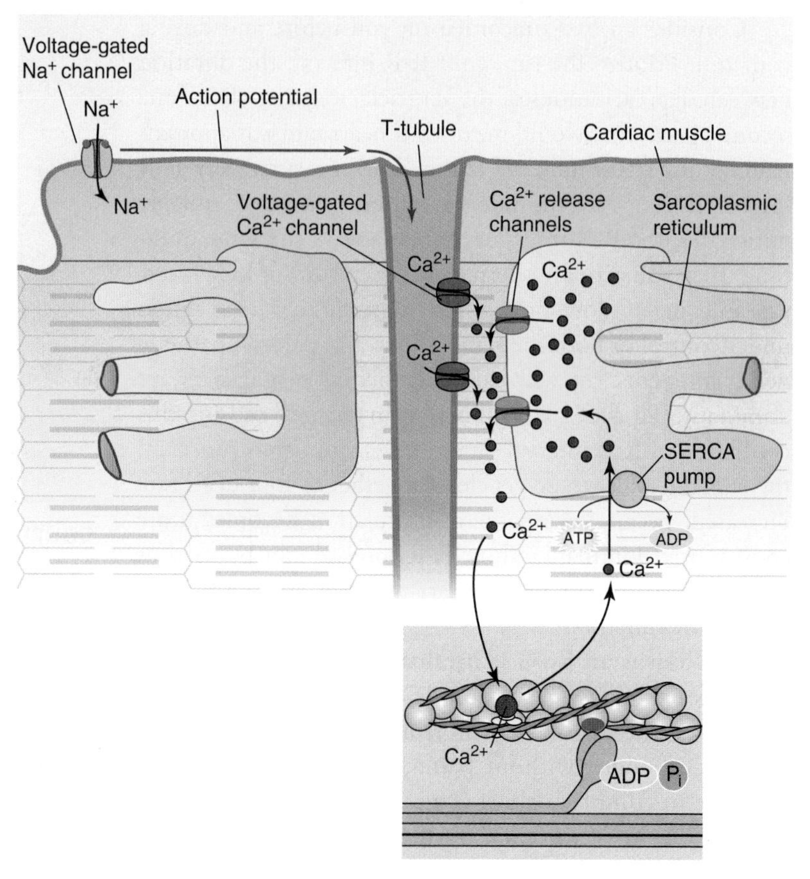

FIGURE 7.15 E-C coupling in cardiac muscle is dependent upon extracellular Ca^{2+} to begin release of Ca^{2+} from the sarcoplasmic reticulum.

What Makes My Heart Pound When I Run?

During your run, your heart not only depolarizes more quickly but also there is an increase in contractility, the force of contraction, which you experience as your heart pounding. The same sympathetic nerve fibers that innervate the pacemaker cells also branch off to the ventricle. Norepinephrine released from the nerve terminals will bind to β1 receptors, phosphorylate I_{Ca} channels, and increase the amount of extracellular calcium that enters the cardiac myocytes. More trigger Ca^{2+} will increase the local Ca^{2+} concentration at the calcium release channels and stimulate their opening. The calcium release channel is part of a larger protein complex known as the ryanodine receptor. The release channel is made up of four protein tetramers that open in proportion to the amount of calcium they sense. Therefore, increased trigger Ca^{2+} is directly responsible for enhanced release of Ca^{2+} from the SR.

Sympathetic stimulation of the ventricular cells through β1 receptors also increases the rate of the SERCA pump, by phosphorylating an associated regulatory protein, phospholamban (**FIGURE 7.16**). Normally, phospholamban partially inhibits the SERCA pump, but when it is phosphorylated, this inhibition is relieved. SERCA moves Ca^{2+} from the cytosol back into the SR at an accelerated rate, Ca^{2+} is removed from troponin, and cardiac myocyte relaxation can occur more quickly, a lusitropic, effect. In addition, the increased sequestering of Ca^{2+} into the SR loads it, so that more Ca^{2+} is available for release following the next release of trigger Ca^{2+}. Increased intracellular Ca^{2+} will allow greater force of contraction. At the whole organ level, phosphorylation via β1 receptors and protein kinase A will allow the ventricle to relax and fill completely, so that when depolarization begins, the force of contraction, that is, contractility, will be greater. More blood will be forced out of the heart at each beat, and stroke volume will increase.

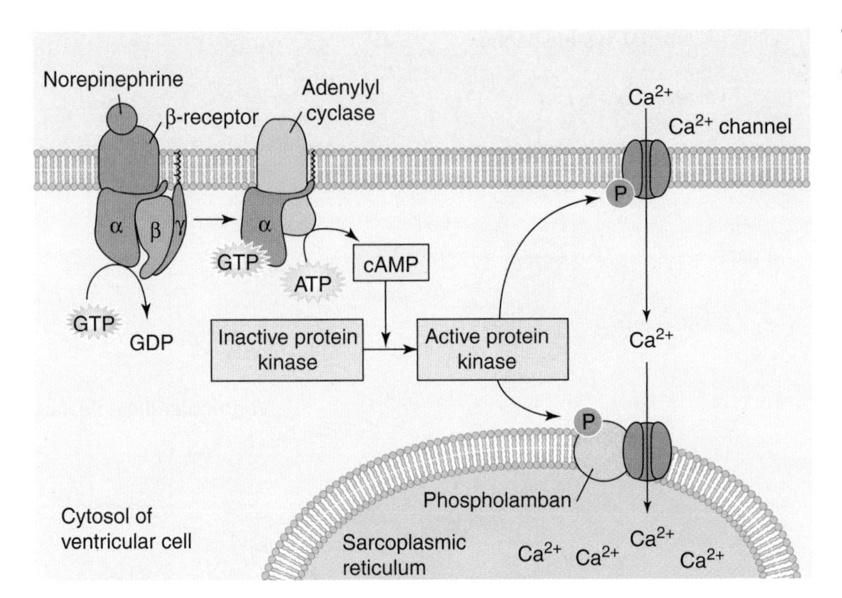

FIGURE 7.16 β-adrenergic stimulation increases contractility and stimulates the SERCA pump.

Cardiac Cycle: The Mechanics of Moving Blood Through the Heart

The heart is a living pump doing work against an incompressible fluid, blood. Forces act upon the heart, and it in turn generates forces. Let's look at the cardiac cycle in detail (**FIGURE 7.17**).

Blood returns to the heart from the superior and inferior vena cavae, emptying into the right atrium. This is a continuous process during diastole. As the right atrium fills, the pressure in the right atrium increases until it exceeds that of the right ventricle, causing the right A-V valve, the tricuspid valve, to open passively, allowing the right ventricle to fill. Simultaneously, oxygenated blood is returning to the left atrium from the lung through the pulmonary vein. As left atrial pressure increases, the left A-V valve, the mitral valve, opens, allowing the left ventricle to fill passively. All this occurs during diastole. Once the SA node depolarizes and begins an action potential, the depolarization of the atria causes atrial muscle contraction, which forces blood from the two atria into their respective ventricles. This increases the volume of blood in the two ventricles. As the depolarization travels down the bundle branch fibers, Purkinje fibers, and ventricular wall, muscle contraction of the ventricles begins. There is a period during which the ventricles contract but have not generated enough force to open the pulmonary semilunar valve, which lies between the right ventricle and the pulmonary trunk, or the aortic semilunar valve, which lies between the left ventricle and the aorta. This phase is called isovolumetric contraction, because the heart is contracting but the volume contained within the heart is not changing. The A-V valves also remain closed during isovolumetric contraction, preventing blood from returning to the atria. This occurs even though the pressure in the ventricles exceeds that of the atria.

How does this happen? The A-V valves are tethered to the ventricular wall by chordae tendineae, which connect to papillary muscles embedded in the ventricular wall. The papillary muscles contract when the ventricle depolarizes, pulling the flaps of the A-V valves closed. Thus, blood is prevented from regurgitating into the atria during ventricular contraction.

When the ventricular muscle contracts strongly enough to generate more pressure on the blood within the ventricle than exists within the arterial tree, the semilunar valves open

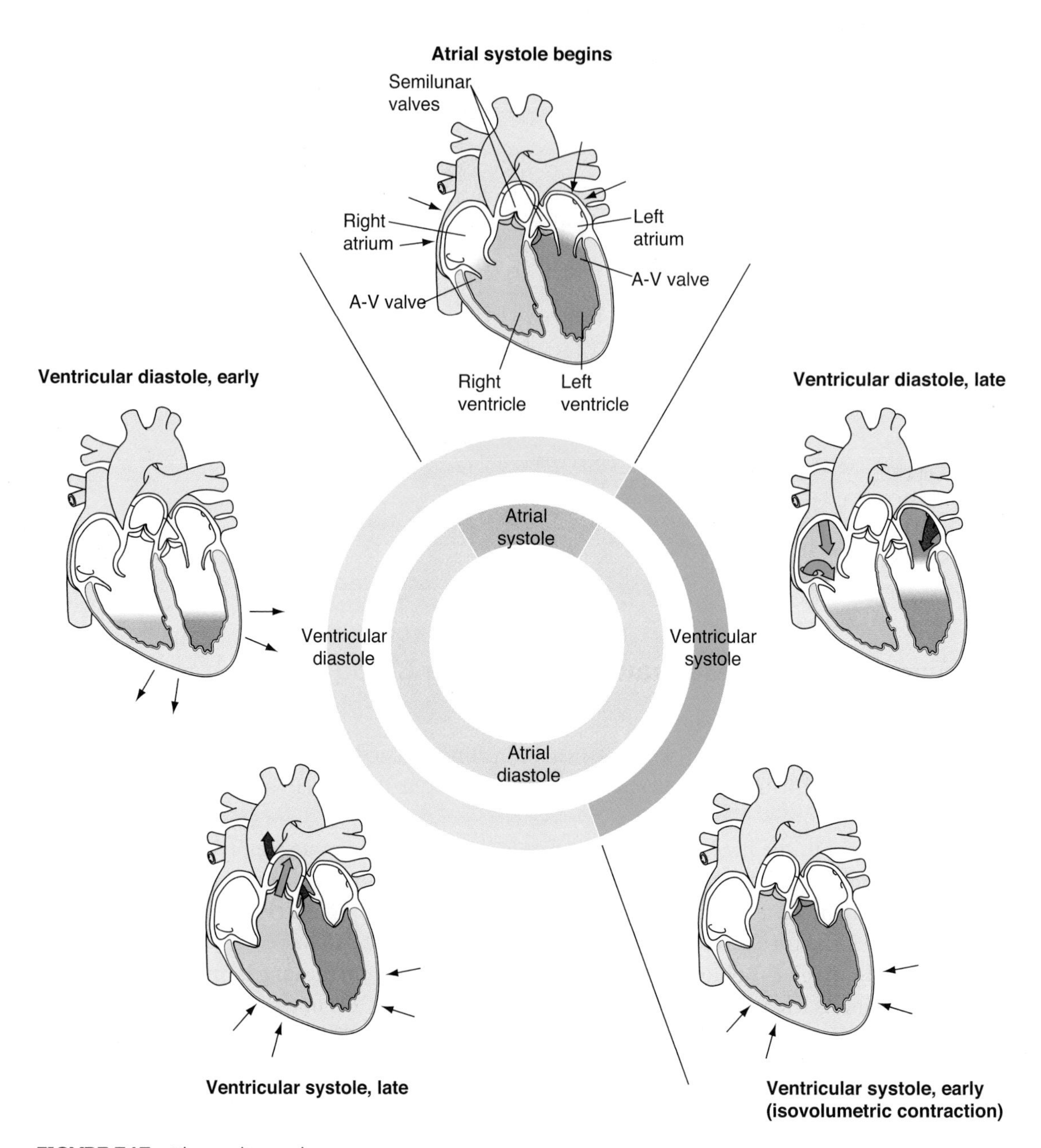

FIGURE 7.17 The cardiac cycle.

and blood is ejected from the ventricles. This is the stroke volume, or ejection volume. It also marks cardiac systole. When repolarization begins and the heart muscle relaxes, diastole and cardiac filling begin again.

What Is Cardiac Output and How Is It Regulated?

Cardiac output is defined as the volume of blood ejected from the heart per minute, but it is most easily summarized by this simple equation: CO = HR × SV—that is, cardiac output = heart rate × stroke volume. Therefore, anything that changes heart rate or stroke volume will change cardiac output. It is actually easiest to think about the components of heart rate and stroke volume separately, especially because in disease, each can be affected uniquely.

Heart Rate

We know that autonomic stimulation is an important controller of heart rate. On your run, sympathetic stimulation will increase your heart rate via β1 receptors, and when you finish your workout, parasympathetic stimulation will slow heart rate via muscarinic receptors. ANS control is the primary controller of heart rate. However, there is another mechanism called the Bainbridge reflex. An increase in atrial pressure, caused by increased return of blood to the heart, causes stretch in receptors located at the junctions of the atria and the returning vessels, the superior and inferior vena cavae and the pulmonary veins. Stretch of these receptors stimulates a selective sympathetic reaction and increase in heart rate without an increase in stroke volume. So, an increase in blood returning to the heart, even momentarily, will reflexively increase heart rate.

Stroke Volume

Stroke volume is dependent upon many factors and adds complexity to the regulation of cardiac output. The absolute volume of blood in the body is an important determiner of venous return, or the amount of blood that returns to the heart during diastole. And because the heart can only pump out as much as is returned, venous return is similarly a vital component of cardiac output. Finally, the amount of time the heart is in diastole will determine the cardiac filling time, that is, the amount of time available for venous return. The longer diastole continues, the greater the amount of blood that can return to the heart and the greater the venous return. These three factors together—blood volume, venous return, and filling time—will determine cardiac preload, or the amount of blood loaded into the heart prior to systole.

It is easy to imagine the consequences of changes in preload. If you were dehydrated during your morning run and had a lower blood volume, there would be less blood returning to the right atrium and less preload. As your heart rate increases during the run, filling time is reduced, and there is less venous return and, therefore, less preload (**FIGURE 7.18**). While this sounds disastrous for the efficiency of the cardiovascular system during exercise, there are several compensatory mechanisms.

Exercise causes sympathetic stimulation, which changes vascular tone. Alpha receptors on the arterioles and veins bind norepinephrine, which will cause vasoconstriction. Venoconstriction, especially, will increase the functional blood volume returning to the heart. Remember that the veins are capacitance vessels that expand when filled. Thus, during constriction, blood is forced toward the heart, increasing venous return. There is another vascular reserve that is recruited by the sympathetic stimulation—constriction of blood vessels serving the stomach and intestines. Vasoconstriction shunts this blood away from those organs and toward skeletal muscle, where blood vessels are dilated due to β2 receptor stimulation by epinephrine. Sympathetic stimulation will also shorten systole due to the increased rate of the SERCA pump. Shortened systole lengthens diastole and filling time, again increasing venous return. All of these effects of sympathetic activation will increase preload and, therefore, increase cardiac output.

In the early 1900s, two physiologists named Otto Frank and Ernest Starling observed that the heart pumped out what was delivered to it, or that stroke volume was dependent upon venous return. This phenomenon was not understood but was well recognized and came to be known as the Frank-Starling law of the heart. The heart contracts against an incompressible fluid, blood, which acts as a rigid component for muscle to contract

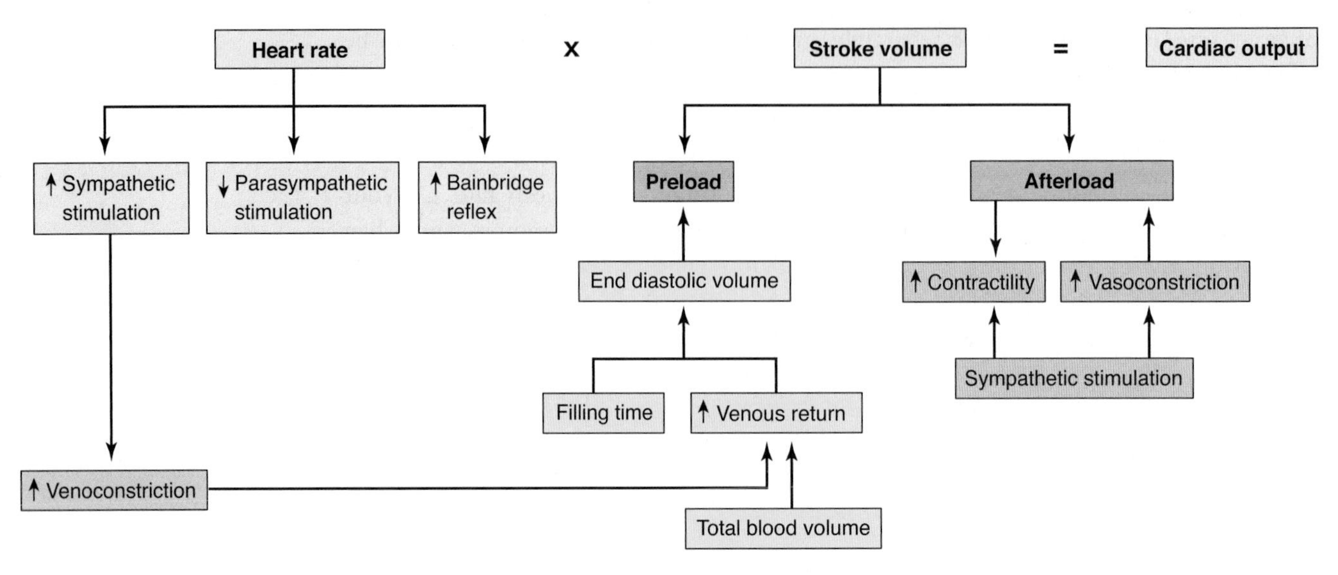

FIGURE 7.18 Factors that change heart rate or stroke volume will change cardiac output.

against. We know that the normal resting length of skeletal muscle is determined by its attachment to bone, and this normal resting length is the point at which actin and myosin are in the ideal anatomical positions for maximal binding. It is different in the heart. While you are sitting quietly, cardiac muscle is stretched minimally by the blood going through it and cross-bridge cycling is inefficient. When more blood fills the heart, as during exercise, the ventricles stretch and actin and myosin are put into improved register. More cross-bridge cycles can be formed, which improves contractility (**FIGURE 7.19**). This is the molecular basis of the Frank-Starling law of the heart, which links increased preload to increased contractility. Therefore, an increase in preload will cause greater contraction of the heart and more blood will be ejected at each beat, that is, a greater stroke volume.

Two final aspects of sympathetic stimulation increase stroke volume. First, phosphorylation of the L-type Ca^{2+} channels in myocytes will increase channel conductance and provide more calcium for cross-bridge cycling, thus improving contractility. Secondly, vasoconstriction of the arterioles will increase the afterload. Remember from the cardiac cycle that the ventricle must generate enough force to overcome the pressure in the vascular tree. When arterioles are constricted, this force increases, so the ventricle must generate more muscle contraction and more force to expel blood. This increase in contractility, during healthy exercise, will increase stroke volume. When this occurs chronically, as in high blood pressure, it places an excessive burden on the heart. However, during exercise, the increase in afterload and contractility is normal and desirable.

All these adjustments to cardiac output are effective only if cardiac rhythm is maintained. If cardiac rhythm is disturbed, all of the regulatory mechanism just discussed will be compromised.

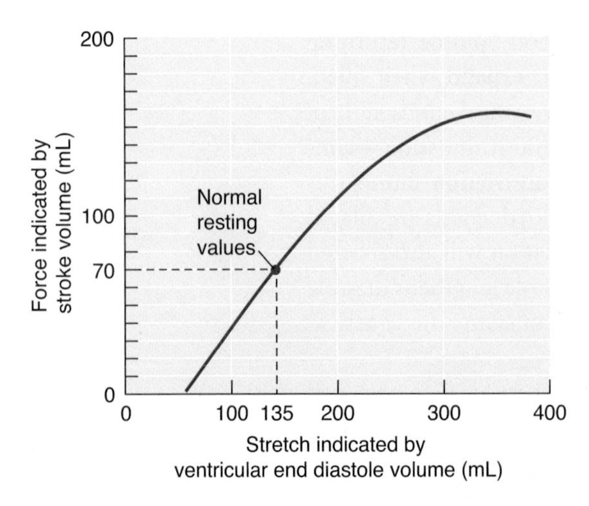

FIGURE 7.19 The Frank-Starling law of the heart reflects changes in myocardial stretch.

Metabolism, O_2 Consumption, and Cardiac Work

Cardiac myocytes are dependent upon O_2 for their metabolic needs. While other animal species maintain glycogen stores within the heart, humans have a very limited supply, available only for minutes of cardiac work. Because the heart must continually be supplied with O_2, there is a lavish coronary circulation within the heart; the myocytes themselves contain the O_2 storage molecule myoglobin, a single-chain protein resembling hemoglobin. Fatty acids, which are metabolized within the mitochondria in an oxygen-dependent process, are the predominant fuel of cardiac muscle. Thus, it is no surprise that mitochondria are also concentrated within cardiac tissue. If fatty acids are not available, the heart uses glucose or lactic acid as an energy source. Note that during exercise, when skeletal muscle is releasing lactic acid into the blood supply, this partially metabolized molecule can be used by the heart as long as O_2 is available. All this means that the heart is very capable of sufficient ATP production, as long as O_2 is present.

What could limit O_2 supply to the heart? Besides the obvious coronary occlusion, which would limit O_2 supply to some portions of the heart, there is another limitation to O_2 delivery. The coronary microcirculation is embedded in the myocardial wall, that is, among the cardiac myocytes themselves. This is necessary if these cells are to be supplied with O_2. However, during systole, when the heart contracts, the muscles often produce enough force to close the capillaries, preventing blood flow and O_2 delivery. Therefore, the heart receives its nutrients primarily during diastole. The longer the period of diastole, the greater nutrient and O_2 availability there is for cardiac myocytes.

Cardiac work, like all physiological work, is measured in ATP usage. We know that cardiac output is HR × SV—but which requires less cardiac work, a faster heart rate and smaller stroke volume or a slower heart rate and a larger stroke volume? Experiments have shown that a slower heart rate and larger stroke volume require less O_2 consumption and, therefore, less cardiac work (**FIGURE 7.20**). It also allows more time for cardiac perfusion. The natural adaptation of training will increase stroke volume and lower heart rate, providing the greatest cardiac output for the least amount of cardiac work.

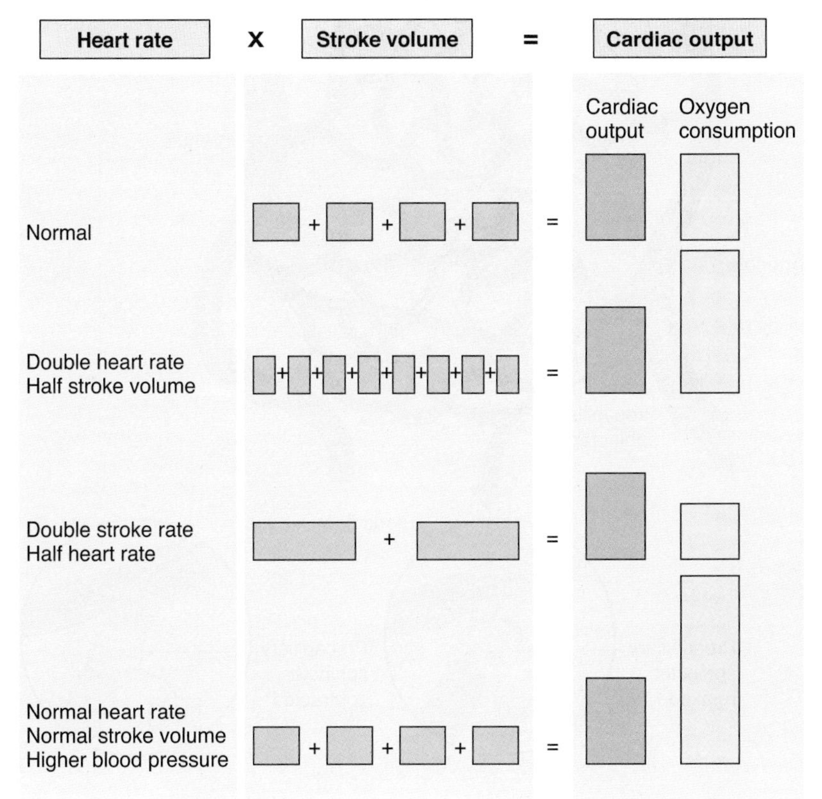

FIGURE 7.20 Cardiac output is a function of heart rate and stroke volume, an increased heart rate uses more ATP than increased stroke volume. Energetically, it is better to increase stroke volume.

Blood Vessels Carry Blood to Tissues—The End Users

If all vessels were completely dilated and fixed at that diameter, then all tissues would be maximally perfused, given a sufficient pressure head. That would mean that nonworking tissues would receive the same amount of blood flow as working tissue. This would be inefficient and wasteful of cardiac work. In fact, blood vessels continually alter their diameter, responding to tissue need and directing blood flow to working tissues.

Anatomically, a layer of smooth muscle surrounds arteries, arterioles, and, to a smaller extent, veins. Vascular smooth muscle responds to local factors that are released by the tissues themselves. For example, CO_2, H^+, K^+, and adenosine (a by-product of ATP hydrolysis) can all cause vascular smooth muscle to relax and dilate, increasing the vascular diameter and increasing blood flow (**FIGURE 7.21**). If you think about the skeletal muscles in your legs during your run, muscle contraction is causing the production of each of these ions, gases, and molecules. Locally, the blood vessels will dilate and increase perfusion to these working muscles. This local control of blood vessel diameter is an important and efficient regulator of blood flow.

The sympathetic nervous system also regulates blood vessel diameter from its central location. Norepinephrine released from sympathetic nerve terminals will bind to α receptors on smooth muscle of vessels in the gut, causing vasoconstriction and reducing blood flow to the stomach and intestines. Blood flow to the kidneys will also be reduced because of vasoconstriction of the afferent arteriole. Constriction of veins will decrease their capacitance and increase venous return to the heart.

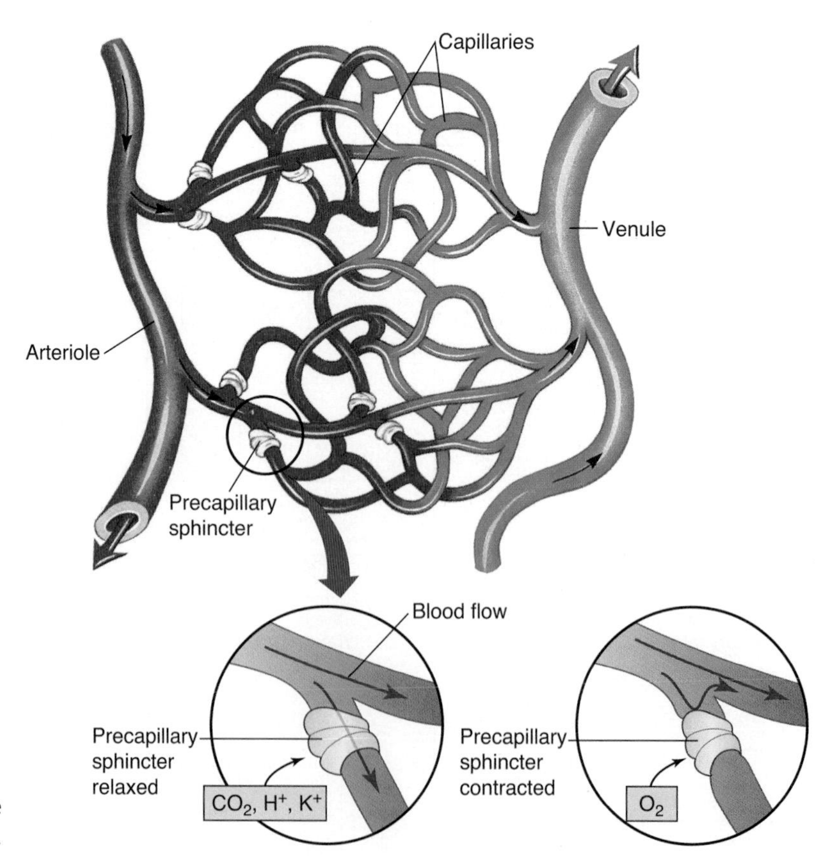

FIGURE 7.21 Metabolites produced by tissue regulate flow through local blood vessels.

Epinephrine released from the adrenal medulla will bind to β receptors on smooth muscle of blood vessels serving skeletal muscle, causing vasodilation and increased blood flow. Once again, the difference in receptor location determines the differential effects of sympathetic stimulation. Even if norepinephrine bound to a β receptor in the skeletal muscle, the effect would be vasodilation. It is the receptor not the ligand that causes the effect.

The physiological signal for nitric oxide production by vascular endothelial cells is shear stress. During exercise, the increased flow of blood through the vessels will increase shear stress and nitric oxide production. Nitric oxide will diffuse through the membranes of smooth muscle, inhibiting myosin binding and activating the SERCA pump, thus reducing [Ca^{2+}]. Smooth muscle contraction is inhibited and blood vessels dilate.

So, during your run, local control, that is, metabolites from working skeletal muscle, will increase vascular diameter, shear stress will dilate vessels in working muscle, and adrenergic receptors located on vascular smooth muscle will reorganize blood flow to favor maximal perfusion of skeletal muscle and increase venous return.

Blood Flow: Behavior of Fluids

There are two primary determinants of flow to the tissue: pressure head generated by the heart and resistance to flow, primarily determined by vessel radius.

A reduction of vascular diameter decreases flow to tissues downstream of these vessels because of increased resistance. Resistance to flow is proportional to vessel radius to the fourth power. Vessel diameter reduction will increase total peripheral resistance (TPR) and reduce flow. Each section of the arterial tree has parallel branching; arteries constitute one set of branches, arterioles another, and capillaries yet another. The site of greatest control of resistance is the arterioles. This occurs because arterioles do not branch as much as other portions of the arterial tree, resulting in fewer parallel arterioles, a smaller relative cross-sectional area, and, therefore, increased resistance. Also, smooth muscle in the arterioles allows for greater constriction, increasing resistance to flow and causing less blood flow.

However, TPR will increase contractility of the heart (remember, it must overcome afterload to eject blood). This will increase the pressure head and flow to areas where there is not vasoconstriction. Flow is proportional to $P_1 - P_2$, or pressure at the pressure head minus pressure at the right atrium. So, increased cardiac work will increase flow (**FIGURE 7.22**).

If Flow Is Essential to Tissues, Why Is Arterial Blood Pressure Important?

Resistance can regulate flow selectively, but flow is irretrievably compromised by a reduced pressure head. Maintaining arterial pressure allows for the greatest flexibility of control of flow, simply by altering resistance, that is, vascular diameter. Remember that increasing TPR will oblige the heart to contract more forcefully to overcome the resistance, thus increasing pressure head, so the heart and vasculature work together on this. The system is designed to maintain pressure head while selectively changing vascular diameter to move blood around to where it is needed.

What is the mean arterial pressure? Mean arterial blood pressure can be summarized by this simple equation:

MAP = pressure during diastole (P_2) + (P during systole – P during diastole)/3

This quantifies the mean pressure in the arterial tree over time, compensating for the difference between systolic and diastolic pressure and the amount of time spent in each part of the cardiac cycle.

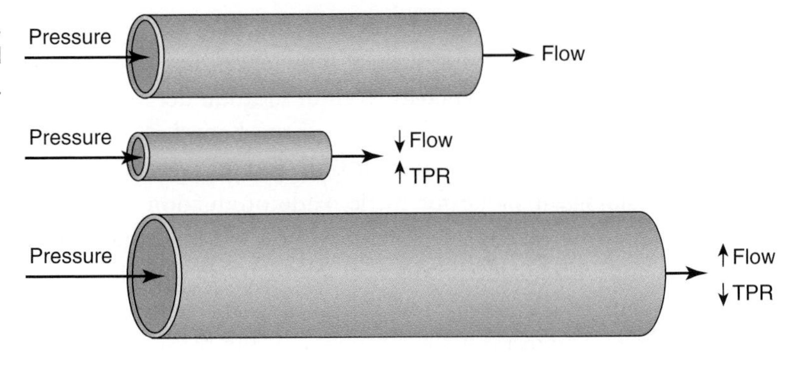

FIGURE 7.22 The pressure head generated by the heart, viscosity of the blood, and resistance of the blood vessels all contribute to flow within the cardiovascular system.

What Regulates Mean Arterial Blood Pressure?

The sympathetic nervous system can work for minutes to hours to days to regulate blood pressure, depending on stress or activity. During the course of your exercise, for example, the sympathetic nervous system increases heart rate and stroke volume, vasoconstriction of some vessels, and vasodilation of others. All this increases your exercising blood pressure slightly.

In addition to the sympathetic nervous system, we also have another short-term regulator of blood pressure—the arterial baroreceptors (**FIGURE 7.23**). These receptors are small stretch organs located in the carotid sinus and the aortic arch. They are sensitive to mechanical deformation, sending action potentials to the cardioacceleratory centers or the cardioinhibitory centers of the brain. They also connect to vasomotor centers in the brain.

What do the baroreceptors do? These receptors are functional during very short-term (seconds) changes in blood pressure, which we experience many times during the day as we make positional changes. For example, when you jumped out of bed this morning, your baroreceptors reacted. As your feet hit the floor and gravity pulled blood from your core to your feet, the baroreceptors sensed a loss of flow past the carotid sinus and the aortic arch. This caused a decrease in baroreceptor action potentials to the brain and a reflex increase in heart rate and increase in vasoconstriction. This transiently increased your blood pressure slightly but, by doing so, maintained your brain blood flow. Tonight, when you lie down again, the opposite will happen. As blood floods the core of your body, the baroreceptors will sense an increased pressure. Your heart rate will decline and your vessels will dilate, thus reducing pressure. These changes are slight, very rapid, and go unnoticed by healthy people. For those suffering malfunction of the baroreceptors, positional changes, like arising from bed, can cause lightheadedness and syncopy.

Baroreceptors are important for regulating very short-term blood pressure changes. They adapt to continue to regulate changes, even in a severely hypertensive person. They have no effect on long-term blood pressure regulation.

Long-term control of blood pressure relies on changes in blood volume, using a variety of hormones you may recognize from studying the endocrine system. During your run, for example, you will lose water as sweat, as a result of your body's process of thermoregulation. Loss of water will increase your blood osmolarity, triggering the release of antidiuretic hormone (ADH) (vasopressin) from the posterior pituitary and reclamation of water at the kidney. This will preserve your blood volume and your blood pressure.

Even without a change in osmolarity, the loss of blood volume would be sensed at the kidney, releasing renin and beginning the conversion of angiotensinogen to angiotensin II.

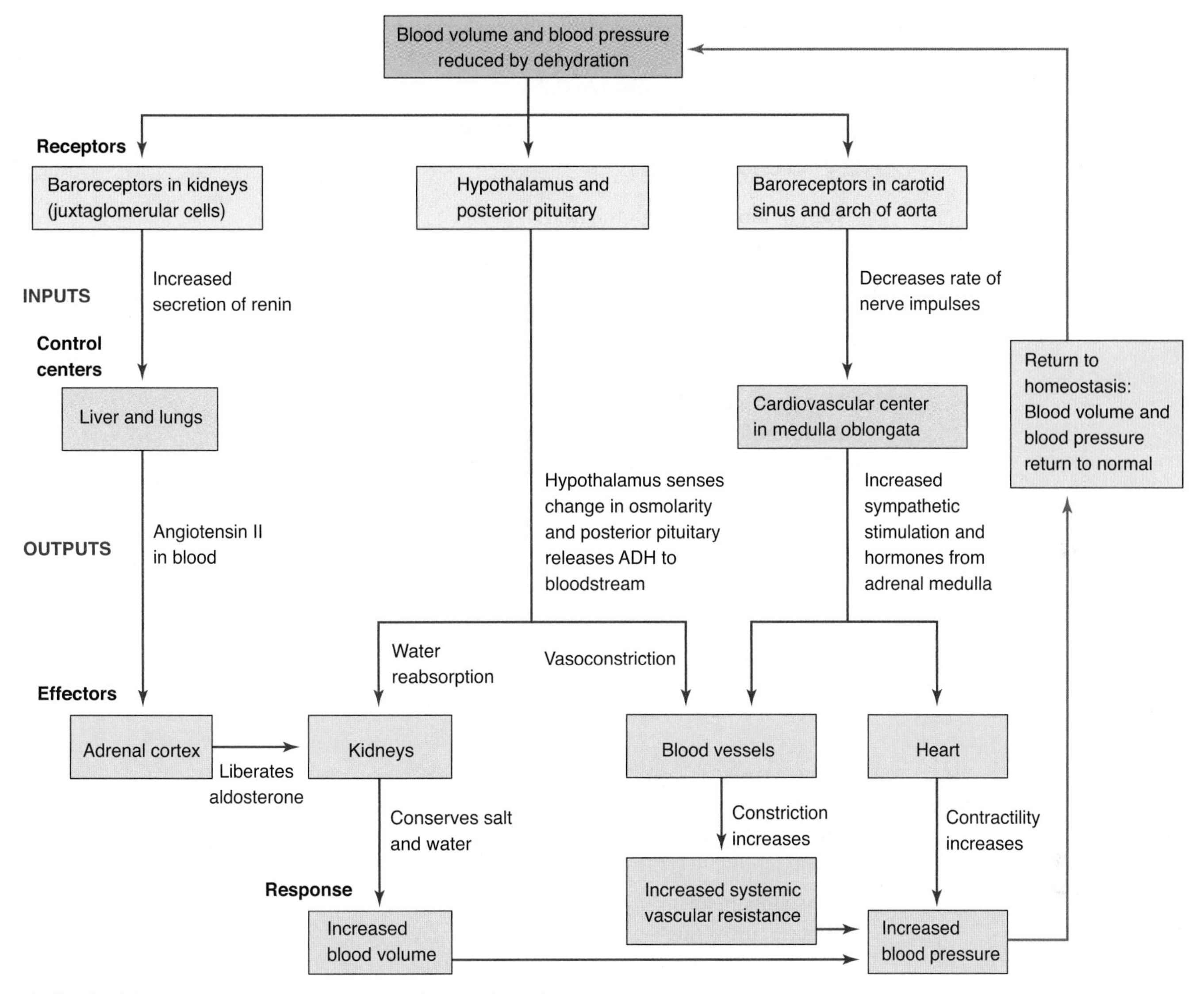

FIGURE 7.23 Short-term and long-term regulators of blood pressure.

Angiotensin II is a powerful vasoconstrictor, which will increase your blood pressure. It also stimulates the release of aldosterone and ADH, both of which work at the kidney to prevent water loss to urine and stimulate thirst centers at the brain. Once you rehydrate, levels of these hormones would go back to resting values.

Finally, if you drink too much water, ANP will increase diuresis at the kidney, bringing your total blood volume back to normal. Regulation of blood volume takes longer than sympathetic stimulation or the barorecptor. It is minutes to hours to days in the final regulation, but it is perhaps the most important determinant of blood pressure regulation.

Summary

The sensation of a beating heart is familiar to us all. While this seems simple, the cardiovascular system, consisting of the heart, blood vessels, and blood, is not a simple system. Each of the components interacts with one another, and the cardiovascular system is intimately tied to fluid volume control and the autonomic nervous system. All this leads to the complexity that is cardiovascular physiology.

Key Concepts

Path of blood through the heart
Cardiac E-C coupling
Anatomy of the vasculature
Fluid exchange at the capillary
Function of hemoglobin
SA nodal action potential
Ventricular action potential
Autonomic regulation of heart rate and contractility
Electrical conduction in the heart
Phases of the cardiac cycle
Regulation of cardiac output
Flow of fluids
Regulation of mean arterial pressure

Key Terms

Aorta
Vena cava
Hemoglobin
Erythropoietin
Sinoatrial node
ECG
Phospholamban
Frank-Starling law
Total peripheral resistance
Mean arterial pressure
Baroreceptors

Application: Pharmacology

1. Your grandfather has been prescribed an angiotensin-converting enzyme inhibitor (ACE inhibitor) for his high blood pressure. How would this drug work, and why would it be effective?
2. Your grandfather has also been prescribed a β-adrenergic antagonist, a blocker of β-receptors, to lower his heart rate and contractility. How will this affect his blood pressure? Why?
3. How might the β-blocker affect cardiac E-C coupling?

Cardiovascular Clinical Case

Type 2 Diabetes Mellitus

BACKGROUND

Type 2 diabetes mellitus (T2DM) not only causes metabolic dysregulation but also is a source of cardiovascular disease. Insulin resistance, hypertension, and coronary artery disease (CAD) have long been associated with diabetes, but it is now thought that insulin resistance may be a major contributor to both hypertension and CAD in patients with T2DM. The physiological mechanisms are different, so we will consider them separately.

Insulin Resistance

Under normal conditions, insulin functions as a vasodilator, increasing vascular diameter and reducing blood pressure. Several smooth muscle cellular ion pumps, including Na^+-K^+ ATPase and the muscular Ca^{2+} ATPase (SERCA), are insulin sensitive, presumably due to ATP availability resulting from glycolysis. Normal glucose transport into a cell allows cytoplasmic glycolysis and ATP production for use by these pumps. Thus, under normal conditions, intracellular $[Na^+]$ remains low as does intracellular $[Ca^{2+}]$.

Insulin resistance disturbs normal glucose uptake, inhibiting the Na^+-K^+ ATPase. Increased cellular Na^+ activates the Na^+/Ca^{2+} exchanger, which moves Na^+ out of the cell and brings Ca^{2+} into it. Raising intracellular Ca^{2+} increases the response of vascular smooth muscle to the vasoconstrictive actions of norephinephrine and angiotensin II, which also increase intracellular Ca^{2+}. At the same time, inhibition of the SERCA pump further increases Ca^{2+} concentrations in smooth muscle, facilitating contraction. Hypertension caused by constriction of arteries results from these ion concentration alterations. Because insulin is also a growth hormone, hyperinsulinemia causes proliferation of smooth muscle cells, so there is a greater mass of smooth muscle with which to constrict arteries. The increase in total peripheral resistance (TPR) is reflected as hypertension.

As TPR increases, afterload increases, requiring more cardiac work to pump blood through the systemic vasculature. A chronically increased workload causes the heart to adapt by increasing its muscle mass, also known as hypertrophy.

Hyperlipidemia

Hyperlipidemia can cause vascular problems. Remember that within the liver, where very-low-density lipoproteins (VLDLs) are made, a high triglyceride content causes the formation of VLDL particles with unusually high triglyceride concentrations. These excess triglycerides are transferred to low-density lipoprotein (LDL) in circulation, and triglyceride-rich LDL particles are formed. This species of LDL is the preferred substrate for hepatic lipase, which removes the triglyceride and converts LDL into a much smaller particle. This smaller particle is much more atherogenic, able to enter the endothelial cell wall and begin plaque formation within arteries. LDLs embedded in the vascular wall are engulfed by macrophages. In addition to the inflammatory pathways that are triggered, the macrophages, filled with lipids, transform into foam cells, which remain resident within the vascular wall. Thus begins the deposition of arterial plaques. Small particle LDLs that remain after triglyceride removal by the liver have been shown to be the most atherogenic particles of all. Through this mechanism, T2DM becomes a serious risk factor for atherosclerosis in general and for CAD in particular.

THE CASE

Your aunt, a type 2 diabetic, has been given medication to lower her triglycerides and to lower her blood pressure, which is 150/90. These medications are in addition to the metformin she takes and the diet to which she is supposed to adhere. It has been a year since her diagnosis, and she hasn't lost weight or been at all successful in keeping her glucose or triglyceride levels under control. She is convinced that the high blood pressure readings are in error and is refusing to take the medication. You are worried that this condition is eroding her health. You decide to talk to her about how T2DM may be responsible for her elevated blood pressure, so she believes there is a problem and will agree to take her medication.

THE QUESTIONS

1. What are the intracellular events that link insulin to hypertension?
2. How would vascular plaques affect blood flow through the vessels? Would plaques increase or decrease vascular resistance? Why?
3. How will you explain the change in blood pressure and the risks that increased blood pressure and triglycerides present? Your aunt is not a scientist, so you must put this into lay terms without diluting the meaning or the seriousness of the discussion.

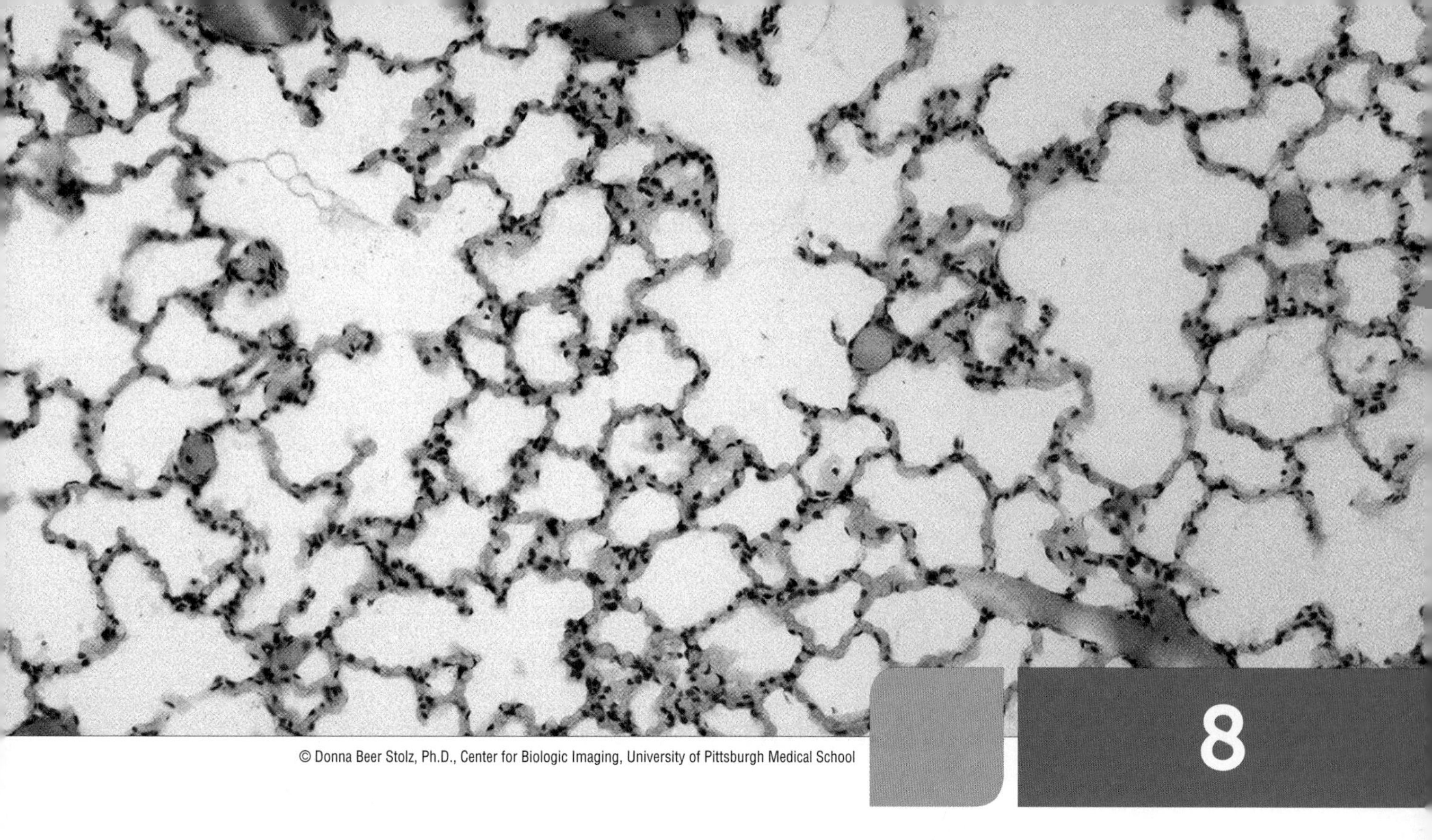

© Donna Beer Stolz, Ph.D., Center for Biologic Imaging, University of Pittsburgh Medical School

8

Respiratory Physiology

Case 1

You have just finished your morning run and are breathing hard. As usual these days, you begin to think about physiology—in particular, the physiology of breathing and what the limits are on your respiratory system. Unfortunately, you suffer from occasional asthma attacks, generally associated with spring pollen release. This year has been particularly troublesome; you find inhalation and exhalation increasingly difficult and you reach for your inhaler. As you do so, you contemplate the mechanics of air movement within the bronchiole tubes. Why is your breathing more difficult, and why does the inhaler help?

Case 2

After years of smoking, one of your mother's friends has been diagnosed with emphysema. He is a barrel-chested man with a peculiar habit of exhaling through his mouth, which is partially closed. This is particularly obvious after he has climbed the stairs. What are the limitations to his breathing? Why does he exhale this way?

Introduction

Inspiration and expiration occur spontaneously throughout our lives, whether we are awake or asleep. But how do we inspire, and where does this air go?

If we were one cell thick, oxygen (O_2) could simply diffuse into the mitochondria and take its place as the final electron acceptor in the electron transport chain. Similarly, carbon dioxide (CO_2), which is produced during the transition reaction and the citric acid cycle, could diffuse out of the cell. Because we are many, many cell layers thick, this cannot happen. So O_2 must be transported within the circulatory system and delivered to individual cells, and CO_2 must be carried away. This simple cellular gas exchange requires a mechanical system to bring atmospheric air into the lung, a gas exchange system that includes a unique protein to carry O_2 and CO_2, and a neuronal regulatory system to drive the mechanical system. We will investigate each aspect of the respiratory system and examine the limitations of each part of the system. The limitations of regulation, mechanics, or gas exchange affect the delivery of O_2 or the elimination of CO_2, which are the ultimate and only functions of this complex system. An understanding of the limitations of the system will help us understand respiration in health and disease.

In healthy individuals, respiration is nearly an invisible function, and we notice a restriction of function only during severe exercise when we may not be able to exchange air fast enough. Our case studies and clinical study involve exercise and some common respiratory diseases: asthma, heart failure, and emphysema. Each of these conditions illustrates an important principle of respiratory function and will help you understand the workings of this system.

How Do We Get Air into the Lung?

During quiet breathing, which you are probably doing as you read this, the primary action that allows air to flow into the lung during inspiration is the contraction of the diaphragm muscle. This slender, plate-like muscle separates the thoracic and abdominal cavities and connects to the parietal pleural membrane that lines the thoracic cavity. Thus, the diaphragm creates a discrete thoracic space sealed off from the abdominal cavity. When the diaphragm contracts, it does what every skeletal muscle does during contraction: it shortens. The unique aspect of the diaphragm is that its resting shape is that of a dome. So, when it contracts, it flattens, which creates a slightly larger thoracic cavity space (**FIGURE 8.1**). How does this result in inspiration?

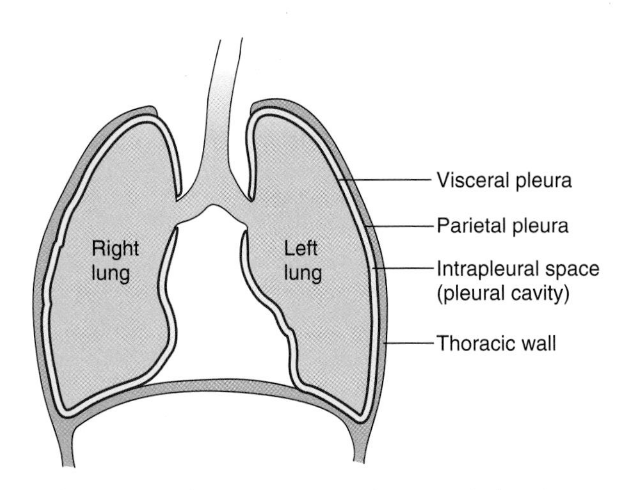

FIGURE 8.1 The diaphragm and the parietal pleural membrane create the thoracic cavity and tether the lung to the body wall. As the diaphragm flattens during contraction, it increases the volume of the thoracic cavity.

As you should know, one portion of the universal gas law relates to volume and pressure. Boyle's law states that volume and pressure are inversely related. That is, if volume increases, then pressure decreases. This is usually stated as $P_1V_1 = P_2V_2$. During a quiet inspiration, the diaphragm contracts, the thoracic cavity enlarges, and the air pressure within the lung falls below normal atmospheric pressure. This allows air to flow down a pressure gradient into the lung. The pressure difference generated during quiet inspiration is very small (approximately 1 cm H_2O, which is equivalent to ~0.74 mmHg or ~0.014 psi), but it is enough to bring 400–500 mL of air into the lung. Because inspiration requires muscle contraction, this is considered an active process requiring adenosine triphosphate (ATP). Notice that while a change in pressure is necessary for inspiration, physiologically, all we can do is change volume. Anything that prevents this change in volume will restrict your ability to inhale. For example, lack of muscle strength, severe curvatures of the spine, or even a heavy weight lying upon your chest will limit thoracic expansion and, thereby, limit inspiration.

Our inspiratory capacity, which measures the maximum amount that you can inhale, is approximately 3,500 mL. The external intercostal muscles, which attach to the ribs, lift the rib cage and increase the anterior-posterior volume of the chest cavity. The sternocleidomastoid muscles attach to the clavicle and the sternum, lifting the rib cage from the top (**FIGURE 8.2**). As we inhale more deeply, we engage these muscles to assist in increasing the thoracic cavity volume and, therefore, the pressure difference. Obviously, the increase in muscle contraction during active or deep inspiration requires more ATP use and increases the physiological work of breathing.

How Do We Get Air Out of the Lung During Quiet Expiration?

Normal, quiet expiration simply requires muscle relaxation of the diaphragm and is passive, that is, it does not require ATP. However, exhalation does rely on a physical property of the lung: elasticity. The normal resting length of the lung is smaller than it is in the chest cavity. That is, if you remove the lung from the chest cavity, it will collapse. The normal elasticity of the lung—that is, the presence of many elastic fibers in the extracellular matrix of the tissue—confers this characteristic. In life, the lung is tethered to the chest wall by the visceral pleura, which lies along the lung and attaches to the parietal pleura, becoming continuous with the chest wall. The envelope formed by the visceral and parietal pleura connects the lung to the chest wall and holds the lung partially open. The pleural membranes also produce pleural fluid, which allows the lungs to expand and contract while suspended in a small amount of fluid, reducing friction on lung tissue (Figure 8.1). Still, during exhalation, as the diaphragm relaxes, elastic recoil makes the lung smaller than it was during inhalation, the decrease in volume increases the pressure, and air flows out of the lung passively (**FIGURE 8.3**).

FIGURE 8.2 (a) Primary and accessory respiratory muscles. (b) During inspiration, the diaphragm contracts and lowers, while the sternocleidomastoid, the scalenes, the pectoralis minor, and the external intercostal muscles raise the ribs and breastbone, increasing the thoracic volume.

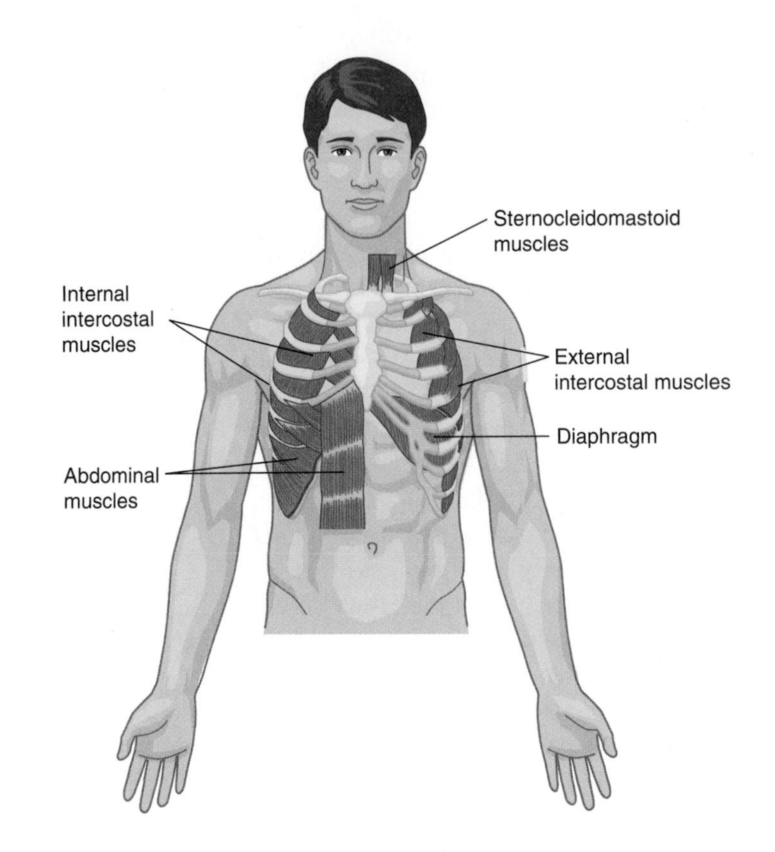

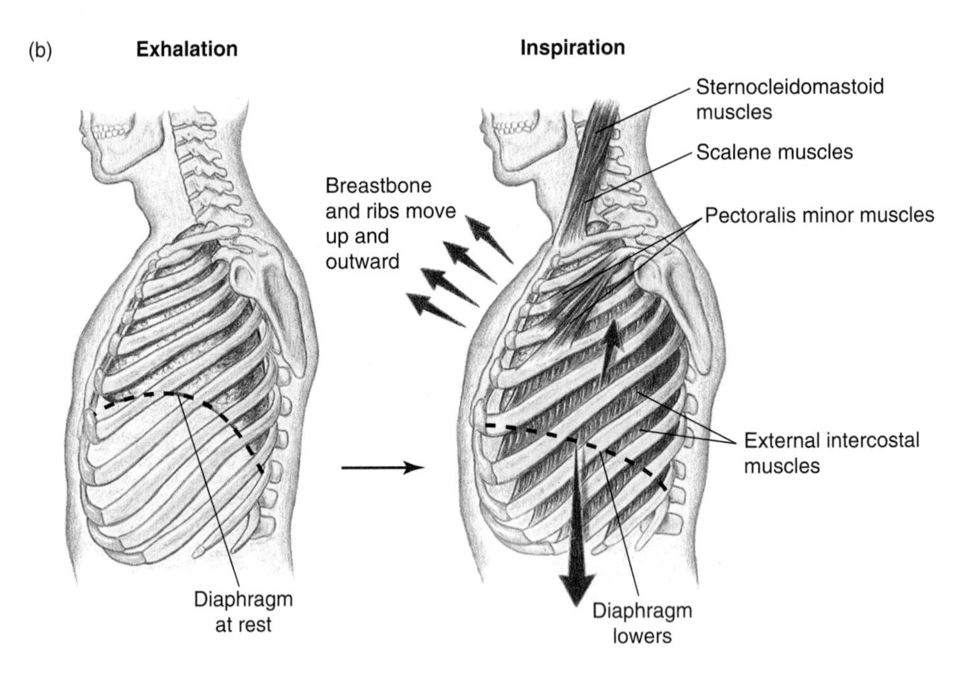

There is another physical component that affects lung expansion and contraction: the chest wall. The resting length of the chest wall structures (connective and muscle tissue) is larger than it appears in life. If the lung is removed from the chest, the thoracic cavity expands. So, the chest wall and the lung are in a dynamic opposition, connected by the pleural membrane. The chest wall pulls outward on the lung, expanding it, while the elastic fibers of the lung pull inward on the chest wall. The balance of these forces gives us a normal chest diameter while retaining the elastic recoil of the lung. It is worth examining

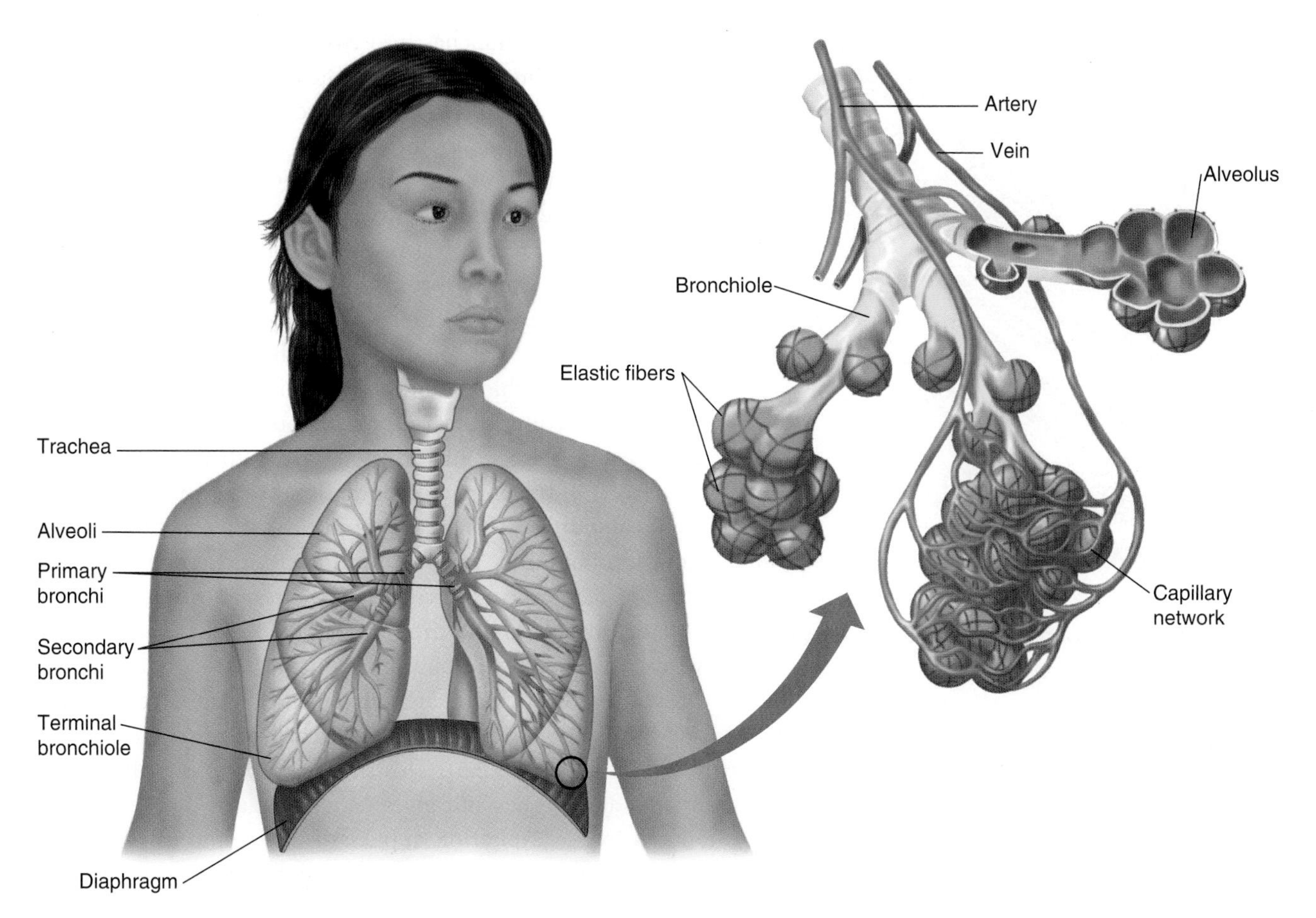

FIGURE 8.3 Elastic fibers surround the alveoli, giving them elastic recoil.

these components that contribute to inhalation and exhalation separately, because disease and age can affect a single component. For example, our friend with emphysema has lost some of the elasticity of the lung. This has two effects: it allows for chest enlargement—hence, his barrel-chest—and it limits lung recoil, so exhalation is difficult. It is easiest to visualize elastic properties by remembering the elastic behavior of balloons. When balloons have been blown up many times, they lose elasticity. This makes them much easier to inflate because you are not working against the force of elastic recoil. However, it also means that a balloon that has lost elasticity deflates less and may remain partially inflated without being tied! A similar effect occurs in an emphysemic lung. Inhalation is not difficult for our friend in case study 2, but exhalation is. The obvious next question is, if passive exhalation is ineffective, is there a way to force expiration?

What Happens During Forced Expiration?

Forced expiration is an active process requiring muscle contraction. The internal intercostals pull the rib cage down and the abdominal muscles contract, displacing the abdominal contents upwards (**FIGURE 8.4**). All of these contractions decrease the thoracic cavity volume, increase the pressure within the cavity, and, therefore, allow air to move down a pressure gradient toward the outside. Again, pressure changes are controlled exclusively by thoracic cavity volume change. Understanding the limitations of expiration requires knowledge of the anatomy of the bronchiole tree.

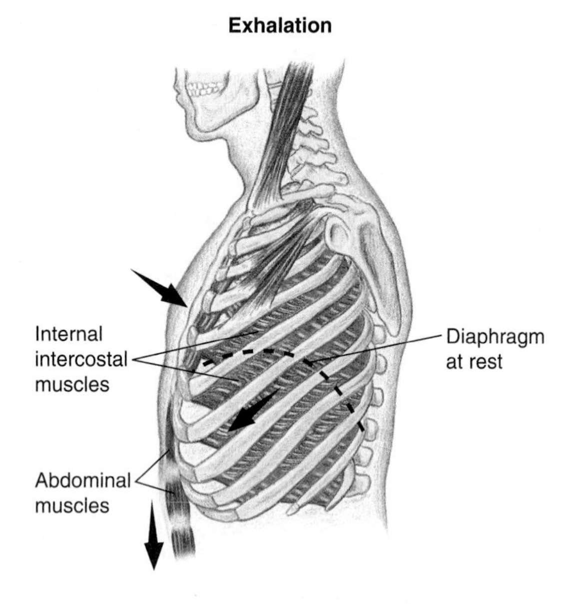

FIGURE 8.4 Accessory expiratory muscles allow us to actively exhale.

The trachea branches into two bronchi, and these bronchi split again into bronchioles. The branching pattern of the airways is similar to that of the cardiovascular system. Each successive generation of bronchioles has a greater number of smaller bronchioles. The trachea, bronchi, and the largest bronchioles are conducting airways and allow no gas exchange. They are supported by cartilaginous plates and have a constant diameter. Smaller bronchioles lack cartilage but are surrounded by smooth muscle, which can control bronchiole diameter. The last portion of the bronchiole tree comprises the respiratory bronchioles and the alveoli. Gas exchange occurs at these structures. The lack of smooth muscle surrounding them promotes ease of diffusion. What this means for the mechanics of breathing, or ventilation, is that these structures are unsupported. This has important consequences for exhalation. During forced exhalation, thoracic and abdominal muscles contract, exerting pressure on the respiratory bronchioles and the alveoli. If the pressure exerted exceeds the pressure within the alveoli or respiratory bronchioles, they will collapse. Respiratory bronchioles are narrow and unsupported and are the first to close. This traps air within the alveoli and prevents their collapse. However, it also prevents air from being expelled (**FIGURE 8.5**). This simple effect of pressure allows our lung to stay partially inflated at all times and gives us a residual volume that cannot be exhaled. This same limitation of expiration applies to our emphysema patient, in whom a forced expiration will simply result in airway closure, trapping air within the alveoli. Because his alveoli have lost elastic recoil, they will hold more air. Thus, less air is exchanged during each breath.

Another phenomenon gives our friend his peculiar form of exhalation. The more rapidly air moves through a tube, the less pressure is exerted on the walls of the tube. In the airways, this means that the more rapidly you exhale, the less pressure is on the respiratory bronchioles, and the more likely they are to close, trapping air. By exhaling more slowly and through pursed lips, which increases resistance and slows flow even further, the small airways can be kept open longer and more air exhaled.

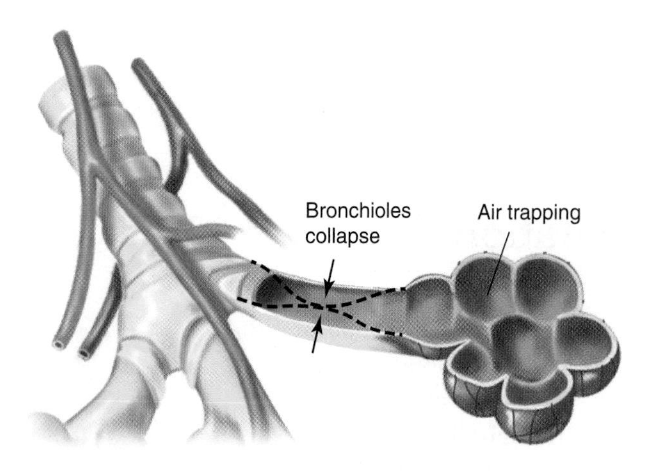

FIGURE 8.5 Mechanism of airway closure.

How Do Flow and Resistance to Flow Affect Ventilation?

The physical laws governing flow of fluids are the same for air in bronchioles as they are for blood in blood vessels. Resistance to flow is a function of viscosity, length of tube, and the diameter of the tube. The narrower the tube, the greater is the resistance. Remember that this is not a linear relationship but an exponential one (resistance is proportional to $1/\text{radius}^4$), so a small change in diameter creates a large change in resistance. Flow will be rapid in the large conducting airways and become slower as the resistance within the smaller bronchioles increases. Autonomic nervous system regulation of smooth muscle surrounding the airways also affects

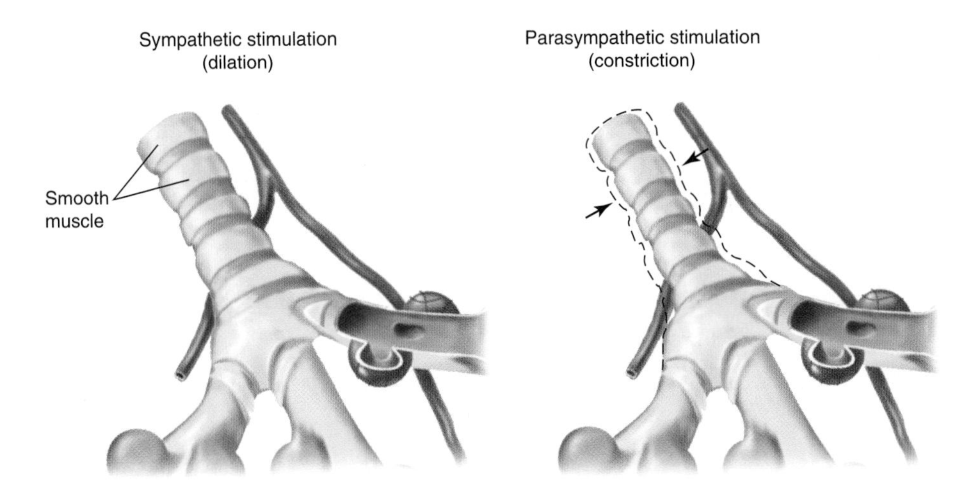

FIGURE 8.6 Sympathetic stimulation causes bronchodilation, while parasympathetic stimulation causes bronchoconstriction.

resistance: parasympathetic stimulation causes smooth muscle contraction, narrows the airways, and increases resistance, while sympathetic stimulation relaxes smooth muscle, widens the airways, and decreases resistance (**FIGURE 8.6**). In a healthy, resting person, the changes in resistance by autonomic stimulation are small and do not limit our air exchange.

However, during an asthma attack, when smooth muscle tension increases because of a reaction to a trigger (an allergen such as pollen, cold air, or even exercise itself can be triggers), there is more resistance to flow and less air reaches the respiratory bronchioles and the alveoli, causing you to gasp for air. The work of breathing increases as more muscles are used for a forced inspiration, to increase the thoracic volume and decrease intrapulmonary pressures. This is the only way to increase the pressure difference, that is, the pressure head and increase the flow down the bronchioles. Your inhaler contains a medication, generally a β-receptor agonist such as albuterol, that dilates the airways, reduces resistance to flow, and increases the amount of air reaching the alveoli.

Ventilation Through Airways Involves Several Resistances

Just as there is laminar and turbulent flow within the blood vessels, so there is within the bronchioles. Air, having less viscosity, can become turbulent at lower velocities than blood. Further, the branching patterns of the bronchial tree promote turbulent flow. Using the same physical principles that we can observe in cardiovascular physiology, we recognize that turbulent flow creates more resistance than laminar flow does. Because the object of ventilation is to get air to the alveoli for gas exchange, the only way to ensure sufficient flow to the alveoli during high resistance to flow is to increase thoracic volume and decrease intrapulmonary pressure (**FIGURE 8.7**). Normally, the increased resistance to flow from turbulence, caused perhaps by rapid breathing during exercise, becomes part of our natural limitation to ventilation.

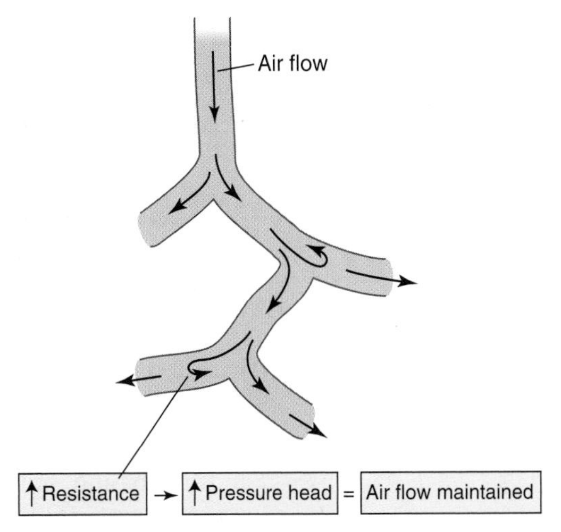

FIGURE 8.7 Narrowed airways increase resistance to flow. Flow volume can only be maintained by increasing the pressure gradient.

While laminar and turbulent flow in arterioles and bronchioles is similar, flow of air through the nasal passages and down to the alveoli differs from the flow of blood within vessels in one respect. In the lung, there is an air-water interface. This interface causes surface tension, not unlike the tension you see on the surface of a drop of water sitting on a table. This layer of fluid along the walls of the bronchioles creates friction as air flows toward

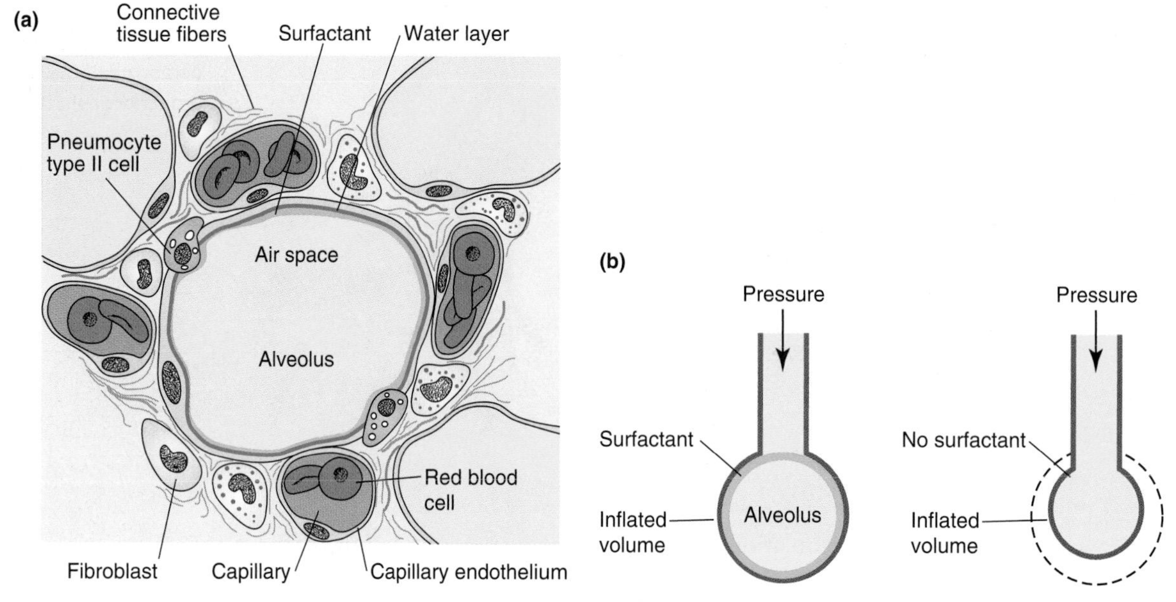

FIGURE 8.8 (a) Anatomical arrangement of the alveolus, type II pneumocyte, surfactant, and blood vessels. (b) Surfactant reduces surface tension and allows greater expansion of alveoli.

the alveoli, and it is there that the major effects of this disparity are felt. The alveoli, surrounded by elastic fibers, expand during an inhalation, as intrapulmonary pressures drop and air flows into them. The expansion of the alveoli requires stretching of elastic fibers and stretching of a very thin fluid layer inside the alveoli. Water, which forms a lattice structure with hydrogen bonds, now has an edge exposed to the air of the alveoli. Physically, this means that alveolar expansion requires force to overcome this surface tension. Why, then, does a normal inspiration require so little ATP? In healthy individuals, we have intra-alveolar cells that create a surfactant molecule—dipalmitoylphosphatidylcholine. This molecule, a phospholipid, relieves the surface tension and allows alveolar expansion to occur more easily. Of course, as the alveoli expand, these surfactant molecules are further apart, and the effort required to expand the alveoli increases. Thus, the surfactant molecule facilitates opening at low alveolar volume and acts as a natural limit to inflation at high volumes (**FIGURE 8.8**). Premature infants, born before the eighth month, have not yet developed the ability to make this surfactant, resulting in extreme difficulty in inhalation. Although this condition used to be fatal, artificial surfactants have become effective treatments for it.

What Happens to Air Once It Is in the Lung? How Is Oxygen Delivered to Tissues?

As air enters the nasal passages, trachea, and bronchi, it is humidified by the water vapor within these moist tubes and is warmed to body temperature. Turbulent flow within the nasal passages and trachea causes particulate matter to settle out and stick to the mucous layer that lines these passages. The trachea is lined with rhythmically beating cilia, which move dirt and other particulates toward the pharynx where they can be eliminated (**FIGURE 8.9**). Thus, the air entering the lung is warmed, humidified, and cleaned before it reaches the alveoli.

Air entering the lungs is 21% oxygen and 78% nitrogen (N_2). This translates to partial pressures of 160 mmHg for O_2 (21% of the sea level barometric pressure of

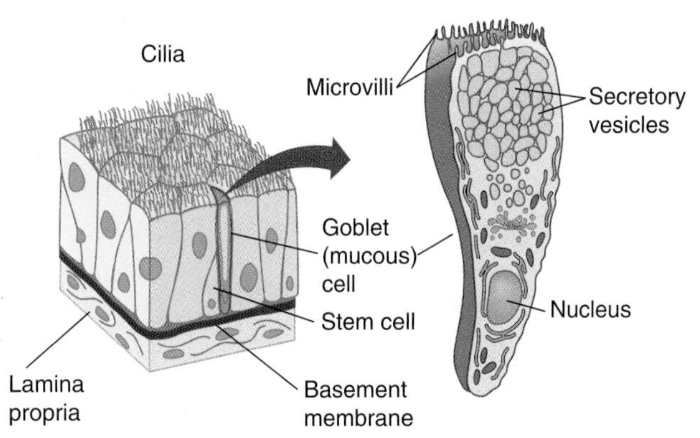

FIGURE 8.9 Ciliated pseudostratified columnar epithelial cells line the trachea and move particulate matter toward the mouth. (Photo © Donna Beer Stolz, Ph.D., Center for Biologic Imaging, University of Pittsburgh Medical School.)

760 mmHg) and 593 mmHg for nitrogen. However, the lungs also contain carbon dioxide, a gaseous product of our own metabolism, and water vapor. Because the total of the partial pressures of gases cannot exceed atmospheric pressure, the partial pressures of both O_2 and N_2 are reduced within the lung. At the alveoli, the partial pressure of O_2 is approximately 100 mmHg. This is the driving pressure of O_2 going into blood (**FIGURE 8.10**).

At this point, once air is in the alveoli, it must diffuse into blood. The factors that govern the rates of diffusion in the lung are surface area, diffusion distance, and the solubility of gas in plasma. Temperature remains constant, as does the viscosity of air, so they generally play no part in gas exchange. The surface area of the alveoli is very large, and the diffusion distance from the interior of the alveoli to the capillaries, which surround those capillaries, is very small. However, O_2 is not very soluble in water. Only about 2% of our O_2 needs are met by dissolved O_2. Instead, O_2 binds to a respiratory pigment, a protein called hemoglobin (Hb) (**FIGURE 8.11**). Red blood cells (erythrocytes) are specialized carriers of the hemoglobin protein and lack a nucleus, mitochondria, and other organelles. Oxygen molecules bind to hemoglobin becoming HbO_2. All this occurs as the pulmonary circulation comes from the right chambers of the heart and splits into a massive capillary bed surrounding the alveoli. During your run, cardiac output may be 20 L/min, which means that 20 L of blood passes over the alveoli for gas exchange each minute. Even at this rapid flow rate, O_2 binds to hemoglobin and is transported back to the heart for delivery to the tissues.

Gases	Inspired air	Alveolar air
H_2O	Variable	47 mmHg
CO_2	000.3 mmHg	40 mmHg
O_2	160 mmHg	105 mmHg
N_2	593 mmHg	568 mmHg
Total pressure	760 mmHg	760 mmHg

FIGURE 8.10 Gas pressures in the lung.

If we look at the behavior of O_2 and Hb binding in whole blood, we will gain an understanding of how metabolism affects O_2 binding to the hemoglobin protein. The oxygen-Hb dissociation curves are shown in (**FIGURE 8.12**). Notice that the curve is sigmoidal. Because O_2 binding to the four subunits of hemoglobin is cooperative, binding of each O_2 increases the ease of binding for the other three molecules of O_2. Normally, hemoglobin loads almost completely (97%–99%) at the lung. Plasma also saturates to the level of O_2 solubility in that fluid. At higher temperatures, higher CO_2 concentrations, and higher H^+ concentrations and lower pH, all of which would accumulate in blood during exercise, the curve is shifted to the right. This does not affect loading very much

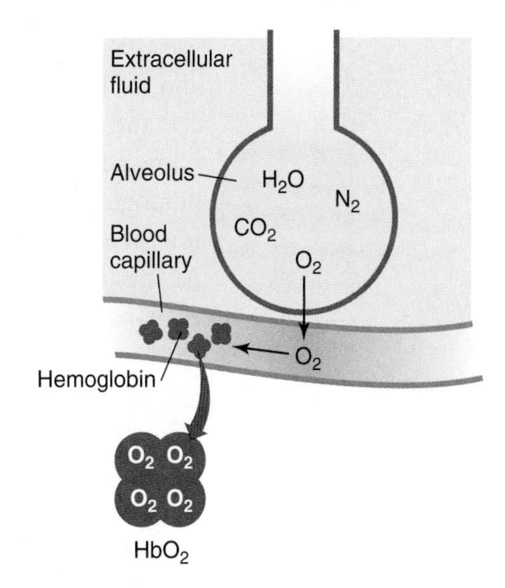

FIGURE 8.11 Path of O_2 from the alveoli to hemoglobin.

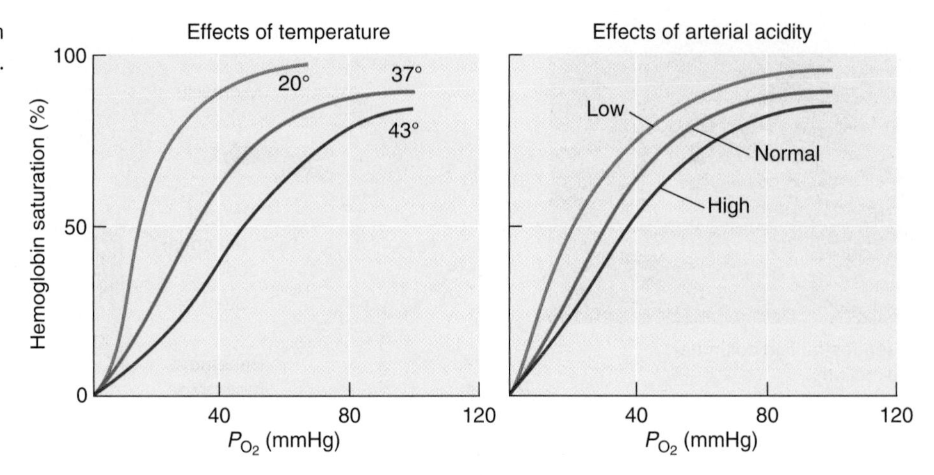

FIGURE 8.12 Oxygen-hemoglobin dissociation curves.

but profoundly affects O_2 delivery to tissues. Hemoglobin has less affinity for O_2 under these conditions, and more O_2 is, therefore, released to tissues, in particular the exercising muscle. As blood travels back to the right atrium and is pumped toward the lung, Hb's binding affinity increases again because of slight reductions in temperature from heavy breathing of cooler air and elevations of pH due to local conditions in the lung. As a result, Hb is fully loaded with O_2 at the lung (Figure 8.12). Thus, hemoglobin, by simple chemical regulation, can function both as an O_2-binding protein and an O_2-releasing protein.

If Only Small Amounts of CO_2 Are Carried on Hemoglobin, What Happens to the Rest of It?

Most CO_2 is carried from the tissue where it is formed to the lungs, where it is eliminated, in the form of the bicarbonate ion. This can be expressed in a simple equation, which will be seen again when we study the kidney and yet again during investigations of acid-base balance:

$$CO_2 + H_2O \leftrightarrow H_2CO_3 \leftrightarrow HCO_3^- + H^+$$

The enzyme carbonic anhydrase facilitates the conversion of CO_2 and H_2O into carbonic acid and the further conversion to the bicarbonate ion and the H^+ ion. Mammals use this bicarbonate-based buffering system as the predominant means of buffering H^+ and CO_2. At the tissue level, CO_2 is formed during metabolism and diffuses out of the cell and into the red blood cells, where it combines with H_2O to transiently become carbonic acid. The carbonic acid is quickly converted to HCO_3^- and H^+. The H^+ can bind to hemoglobin as O_2 is released. HCO_3^- is transported out of the red blood cell in exchange for the Cl^- ion, which allows the conversion of CO_2 and H_2O to H_2CO_3 to continue. At the lung, the opposite occurs. HCO_3^- is transported back into the red blood cell, in exchange for Cl^-; meanwhile, H^+ is released as hemoglobin binds O_2, and the equation runs backwards, creating CO_2, which we exhale (**FIGURE 8.13**).

Although this system may seem unnecessarily complex, it serves two functions: (1) conversion of CO_2 to H^+ creates an ion that can be bound to Hb and, thus, does not contribute to a change in pH, and (2) it produces an ion, H^+, that can be eliminated by the kidney. Both of these processes help us maintain our normal blood pH of 7.4. However, just as there are a finite number of O_2 binding sites on hemoglobin, so there are a

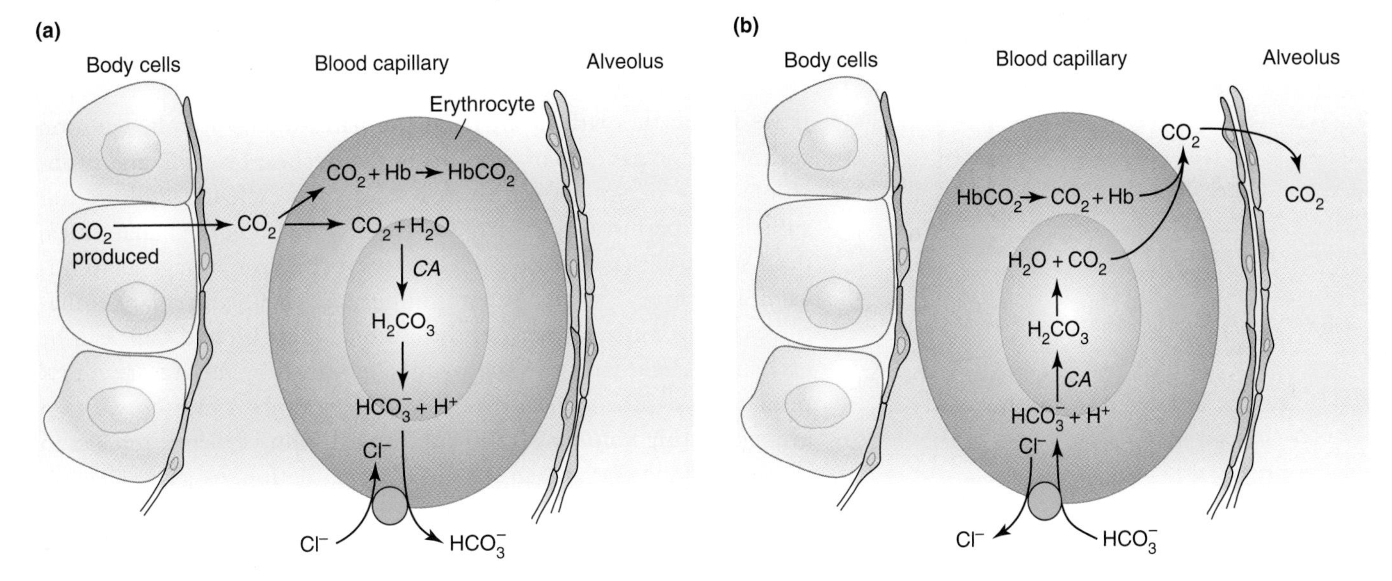

FIGURE 8.13 Path of CO_2 from the tissues to the lung.

finite number of H^+ and CO_2 binding sites. Furthermore, HCO_3^- can become limiting. So, during extreme exercise or in the case of disease, if we cannot eliminate CO_2 as fast as it accumulates, we may become acidotic.

Which Is More Important: O_2 Binding or O_2 Release?

We tend to think that any modulator that increases O_2 binding affinity to Hb is advantageous. However, O_2 delivery is just as important as HbO_2 carrying capacity. For example, let's suppose that the run you took this morning was at 5,000 feet above sea level. The atmosphere surrounding us is thinner at 5,000 feet, so the barometric pressure is reduced to about 703 mmHg. While O_2 still makes up 21% of this atmosphere, there is a lower partial pressure because of the lower barometric pressure. Calculating partial pressures of oxygen (703 × 21%), we discover a pressure of 148 mmHg for O_2 instead of 160 mmHg. The water vapor in the lung stays the same, so now the driving pressure is 89 mmHg instead of 100 mmHg. Your run at this altitude may trigger increased breathing rate and depth, thus increasing gas exchange to match your O_2 needs. It also increases your carbon dioxide excretion from exhalation. In fact, you may eliminate so much CO_2 that you actually reduce the concentration of free H^+ ions in the blood, and raise blood pH above the optimal level of 7.4.

$$CO_2 + H_2O \leftarrow H_2CO_3 \leftarrow HCO_3^- + H^+$$

This can actually reduce O_2 delivery to tissues, because O_2 has a greater binding affinity to Hb at higher pH values. Over hours to days, we adapt to these potential mismatches of O_2 binding and delivery by increasing production of bisphosphoglycerate (BPG). As red blood cells metabolize more, they produce more of the precursor to BPG. BPG binding to Hb reduces O_2 binding affinity and increases O_2 delivery. Your run on the third day at altitude will be much more comfortable than it was on the first day.

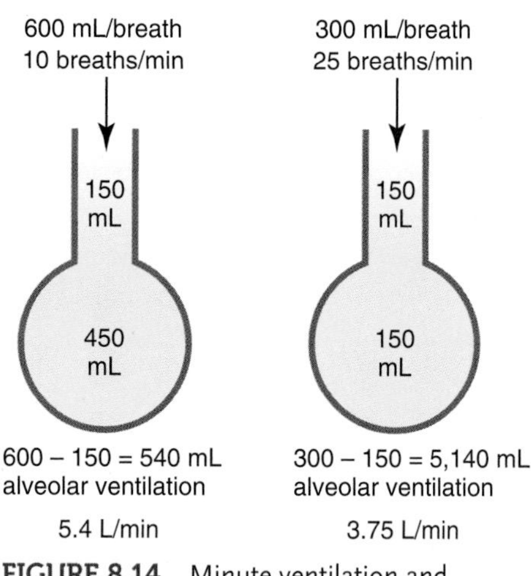

FIGURE 8.14 Minute ventilation and anatomical dead space.

Ventilation Must Reach the Alveoli for Gas Exchange to Occur

If we follow the path of air from the nares to the alveoli, we soon realize why no gas exchange occurs in the trachea, bronchi, and bronchioles. The tissues are too thick to allow gas exchange and are not sufficiently perfused by capillaries. These spaces are called anatomical dead space, because no gas exchange can occur. An average anatomical dead space is 150 mL. What is important, then, is alveolar ventilation, or how much air reaches the alveoli, where O_2 and CO_2 can be exchanged. Ventilation volumes are determined by respiratory rate × respiratory depth. Typically, we breathe 16 times a minute with an average resting volume of 500 mL, or 8 L/min; 150 mL remain in the conducting airways, so the alveolar ventilation, or gas exchangeable air, is only 350 mL per breath, or 5.6 L/min. Notice that the 150 mL is a fixed volume. So, rapid shallow breathing of 300 mL per breath 25 times/minute only produces an alveolar ventilation of 3.75 L/min, whereas 10 breaths/min at 600 mL produce 5.4 L/min (**FIGURE 8.14**). Clearly, increasing respiratory depth is more effective at improving alveolar ventilation. Additionally, it requires less work of breathing because there are fewer inspirations, and as you recall, it is inspiration that requires ATP.

Pressures Within the Lung Itself Can Cause Ventilation-Perfusion Mismatch

For O_2 to enter the circulatory system and reach tissues, blood must flow uninterruptedly through the capillaries, which cover the alveoli. Pressures within the intrapulmonary space continuously change during the respiratory cycle. These pressures can affect blood flow within the extra-alveolar vessels, as well as the intrapulmonary ones. The extra-alveolar vessels lie away from the alveoli and are pulled open as the lung inflates. This promotes blood flow toward the capillary beds, which surround the alveoli and participate in gas exchange. These capillaries are, in turn, subjected to pressures from the alveoli themselves. If the pressure exerted by the alveolus exceeds that within the pulmonary capillary, it will tend to close that capillary, limiting blood supply to that alveolus and, therefore, gas exchange. Notice that this time the limitation is not ventilation, but perfusion! In fact, blood flow within the pulmonary capillaries is subject to gravity. So, when we are in an upright position, the capillary blood pressure in the upper portion of the lung is lower than it is in the lower portion of the lung. Furthermore, it is lower during cardiac diastole than it is during systole. In the upper lobes of the lung, during diastole, the pulmonary capillaries tend to collapse because the pressure of the air in the alveoli is greater than the blood pressure in the capillary. The alveoli are well ventilated, but not well perfused. In the lowest part of the lung, capillary blood pressure almost always exceeds alveolar pressures, so perfusion continues uninterrupted during both systole and diastole. However, less air reaches these alveoli, so there is more perfusion than ventilation. Thus, a gradient is established in the upright lung, where ventilation exceeds perfusion in the top portion while perfusion exceeds ventilation in the bottom portion (**FIGURE 8.15**). Both conditions reduce the efficiency of gas exchange and limit both the amount of O_2 that is delivered to tissues and the amount of CO_2 that is exhaled.

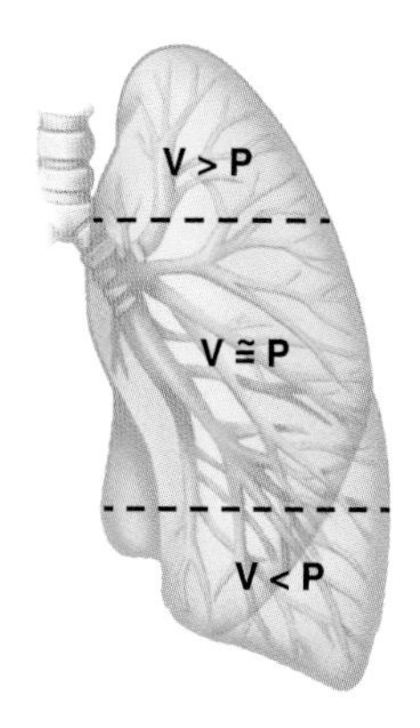

FIGURE 8.15 Ventilation-perfusion inequalities in the upright lung.

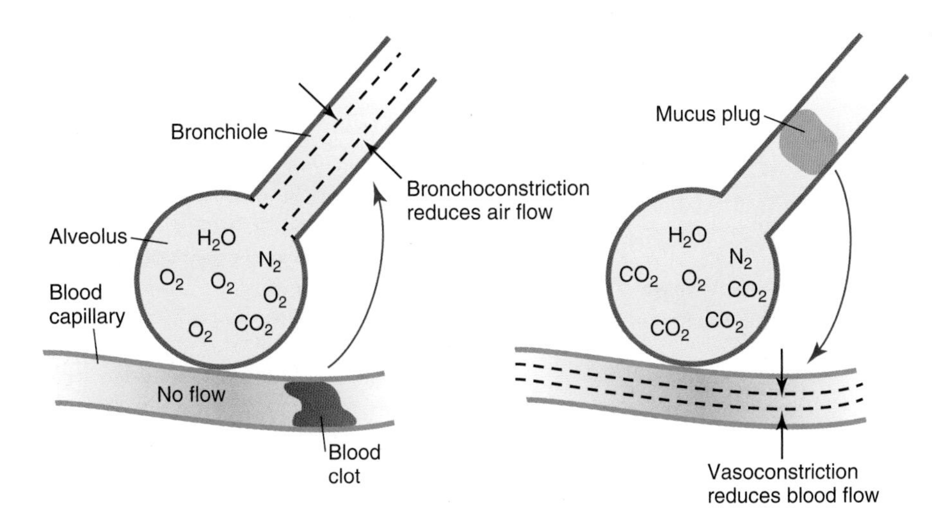

FIGURE 8.16 Bronchoconstriction and pulmonary hypoxic vasoconstriction serve to minimize ventilation-perfusion inequalities.

There are elegant compensations for ventilation-perfusion mismatch that tend to equalize airflow and blood flow within the lung. The first occurs when alveoli are ventilated, but not perfused. Imagine an air-filled alveolus with no blood flow to it. A small blood clot could cause this. Because no gas exchange occurs, this alveolus is filled with O_2 and no CO_2. This results in an alkaline environment, which causes local bronchoconstriction. Narrowing the airway increases the resistance and reduces the airflow, or ventilation, of the nonperfused alveolus, thus improving the match between blood flow and airflow.

The second mechanism, pulmonary hypoxic vasoconstriction, causes vasoconstriction of pulmonary vessels whenever alveoli are hypoxic, that is, poorly ventilated. Constriction of the vessels reduces blood flow to this area and shunts it to a well-ventilated area (**FIGURE 8.16**). Notice that hypoxic vasoconstriction in the lung is the *opposite* of what happens in the systemic vasculature during hypoxia! Both of these mechanisms equalize ventilation and perfusion in the normal lung.

What Causes Me to Breathe While Sleeping?

Breathing is an autonomic function as well as a voluntary one. It is unique as an autonomically controlled system. We can breathe at will, voluntarily, or we can forget about it and we will breathe as needed. The primary drive to breathe is blood CO_2 concentration. The sensor of carbon dioxide concentration $[CO_2]$ is the central chemoreceptor, several clusters of specialized cells located in the medulla of the brain stem. As CO_2 accumulates in arterial blood, it can freely diffuse through the blood-brain barrier and enter the chemoreceptor cells. Here, CO_2 is converted to H^+ and the pH within the chemoreceptor cells declines. Remember the Henderson-Hasselbach equation:

$$CO_2 + H_2O \leftrightarrow H_2CO_3 \leftrightarrow HCO_3^- + H^+$$

This increases neuronal firing to the dorsal respiratory group, also located in the medulla. Neurons from the dorsal respiratory group project to the phrenic nerve, which causes diaphragmatic contraction, and we inspire. Thus, regulation of breathing is directly linked to blood CO_2 concentrations (**FIGURE 8.17**). Notice that H^+ concentrations do not affect breathing rate because H^+ cannot cross the blood-brain barrier. An increase in CO_2 production, from metabolism, during a run, increases your respiratory rate and depth through this same mechanism.

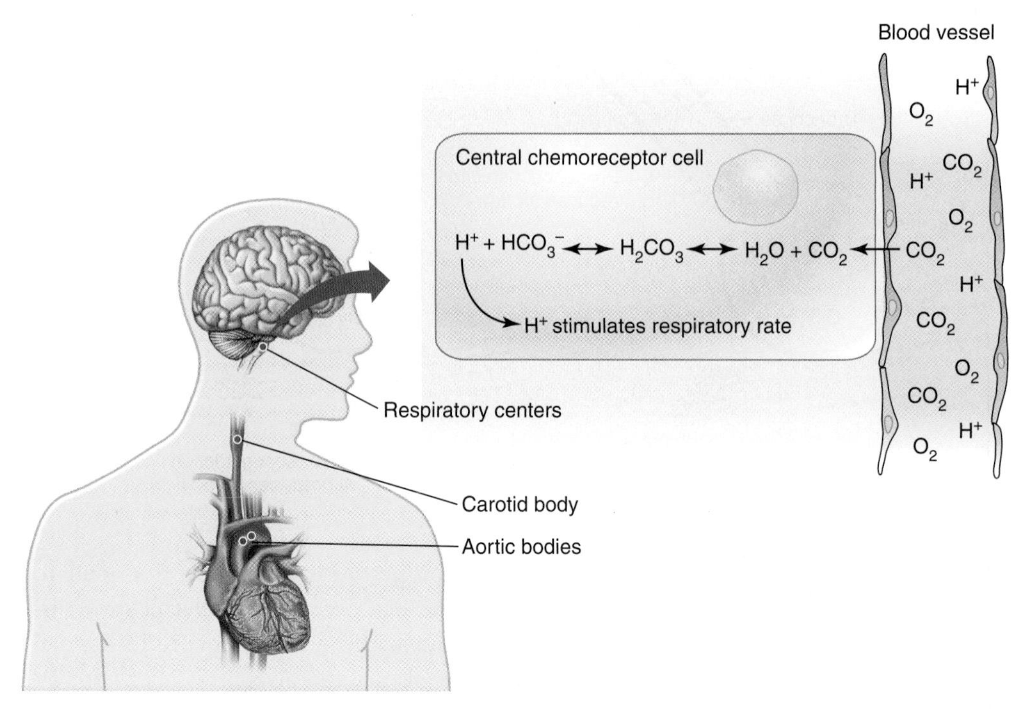

FIGURE 8.17 Regulation of respiration by the central chemoreceptors is based on CO_2. Peripheral chemoreceptors, located in the carotid sinus and aortic arch, monitor H^+, CO_2, and O_2 concentrations in the blood.

There are also peripheral chemoreceptors, located in the carotid sinus and the aortic arch, which sense changes in blood H^+, O_2, and CO_2. These peripheral chemoreceptors are generally inactive, because the central chemoreceptor is the most sensitive and generally maintains blood CO_2 and, therefore, H^+ and O_2 within normal parameters (Figure 8.17). However, under some circumstances, the peripheral chemoreceptors are essential. For example, during your exercise at altitude, you may actually breathe enough to lower blood CO_2. However, your respiration will not cease because of the peripheral chemoreceptors, which, through their own afferent neurons, will initiate inspiration in response to low O_2.

Our emphysemic friend, who has such a difficult time exhaling, will retain more CO_2 and, thus, have a higher blood CO_2 concentration. While this would normally increase respiration, over time, the central chemoreceptors adapt to higher blood [CO_2] and do not respond. The peripheral chemoreceptors do respond to low O_2 within the blood, and our friend breathes in response to low O_2, not high CO_2. The odd result could be that giving O_2 to this person could actually impair his breathing!

There are other neuronal inputs to breathing that control the relative depth of breathing, the duration and speed of inspiration, and the initiation of expiration. There are stretch receptors within the lung itself that prevent overinflation of the alveoli and inputs from the pons, cranial nerves, and thalamus that affect the rate and depth of expiration, especially during forced expiration. However, the primary determiners of respiratory drive are the central and peripheral chemoreceptors.

Summary

We have now seen the limitations of respiration imposed by the mechanics of ventilation and the chest wall, gas exchange and hemoglobin carrying capacity, ventilation and perfusion ratios, and neural control of regulation. In the asthma case, increased resistance to flow resulted in limited ventilation, while exercise increased O_2 demand and CO_2 elimination requirements. In the emphysema case, the loss of elastic recoil and airway closure affected expiration and CO_2 elimination. This, in turn, changed the neural control of breathing, switching it from the central chemoreceptors to the peripheral ones.

Key Concepts

Mechanics of ventilation
Autonomic control of airway diameter
Gas exchange in the lung
Bicarbonate equation
O_2-hemoglobin dissociation curves
Ventilation-perfusion inequalities
Neural control of ventilation
Pulmonary volumes and capacities

Key Terms

Visceral pleura
Parietal pleura
Diaphragm
Bronchioles
Surfactant
Minute volume
Minute ventilation
Pulmonary hypoxic vasoconstriction

Application: Pharmacology

1. If you are prescribed an epinephrine-based inhaler for your asthma, which is related to seasonal allergies, how will this help?
2. Your friend with emphysema is so impressed by the improvement in your breathing after you use your inhaler that he asks to borrow the device. Would it be beneficial for your friend with emphysema? What would limit its effectiveness?
3. A new type of inhaler has become available in the last few years. It is an anti-cholinergic agent. Will this help your asthma? Why or why not?
4. Smoking not only introduces particulate matter into the airways; it also impairs the movement of the cilia on the epithelium. What might be some of the consequences of this on cleaning of the airways? What would it do to airway diameter? What effect would that have on breathing?

Respiratory Clinical Case Study

Heart Failure

BACKGROUND

Your grandfather has heart failure and complains of shortness of breath. In fact, that was one of the problems that caused him to seek medical attention in the first place. He has been diagnosed with left ventricular heart failure. He wonders why, if it's his *heart* that's failing, he has such trouble with his breathing.

Heart failure is a multiple organ system disease. Even though the actual problem is with cardiac contractility, this causes real difficulty with gas exchange in the lung. Keep in mind that stroke volume is compromised in left ventricular heart failure because of a decline in contractility. Since the heart is not ejecting as much, the end systolic volume increases. This results in less atrial filling during the next diastole. Blood that would have come back from the pulmonary circulation to the heart remains in the lung, increasing the hydrostatic pressure at the capillaries. Just as in systemic capillaries, an increase in hydrostatic pressure promotes fluid loss from the capillaries into the extracellular space. Normally, the exudate in the lung is picked up by an abundant lymphatic capillary circulation, and the alveoli remain relatively dry. However, when too much fluid leaves the pulmonary circulation, it can overwhelm the ability of the lymphatic system to absorb it and fluid can enter the alveoli. This is the cause of pulmonary edema. Even a small amount of fluid will increase the diffusion distance for air from alveolar space to the red blood cells within the pulmonary circulation and compromise O_2 delivery. So, when your grandfather exerts himself, he experiences shortness of breath and must limit his activity to match respiratory efficiency. This limitation of activity and the discomfort that impaired gas exchange causes are important clinical signs of heart failure. Please note, however, that our patient will complain of breathing problems, and, if he is also asthmatic, may request a new inhaler from his physician!

THE CASE

Your grandfather has had some problems with asthma triggered by seasonal allergies all of his life, as do you. So, in the springtime, when the trees are heavy with pollen, you and he both have some trouble breathing. This year is different, however. Despite your efforts, your grandfather is still not taking his medication as prescribed, and the pollen load has made breathing more difficult this spring than ever before. He decides to discuss his breathing problems with his doctor. Unfortunately, he fails to tell his doctor that he is not taking the medications as prescribed and discusses only his lifelong history of seasonal allergies and breathing difficulties. He comes home with an epinephrine-based inhaler, which does help his breathing somewhat. Because the inhaler has had some benefit, he is even more resistant to your arguments that he should take his medications.

THE QUESTIONS

1. Your grandfather has been diagnosed with left ventricular heart failure. Explain in simple terms why he finds breathing so difficult as a result of his heart failure.
2. Why is fluid in the alveoli a hindrance to breathing? Be specific in your answer.
3. How would the inhaler improve his breathing?
4. What limits the effectiveness of the inhaler in this situation?
5. If the epinephrine gets into circulation, what effect might it have on the heart?

© Image Source/age fotostock

9

Renal Physiology and Acid-Base Balance

Case 1

You come home after a busy day, very thirsty, because you have had nothing to drink all day. You collapse on the couch with a bag of potato chips and a very large cola. Halfway through the bag of chips you begin to think about the consequences of your snack on fluid balance and how it will be regulated at the nephron. Where will the cola go after it enters your stomach? On reflection, you realize you haven't urinated in hours. Has your urine output been hormonally influenced? How do we conserve water during times of scarcity, like today? How are sodium and glucose regulated at the nephron?

Case 2

In a mad rush to lose weight before her wedding, a woman experiments with induced vomiting after meals. After three days of this, she is taken to the hospital in distress and is diagnosed with metabolic alkalosis. How did this happen, and what would you expect her partial pressure of carbon dioxide (pCO_2) and blood bicarbonate (HCO_3^-) levels to be?

Case 3

Just before a big exam, you find yourself hyperventilating and start to black out. Your friend suggests you breathe into a bag. Describe what is happening and why.

Introduction

Most discussions of physiology, in addition to the description of cells and organs, emphasize how important blood supply is to tissues and the role that blood plays in the nourishing of cells and removal of waste products. But where do all these waste products go? What prevents them from traveling endlessly through our closed circulatory system?

The kidneys, which receive 25% of the cardiac output, filter blood continuously, forming a filtrate of blood, urine, as a waste product. In this discussion, we will learn how (1) blood is filtered; (2) urine is formed; (3) urine volume is regulated to control total body water; (4) the kidney contributes to ion homeostasis and drug removal; (5) the kidney contributes to acid-base balance; and (6) urination (micturition) is regulated.

Structure of the Kidney, the Nephron, and Associated Blood Vessels

The kidneys are paired, bean-shaped organs lying on either side of the body and are served by renal arteries that arise from the descending aorta. Outflow from the kidneys is via the renal veins. In cross section we can see the renal pyramids, which are an orderly arrangement of nephrons (**FIGURE 9.1**). Just as a symmetrical arrangement of proteins in skeletal muscle appeared as striations, so the parallel alignment of nephron tubules in the kidney impart a pleated appearance.

The nephron is the filtration apparatus of the kidney and will be the focus of most of this discussion. The nephron is served by the afferent arteriole that brings blood to the glomerulus, a network of capillaries within an enclosed capsule, the Bowman's capsule, that will hold the blood filtrate. The blood remaining after filtration at the glomerular capillaries exits the nephron through the efferent arteriole (not to be confused with the afferent arteriole—think **a**fferent = **a**rriving, **e**fferent = **e**xiting). The efferent arteriole then descends and loops around the nephron in yet another capillary bed known as the

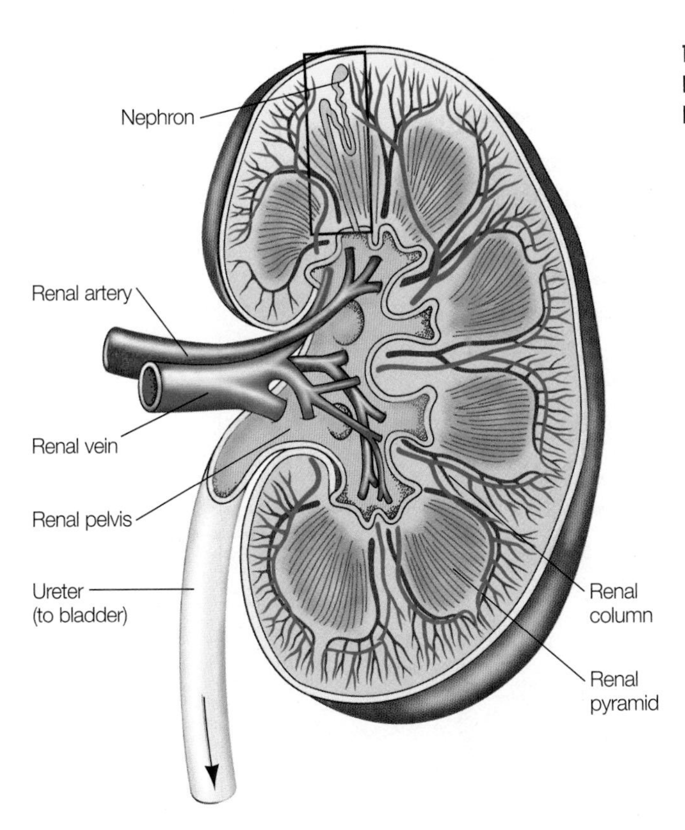

FIGURE 9.1 In cross section, the nephrons of the kidney appear as linear striations. The renal artery divides and spreads throughout the kidney to serve each of the pyramids, which are clusters of nephrons.

peritubular capillaries. The peritubular capillaries and their extensions, the vasa recta, join to the interlobular vein, which eventually culminates in the renal vein (**FIGURE 9.2**). Note that the efferent arteriole lies between two capillary beds: the glomerular capillaries and the peritubular capillaries. This anatomical feature will be important as we discuss urine formation in later sections.

How Does Blood Become Urine?

Urine formation begins at the glomerular capillaries, which are fenestrated, meaning that they have numerous tiny spaces between the union of the endothelial cells that comprise the capillaries. These spaces allow for free passage of water, ions, glucose, and amino acids from the blood space into the Bowman's capsule space that surrounds the glomerular capillaries (**FIGURE 9.3**). Just as we see in systemic capillaries, hydrostatic pressure of the blood within the afferent arteriole provides the force of filtration out of the capillary and into the Bowman's space. Hydrostatic pressure is related to blood pressure, so in general, the greater the blood pressure in the afferent arteriole, the greater is the hydrostatic pressure and the pressure for filtration. As fluid leaves blood, pressure in Bowman's capsule acts as a counter force to filtration.

While they are leaky, the glomerular capillaries are not permeable to proteins or red blood cells. This is partially a size restriction, but another important barrier to protein loss is the negatively charged basement membrane that covers the capillaries. Because proteins are also predominantly negatively charged, the charge repulsion prevents filtration of proteins. There is another barrier to filtration, the podocytes, which lie interior to the basement membrane, facing Bowman's space. The podocytes do not cover the capillary endothelium entirely but contain filtration slits through which small molecules can pass.

Plasma proteins that remain in the capillaries provide the oncotic pressure that draws water back from Bowman's capsule into the blood, thus completing the balance of filtration

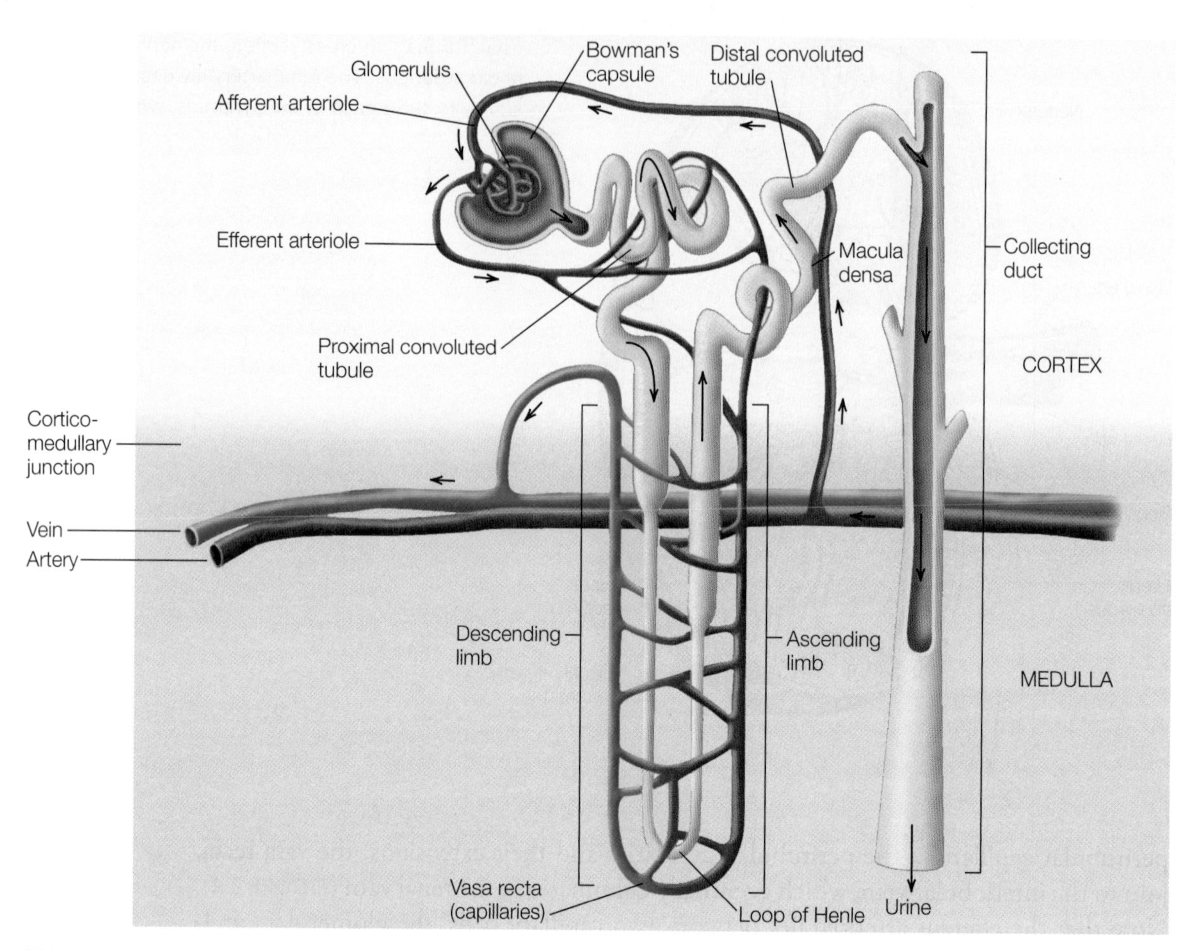

FIGURE 9.2 The anatomy of an individual nephron.

forces, similar to what we see in systemic capillaries. Clearly, an increase in hydrostatic pressure will increase filtration, and a decrease in hydrostatic pressure will decrease filtration. A decrease in plasma proteins, as might be seen in liver disease, will cause a decrease in reabsorption and therefore an increase in filtration.

The Diameters of the Afferent and Efferent Arterioles Regulate Glomerular Filtration Rate and Renal Blood Flow

The diameters of both the afferent and efferent arterioles are subject to hormonal control. Changing the diameter of either the afferent or the efferent arterioles will alter the hydrostatic pressure and, therefore, filtration. At the same time it will affect the amount of blood remaining after filtration, or the renal blood flow (RBF). Glomerular filtration rate (GFR) is a measure of the volume of blood that is being "cleaned" by all of the nephrons collectively, each minute, an important homeostatic function. Renal blood flow provides O_2 and nutrition to the kidney tubule cells. These cells are very active metabolically, so a relatively high blood flow to the tubule cells is necessary for their function. Clearly, both GFR and RBF are vital functions, and the balance of these two processes is regulated by a variety of hormones (Figure 9.3).

Epinephrine, released during sympathetic stimulation and acting at G-protein coupled receptors on the afferent arteriole, causes constriction of this vessel. Constriction reduces the hydrostatic pressure and GFR in the glomeruli and decreases RBF. During sympathetic stimulation of the autonomic nervous system, this reduces filtration and flow to the kidney but allows blood to be shunted to skeletal muscle as part of our "fight or flight" state.

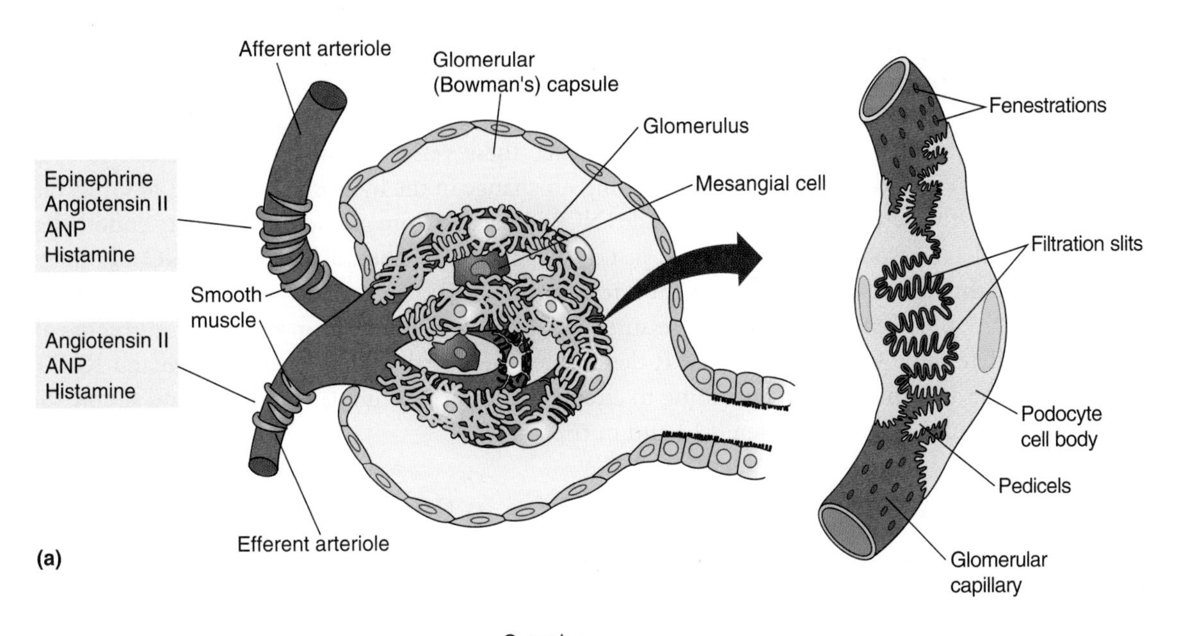

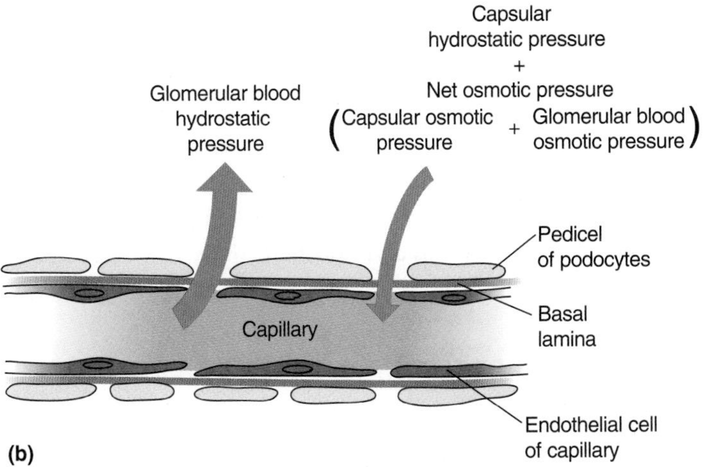

FIGURE 9.3 (a) The Bowman's capsule is designed for filtration, allowing water, ions, and small molecules to filter into the Bowman's space and then the proximal convoluted tubule. Blood flow into the glomerular capillaries via the afferent arteriole and out of the capillaries by the efferent arteriole is regulated by hormones. (b) Hydrostatic pressure favors filtration, and osmotic pressure of the blood favors retention of fluid within the capillaries. Osmotic force within the Bowman's capsule will increase filtration.

As a temporary measure this is useful, but over an extended period this could result in ion imbalances in blood and inadequate blood supply to the nephrons. This occasionally occurs in athletes engaged in extended events.

Angiotensin II has dual effects on GFR and RBF. At low concentrations, angiotensin II causes constriction of the efferent arteriole. Reducing the diameter of the efferent arteriole increases resistance to blood flow and produces a back pressure on blood leaving the afferent arteriole. This increases hydrostatic pressure and increases GFR while reducing RBF. At higher concentrations—as we would see in the case dehydration, for example—angiotensin II also binds to less-sensitive receptors on the afferent arteriole, reducing both GFR and RBF. This prevents water loss to urine during dehydration but has the same potentially negative effects that we saw with epinephrine.

Atrial natriuretic peptide (ANP), which is released from the right atrium in response to an increase in blood volume, targets both the afferent and efferent arterioles as well but has different effects at each. ANP causes dilation of the afferent arteriole and constriction of the efferent arteriole. Each of these actions increases GFR and reduces RBF. However, because ANP is released when there is an increase in blood volume, there is still sufficient RBF to nourish tubular cells.

Several other hormones can influence these vessels. Histamine release dilates both afferent and efferent arterioles, causing no change in the hydrostatic pressure or GFR, but greatly increasing RBF. Nitric oxide (NO) is released from glomerular capillary endothelial cells in response to shear stress, just as we saw in the systemic capillaries. NO can dilate either the afferent or the efferent arteriole.

Blood pressure, oncotic pressure, and arteriole diameter as mediated by hormones all determine GFR and RBF. GFR determines the initial volume of the filtrate and RBF is the amount of blood remaining to circulate into the peritubular capillaries and supply the tubular cells with oxygen (O_2) and nutrients.

What Is in the Filtrate That Lies in the Bowman's Capsule?

The leakiness of the glomerular capillaries allows water, ions including Na^+, K^+, Ca^{2+}, HCO_3^-, H^+, and Cl^-, glucose, amino acids, metabolites, toxins, and drugs to all be filtered into Bowman's space. You will notice that many of the molecules that are freely filtered may be important for us to retain, and yet they are not retained at this stage in filtration. Only proteins are retained within the glomerular capillaries. Why does this happen?

The glomerulus is nonselective in filtering but is very selective in reabsorption. (We will focus on the mechanisms of nutrient and ion reabsorption along the nephron in this discussion.) Selectivity in reabsorption allows the nephron to eliminate foreign substances or metabolites because they are not selectively reabsorbed, remain in the filtrate, and are excreted in urine. Thus, the kidney serves as an important route for waste removal from the blood and drug elimination.

What Happens to the Filtrate at the Proximal Convoluted Tubule?

The filtrate that is collected in the Bowman's capsule moves by simple pressure gradient into the proximal convoluted tubule, the next portion of the nephron. While the nephron appears continuous, there are histological differences in each named section that reflect their dissimilar functions. The proximal convoluted tubule is a workhorse of transport, evidenced by the brush border on the apical (tubular or filtrate) side of the cell and the extensive invaginations on the basolateral (blood) side of the cell (**FIGURE 9.4**). The invaginations on the basolateral side are populated by mitochondria, because this section of the tubule is very metabolically active, transporting ions from the filtrate into the extracellular fluid (ECF). It is important to remember the direction of transport in the PCT. Water and ions will move from the filtrate side to the ECF, where they are transported into the blood via the peritubular capillaries. Traditionally, the ECF is called the blood side, because molecules transported from the filtrate into the ECF will quickly enter the capillaries and become part of the blood.

Sodium can be transported in two different ways in the PCT: the transcellular route and the paracellular route. We will start with the more complex pathway, the transcellular route, which crosses the proximal tubular cell. Na^+ concentration in the filtrate is high (~145 mM) because Na^+ is freely filtered at the glomerular capillaries. However, Na^+ concentration inside of the proximal tubular cell is low (~12 mM), as in all cells of the body. Between the filtrate and the intracellular space of the proximal tubule cell, there is a very favorable driving gradient for Na^+ that can be used to drive cotransport or secondary

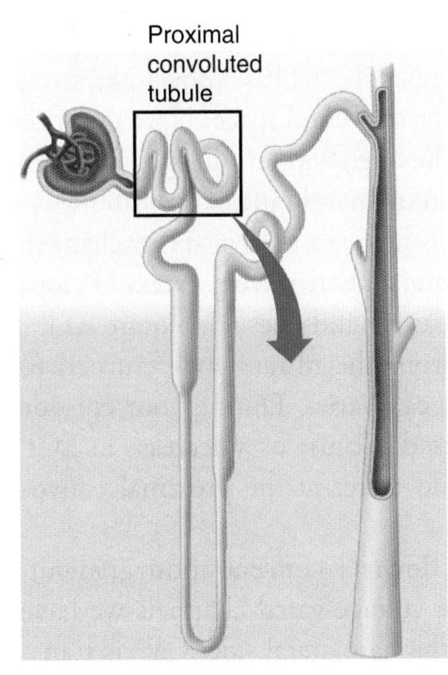

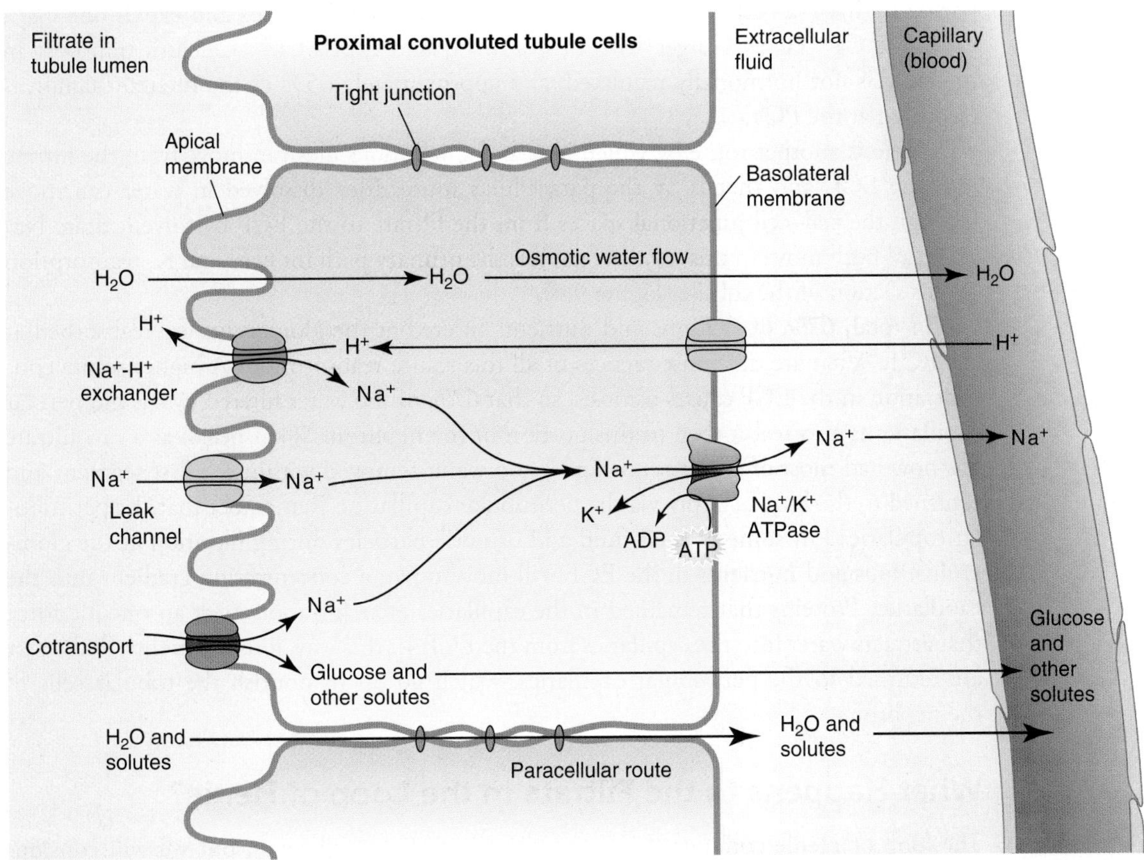

FIGURE 9.4 Reabsorption of ions and nutrients at the proximal convoluted tubule drives the osmotic reabsorption of water. Movement of ions and water occurs through the cell in the transcellular route and between cells, in the paracellular route.

active transport. Na^+ binds to one site on a transport protein while another ion or small molecule binds to another site. Na^+ moves down its concentration gradient, passively, and the other molecule is transported along with the Na^+ as a passenger. In the PCT, in the transcellular pathway, Na^+ transport is coupled to amino acids, glucose, phosphate, sulfate, lactate, and other small molecules. Once inside the proximal tubular cell, Na^+ is transported to the ECF by the Na^+/K^+ ATPase, which maintains the gradient and moves Na^+ to the blood side, reclaiming it from the filtrate. The Na^+/K^+ ATPase moves Na^+ from the proximal tubular cell into the ECF, where it will enter the blood space. Note that this is a process that requires energy, which is provided by the adenosine triphosphate (ATP) manufactured by the mitochondria located at the basolateral membrane of the tubule.

Na^+ is also transported into the proximal tubular cells through the Na^+-H^+ exchanger, which transports Na^+ into the cell and H^+ into the filtrate, thus removing excess H^+ ions while salvaging Na^+. This electrically neutral exchange is passive and does not require ATP.

Through all these mechanisms, Na^+ is recovered from the filtrate and returned to the ECF, where it will be picked up by the peritubular capillaries. During your episode of dehydration in the opening case, angiotensin is released because of a decrease in ECF volume; it will increase the transport of both NaCl and water at the proximal convoluted tubule.

Ca^{2+} enters the proximal tubular cell and passes down a concentration gradient, through epithelial calcium channels, different from the voltage-gated channels we have seen previously. However, the transport is passive. On the basolateral side, Ca^{2+} is transported by the Na^+-Ca^{2+} exchanger, which brings in three Na^+ ions and expels one Ca^{2+}, or by the Ca^{2+}-H^+ exchanger. This brings in H^+ and expels Ca^{2+}. Calcium transport in the PCT is not hormonally regulated, and approximately 65% of the filtered calcium is recovered at the PCT.

There is another route by which ions and small molecules can move from the filtrate into the ECF, and that is by the paracellular route. Ions dissolved in water can move through the cell–cell junctional spaces from the filtrate to the ECF by solvent drag. Na^+ and Ca^{2+} both move in this way. In fact, it is the primary path for Ca^{2+} and K^+ reabsorption in this section of the tubule (Figure 9.4).

In total, 67% of the ions and nutrients filtered at the glomerulus are reabsorbed at the PCT. What are the consequences of all this solute reabsorption? A higher solute concentration in the ECF causes osmosis, so that 67% of the water filtered by the glomerular capillaries is also reabsorbed in this portion of the nephron. What began as a raw filtrate has now had most of its ions, nutrients, and water removed within the first segment and returned to the blood supply via the peritubular capillaries. Remember that the peritubular capillaries lost some of their fluid and osmotic particles during filtration at the glomerulus. Ions and nutrients in the ECF will move down a concentration gradient into the capillaries. Proteins that remained in the capillaries after filtration exert an oncotic force that attracts water into the capillaries from the ECF. In this way, ions, nutrients, and water are returned to the peritubular capillaries, which go on to nourish the tubular cells of the nephron.

What Happens to the Filtrate in the Loop of Henle?

The loop of Henle contains three anatomically distinct cell types, but we will consider them all in one section because their functions are so closely interdependent. At the beginning of the loop of Henle, the filtrate has the same osmolality as blood and the original filtrate, 300 mOsm. Remember that ions and water were filtered and reabsorbed equally, so the volume changed in the proximal convoluted tubule, but not the osmolality. In the loop of Henle, both the volume and the osmolality will change.

The tubular cells of the thin descending limb of the loop of Henle are poorly invaginated and have few mitochondria. These cells are impermeable to Na^+. However, they are well-equipped with aquaporins, or water channels, that allow water from the filtrate to flow passively into the ECF and be absorbed into the peritubular capillaries. Thus, as the filtrate travels down the thin descending limb, the volume is reduced and the osmolality increases (**FIGURE 9.5**). The maximum osmolality achieved during dehydration is 1,200 mOsm, primarily as a result of the high concentration of NaCl at the base of the loop of Henle.

In the thin ascending portion of the loop of Henle, NaCl moves passively down a concentration gradient through transporters into the ECF. This section of the loop is impermeable to water but very permeable to NaCl. Thus, as the filtrate moves toward the thick ascending limb, the osmolality drops, approaching 400 mOsm by the end of the thin ascending limb.

Cells of the thick ascending limb have plentiful mitochondria and are active metabolically. On the apical or filtrate side lies the Na^+/K^+-$2Cl^-$ transporter, which brings all three ions into the tubular cells in an electroneutral transport that does not require ATP. Also on the filtrate side is the Na^+-H^+ exchanger, which brings in Na^+ in exchange for an excluded H^+. This salvages Na^+ from the filtrate and adds H^+ to the filtrate, thus eliminating a metabolic waste product. Both of these exchange proteins operate passively, but on the basolateral surface of the cell, the blood side, Na^+ is actively pumped into the ECF by the Na^+/K^+ ATPase. K^+ and Cl^- are transported passively out of the thick ascending limb tubule cells by a K^+-Cl^- symport protein. The filtrate in the thick ascending limb is very positively charged because of the concentration of positive ions. This charge differential causes a driving gradient across the paracellular pathway for Na^+, K^+, Ca^{2+}, and Mg^{2+} to travel to the ECF. This segment is very permeable to ions but is impermeable to water, so by the end of the thick ascending loop of Henle, the osmolality has decreased to 120 mOsm, less than half the osmolality of blood (Figure 9.5).

Now we can see the interdependence of the different segments of the loop of Henle. Water moves by osmosis through water channels in the thin descending limb because of the high osmolality of the ECF established by the thick ascending limb. The volume of the filtrate is thus reduced during its travel down the thin descending limb, while the osmolality of the filtrate is reduced and ions salvaged in the thin and thick ascending limbs.

How important are these ion gradients? Their significance can be illustrated by the use of loop diuretics, drugs that are given to decrease total blood volume. Loop diuretics, including the well-known drug furosemide, inhibit the Na^+/K^+-$2Cl^-$ transporter in the thick ascending limb. Without that transporter, there is less Na^+ and K^+ to be transported into the ECF by the basolateral Na^+/K^+ ATPase and the K^+-Cl^- symporter. The osmolality of the ECF is decreased, removing the driving gradient for water from the thin descending limb. The result is a greater volume of filtrate with a higher ion concentration and osmolality entering the distal convoluted tubule (DCT). Ultimately, it causes a greater diuresis and a loss of total body water.

What Happens to the Filtrate in the Distal Convoluted Tubule and the Collecting Duct?

The final processing of the filtrate comes in these two sections. Because both are sensitive to volume-regulating hormones, we will consider them together.

The early portion of the DCT is much like the thick ascending limb of the loop of Henle: impermeable to water and an active transporter of NaCl out of the filtrate. Thus, early in the DCT, the filtrate continues to be diluted, but not reduced in volume. Cells in the latter portion of the DCT and the cells of the collecting duct are similar. There are two cell types, the principal cells and the intercalated cells, each of which has unique functions.

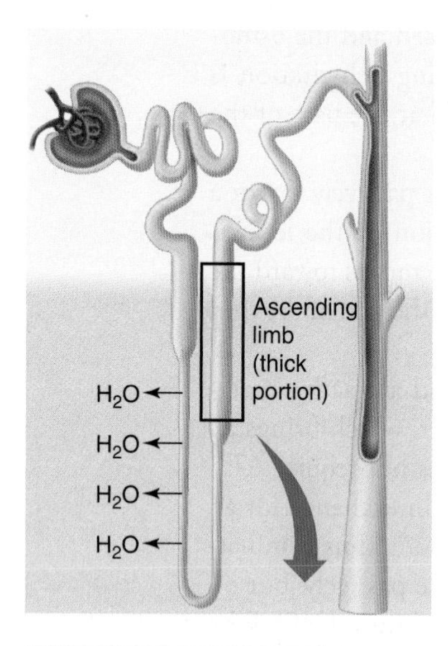

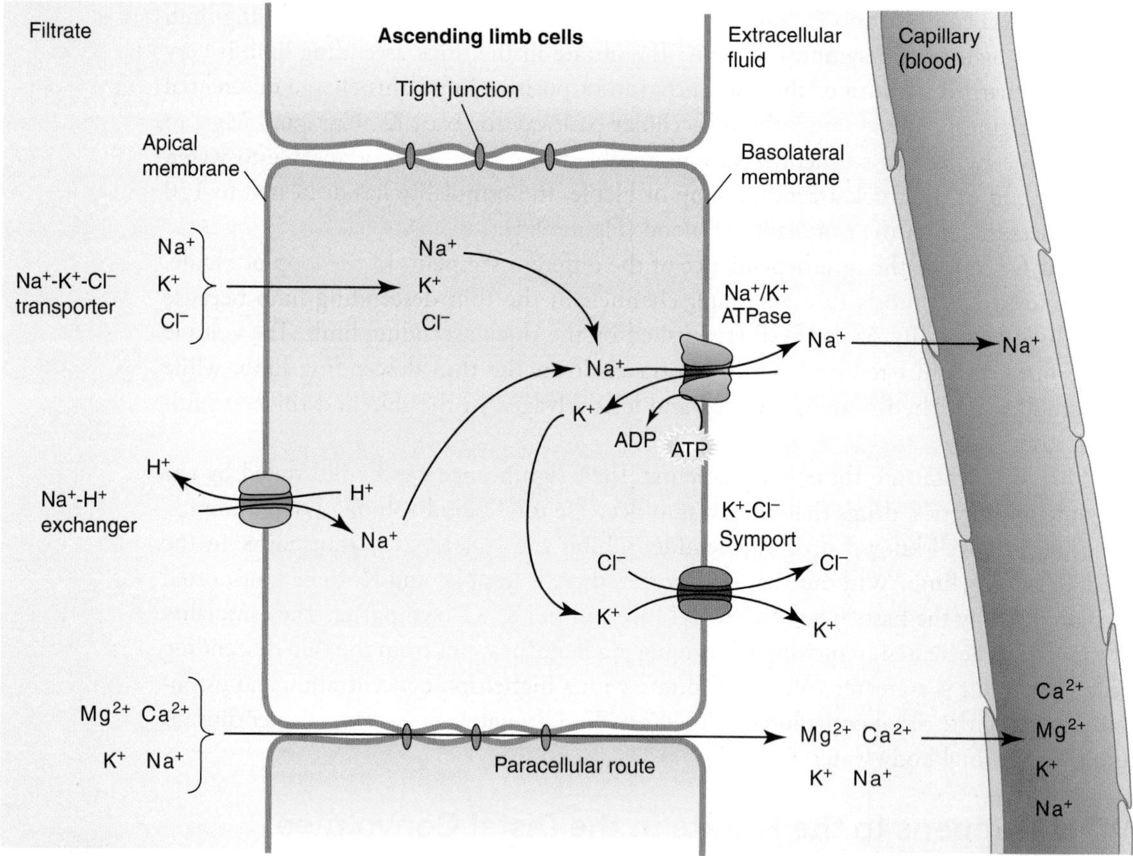

FIGURE 9.5 The descending loop of Henle and the ascending loop of Henle contain very different transporters. Together they set up osmotic gradients for water and ion reabsorption.

The principal cells contain epithelial Na^+ channels (ENaC), which passively allow Na^+ to move down a concentration gradient from the filtrate, into the principal cells, where it is transported to the ECF by the Na^+-K^+ ATPase. K^+ moves down its concentration gradient from the intracellular space of the principal cell into the filtrate, thus providing a good driving gradient for the Na^+K^+ ATPase. The intercalated cells are important for acid-base balance, specifically for bicarbonate ion salvage and H^+ ion excretion. On the filtrate side, intercalated cells have a K^+-H^+ ATPase that moves K^+ into the cell and H^+ out.

Another active process, a H^+ ATPase moves H^+ into the filtrate. Both of these mechanisms use the formation of carbonic acid from water and carbon dioxide (CO_2) to eliminate H^+ and create new bicarbonate buffering ions for the blood (**FIGURE 9.6**).

Both the late DCT and the collecting duct are sensitive to hormones. In our opening case, you are dehydrated. Where the thin ascending limb of the loop of Henle meets

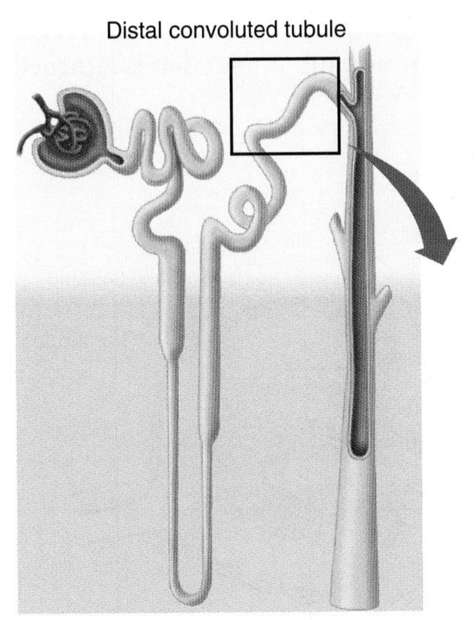

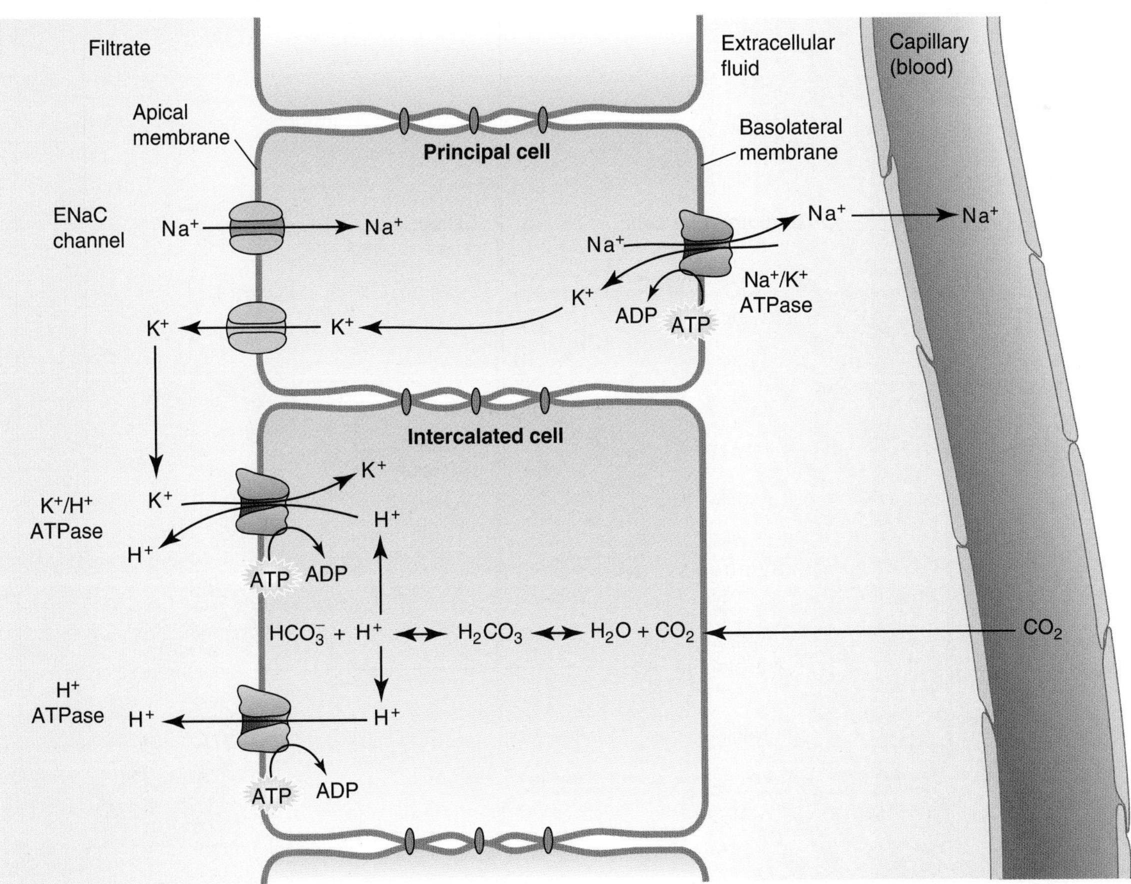

FIGURE 9.6 Different ions are transported by the principal cells and intercalated cells of the distal convoluted tubule and the collecting duct.

the DCT, a cluster of cells (known as the juxtaglomerular apparatus) lies between the tubule and the afferent and efferent arterioles. These cells serve as a sensor of filtrate volume, which is proportional to blood volume. Specifically, a reduced flow of Na^+ past the cells of the juxtaglomerular apparatus causes a release of the enzyme renin, which initiates the production of angiotensin II and the release of aldosterone from the adrenal glands. Aldosterone acts at the cells of the thick ascending limb and the DCT to increase the ENaC in the apical membrane, increase the Na^+/K^+ ATPase pumps in the basolateral membrane, and increase key proteins in the metabolic pathways to facilitate ATP production (**FIGURE 9.7**). Thus, aldosterone will increase the amount of Na^+ that is returned

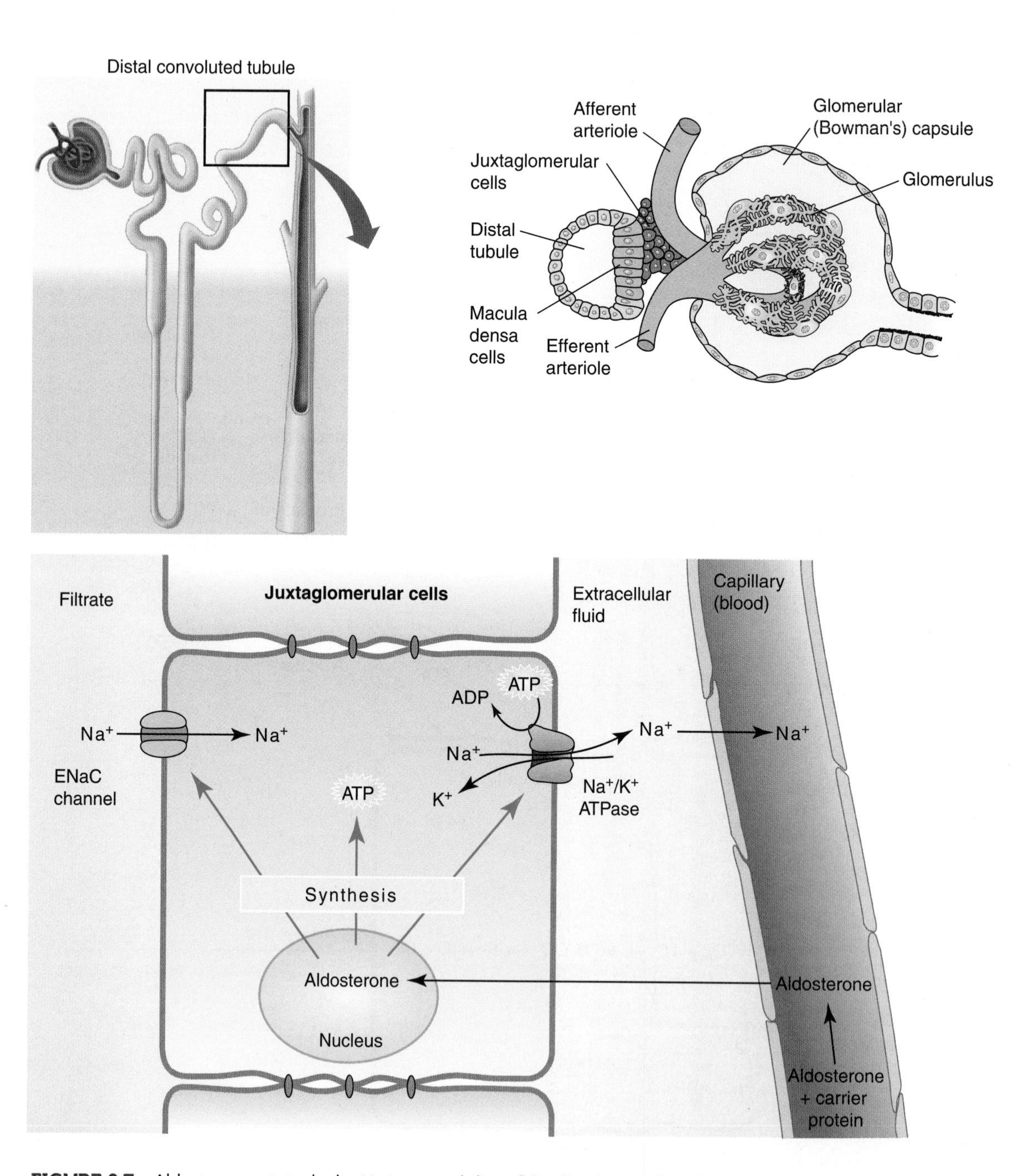

FIGURE 9.7 Aldosterone controls the Na^+ permeability of the distal convoluted tubule and the collecting duct.

to the ECF and, therefore, the water that will follow by osmosis, increasing blood volume and reducing urine volume. Because all of aldosterone's actions involve protein production, this hormone takes hours to days to affect the kidney.

Antidiuretic hormone (ADH) acts to increase the number of aquaporins, or water channels, in the collecting duct. The source of ADH is the posterior pituitary, released when osmoreceptor cells in the hypothalamus sense an increase in blood osmolality, as would occur during dehydration. ADH acts through a G-protein coupled receptor linked to cyclic adenosine monophosphate (cAMP) and protein kinase A to fuse vesicles filled with aquaporins into the membrane of the collecting duct (**FIGURE 9.8**). These channels will allow rapid osmosis of water into the ECF. Note the interdependence of aldosterone action and ADH. Aldosterone causes Na^+ to increase in the ECF, which increases the osmolality of that space. Close by, aquaporins provide an easy route for water to follow this osmotic gradient, thus reducing the water content of the final filtrate. Thus, during dehydration, the volume of body water that is lost as urine is decreased.

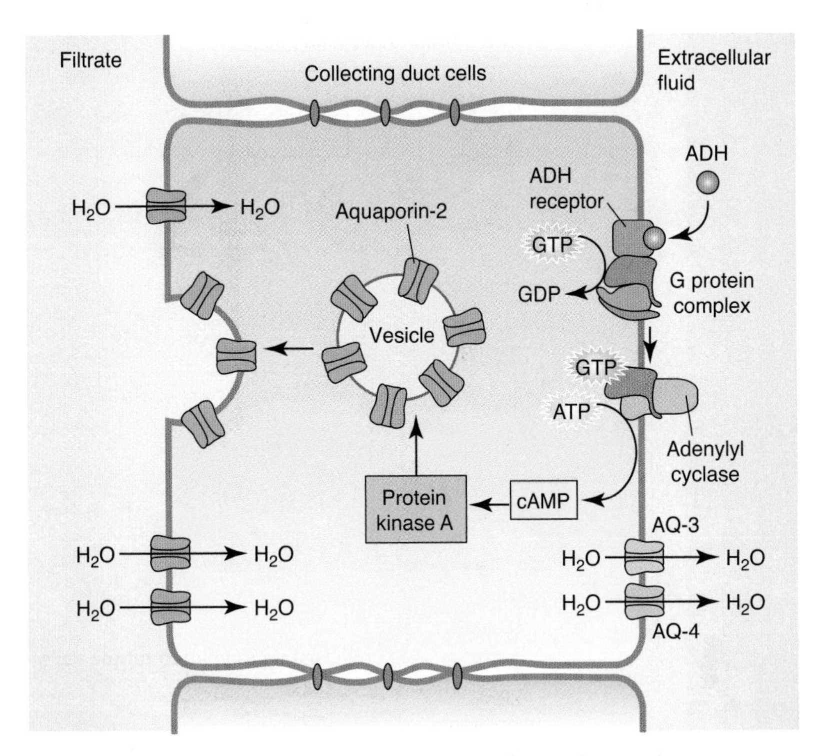

FIGURE 9.8 ADH controls water permeability of the collecting duct.

What happens if you are not dehydrated? Without these hormones, the filtrate at the DCT is dilute, with an osmolality of approximately 120 mOsm. Most of the ions and water have already been reabsorbed. Without the aquaporins, the collecting duct is water impermeable, so the final urine volume and osmolality will approximate that of the DCT. More water will be lost in urine, but it was excess water anyway!

Ca^{2+} balance is also managed at the DCT. Parathyroid hormone, released by parathyroid glands in response to a decrease in plasma Ca^{2+}, has one of its targets in the DCT. Parathyroid hormone causes an increase in Ca^{2+} transport across the apical membrane through Ca^{2+} channels and an increase in transport across the basolateral membrane by the Na^+-Ca^{2+} exchanger and a Ca^{2+} ATPase. Thus, when Ca^{2+} is in short supply, its reclamation from the filtrate is enhanced by parathyroid hormone at the DCT (**FIGURE 9.9**).

There is one more important molecule that contributes to the high osmolality of the ECF in the interstitium near the collecting duct: urea. Ammonia is produced by the liver during the deamination of amino acids. However, because ammonia is toxic, even in small quantities, it is converted to urea within the liver. Urea is filtered in the glomerulus. All of the nephron is impermeable to this molecule until it reaches the collecting duct. The collecting duct is permeable to urea, and this permeability is increased by the hormone ADH. Urea is osmotically active, so when released into the ECF, it increases the driving gradient for water to leave the collecting duct (**FIGURE 9.10**). Even a waste product is utilized for the concentration of urine!

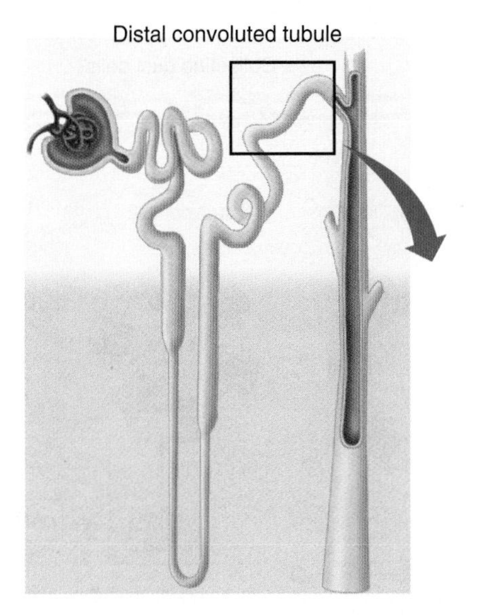

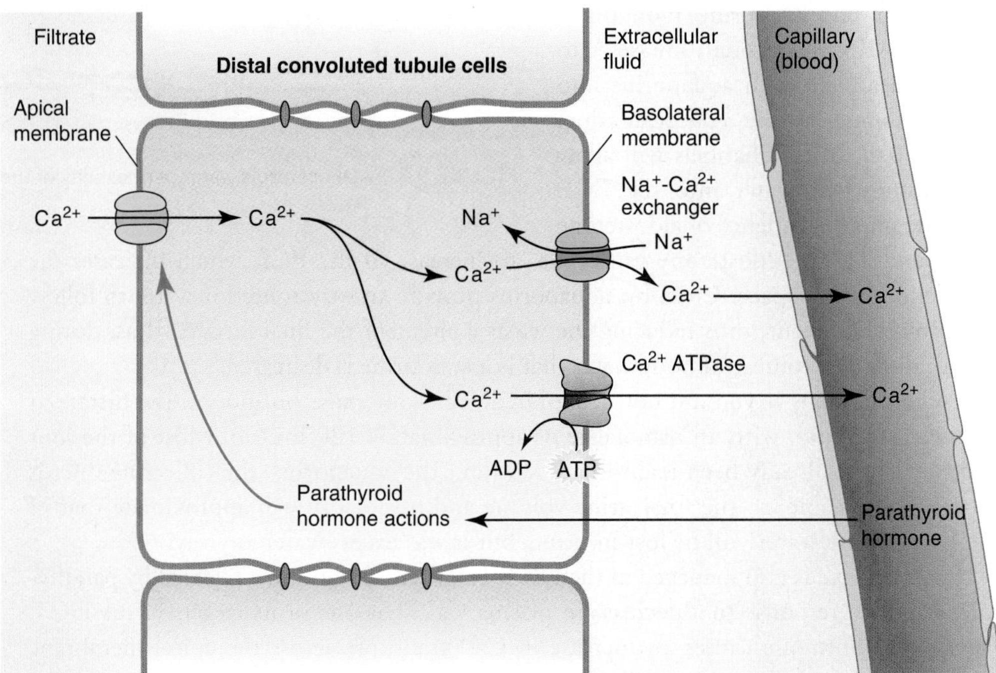

FIGURE 9.9 Parathyroid hormone regulates Ca^{2+} uptake in the distal convoluted tubule.

Summary of Urine Formation by Filtration

The process we have just followed—filtering of blood, reclamation of water and ions by the nephron, and the production of urine waste—continues every minute of your life. With each heartbeat, 25% of your cardiac output goes to the kidney, which then systematically reorganizes the ion composition of blood and salvages the nutrients while allowing metabolites and other "unrecognized" molecules to be removed with excess water in urine.

Secretory Actions of the Kidney

So far, the movement of molecules has always been from the filtrate and back to the blood. However, the kidney is also capable of a secretion, or movement, of molecules from the blood to the filtrate. Because potassium is in high concentration in many foodstuffs, we

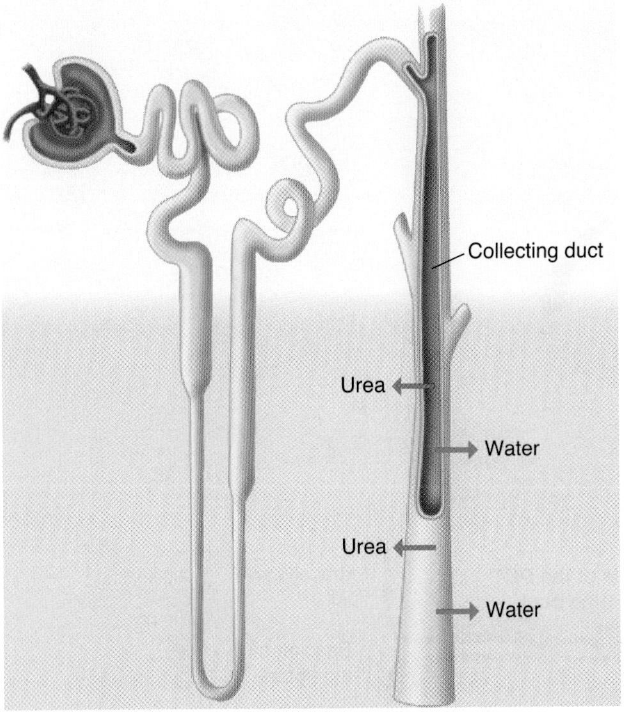

FIGURE 9.10 Urea is an important osmotic force for the reabsorption of water.

generally take in more K^+ than we need. The principal cells of the DCT and the collecting duct can secrete K^+ from the tubular cells into the filtrate, thus causing the elimination of K^+. How does this happen?

As we have seen, Na^+ is actively reabsorbed using the Na^+/K^+ ATPase, a protein that is upregulated by aldosterone. The Na^+/K^+ ATPase will increase intracellular K^+ concentrations, which will then cause K^+ to diffuse through K^+ channels into the filtrate. K^+ loss to the filtrate will increase when aldosterone is active (**FIGURE 9.11**).

Using a similar cellular mechanism, H^+ is secreted into the filtrate in the proximal convoluted tubule, when the Na^+-H^+ exchanger removes Na^+ from the filtrate and returns it to the ECF. H^+ is lost to the filtrate, that is, secreted. Thus, reclamation of valuable ions is linked to loss of metabolites such as H^+. While the pumps and exchangers evolved to handle ions, they can also be used by some drugs or drug metabolites, which are also secreted into the filtrate. Some drugs are simply lost in the filtrate because they were not reabsorbed, but others are actively secreted into the filtrate. In either case, the kidney becomes an important route for drug and normal metabolite elimination. This is why urine has traditionally been a diagnostic tool for medicine.

Where Does Urine Go After the Collecting Duct, and What Regulates Elimination?

The filtrate at the end of the collecting duct is finally the finished product—urine. Urine collects in the minor and major calyces of the kidney and finally is collected in the ureters that drain into the bladder. The bladder serves as a reservoir for urine, to be eliminated at our discretion. Urination, or more formally, micturition, is regulated by an interaction of the autonomic nervous system and the somatic nervous system.

The urinary bladder is surrounded by smooth muscle, innervated by both branches of the autonomic nervous system. Sympathetic stimulation to the bladder causes relaxation

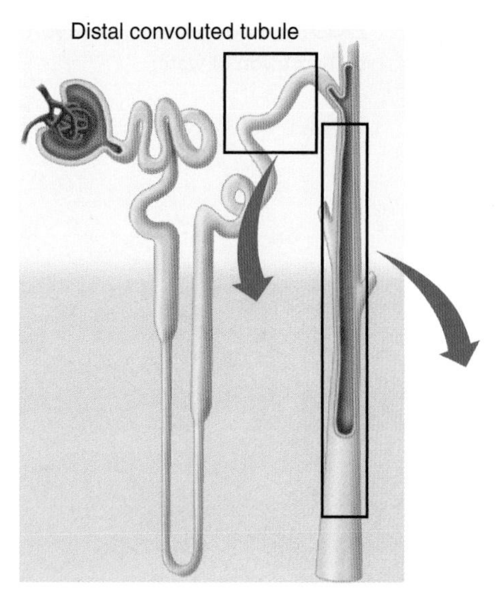

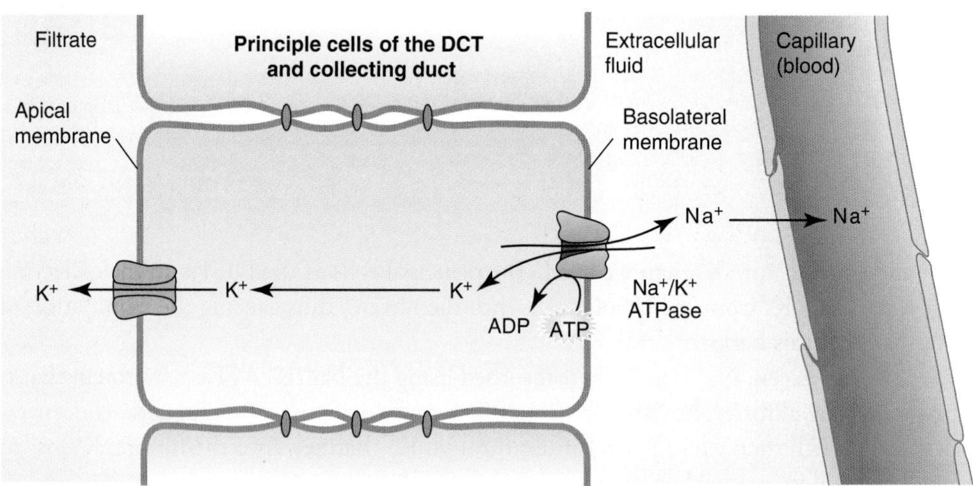

FIGURE 9.11 Ions can be secreted into the filtrate in several parts of the nephron.

of the primary muscle, the detrusor muscle, allowing the bladder to relax and fill. Parasympathetic stimulation of the same muscle causes contraction, which will reduce the diameter of the bladder and increase the pressure for voiding. How do we know when our bladder is full? Stretch receptors in the wall of the bladder respond to increasing stretch of the bladder and send this information to the cortex via pelvic afferent sensory nerves (**FIGURE 9.12**). We now know it is time to void. When the opportunity arises, we can voluntarily relax the skeletal muscle of the external sphincter and allow urine to pass down a pressure gradient from inside the bladder to the receptacle of our choice!

Acid-Base Balance Is Primarily Regulated by Bicarbonate

Most of us learned in our first biology course that enzymes are pH dependent and that under extremes of acidity or alkalinity, proteins will lose their functional structure. In health, we assume that our cellular pH is optimal and rarely give a thought to its regulation. Our normal extracellular pH is 7.4, and a drop of as little as 0.2 pH units will cause serious health disturbances. Clearly, pH is a well-regulated parameter. Regulation of pH is

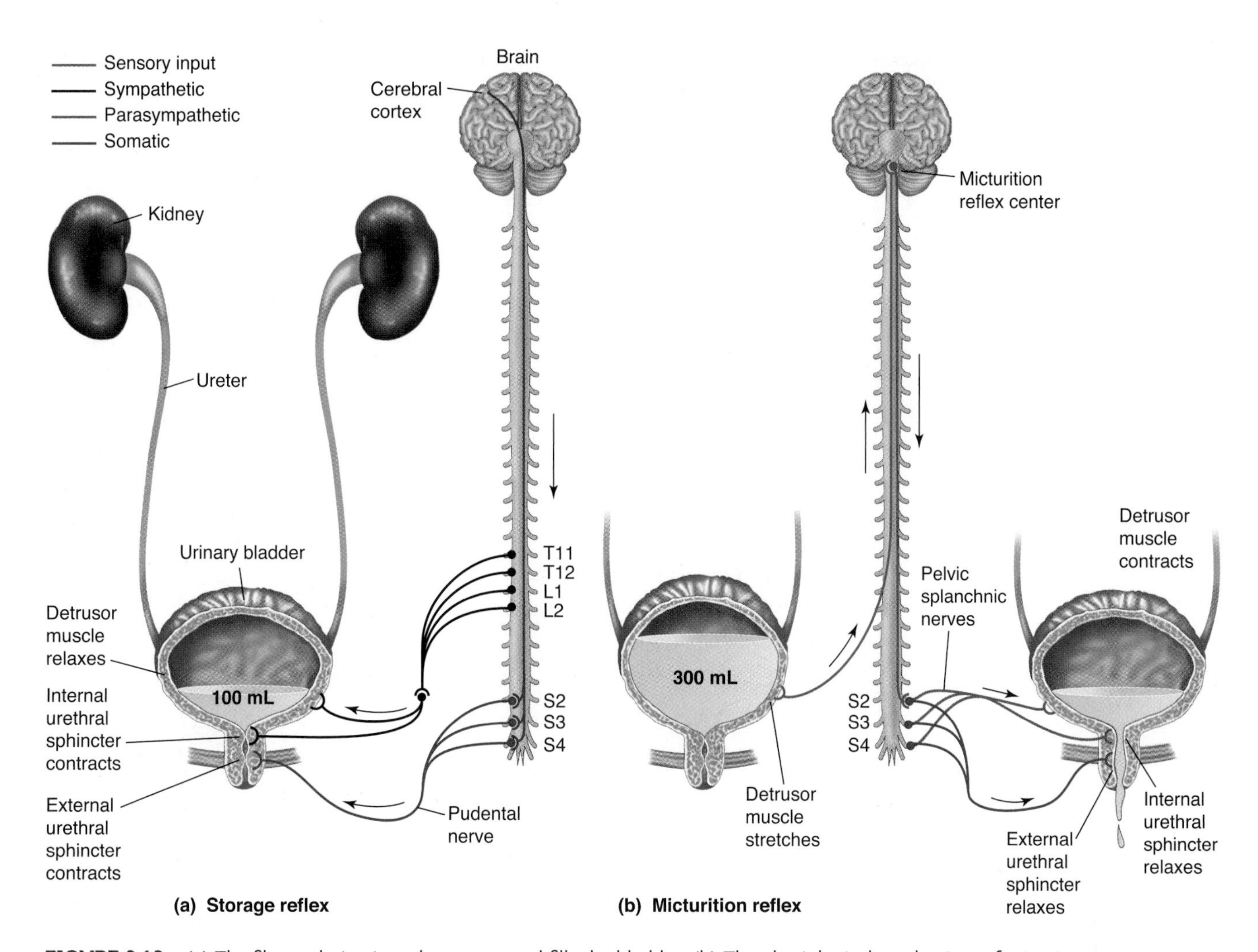

FIGURE 9.12 (a) The filtrate drains into the ureters and fills the bladder. (b) The physiological mechanism of urination, or micturition, requires coordination of the autonomic and voluntary nervous systems.

by the respiratory system and the renal system acting in a coordinated manner to maintain our normal pH, or H^+ concentration.

The primary buffering molecule of the blood is the bicarbonate ion, which is derived from CO_2 and H_2O by the enzyme carbonic anhydrase. The bicarbonate equation shows the relationship between CO_2, H^+, and HCO_3^-:

$$CO_2 + H_2O \leftrightarrow H_2CO_3 \leftrightarrow HCO_3^- + H^+$$

We know that CO_2 is eliminated at the lung during exhalation. The kidney is a powerful eliminator of H^+ in several sections along the nephron, including the proximal convoluted tubule, the thick ascending limb of the loop of Henle, and to a lesser extent the DCT and the collecting duct. The kidney not only eliminates H^+ but also reabsorbs HCO_3^-, a basic molecule, which is the primary acid-base buffer in the blood. H^+ elimination and HCO_3^- reabsorption are generally linked.

H^+ Ions Are Lost and HCO_3^- Is Reabsorbed by Multiple Transport Mechanisms

Both H^+ and HCO_3^- are freely filtered at the proximal convoluted tubule. If HCO_3^- were not reabsorbed, it would be devastating for our acid-base balance, because we would lose

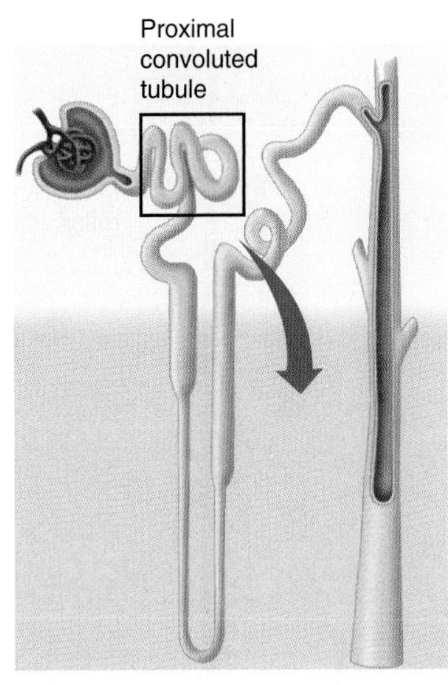

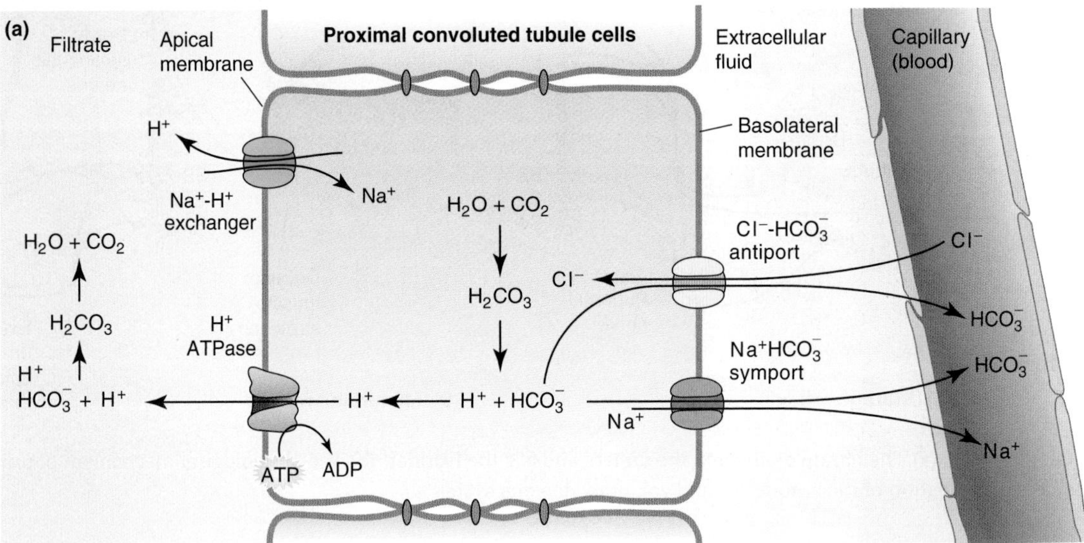

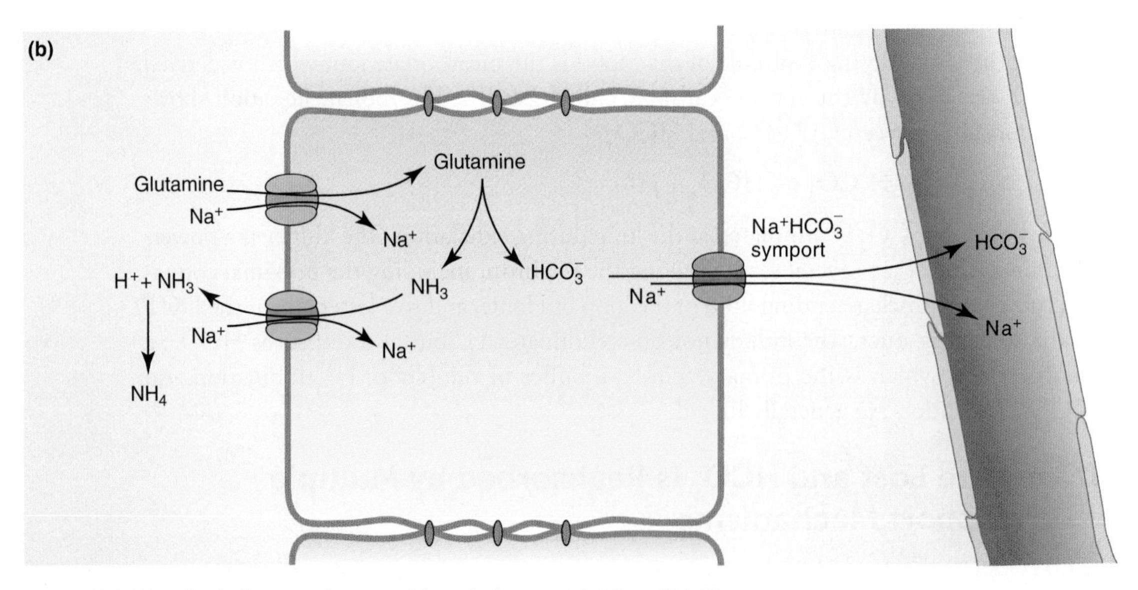

FIGURE 9.13 The kidney regulates acid-base balance with H^+ and HCO_3^-.

most of our buffering capacity in urine! However, the driving gradient for Na^+, established by the Na^+/K^+ ATPase, allows Na^+ to come into the PCT tubular cell and H^+ to be eliminated into the filtrate, by exchanger. H^+ produced by the breakdown of carbonic acid is pumped into the filtrate by the H^+ ATPase, an energy-using process. H^+ in the filtrate can combine with HCO_3^- to become H_2O and CO_2. Water will be reabsorbed later in the nephron, and CO_2, a gas, can diffuse back into the kidney tubule to recombine with intracellular H_2O to generate more intracellular HCO_3^-. Thus, H^+ is put into the filtrate and HCO_3^- is recovered from the filtrate (**FIGURE 9.13**).

On the basal side, the blood side, HCO_3^- is returned to the blood by two proteins: a $Na^+HCO_3^-$ symporter, which transports both Na^+ and HCO_3^- to the blood, and a Cl^--HCO_3^- antiporter. Both of these proteins return HCO_3^- to the blood. Transport mechanisms are similar in the thick ascending limb of the loop of Henle.

The intercalated cells of the DCT and collecting duct transport H^+ out of the tubular cell and into the filtrate through the K^+/H^+ ATPase and the H^+ ATPase. On the basal side, the HCO_3^--Cl^- antiporter moves HCO_3^- into the blood (Figure 9.13)

Once in the filtrate, H^+ ions can combine with phosphate, an important urinary buffer, or NH_3, a product of glutamine deamination in the tubular cells. NH_3 is membrane permeant and, therefore, can diffuse into the filtrate, where it binds to H^+, becomes NH_4, and is trapped within the filtrate, no longer able to cross a plasma membrane. As you notice, the primary excretion product of the kidney is H^+, while the primary excretion product of the lung is CO_2. However, because of carbonic anhydrase, the enzyme that facilitates the combination of CO_2 and H_2O into carbonic acid, and the breakdown into HCO_3^- and H^+, both lung and kidney contribute significantly to maintaining our acid-base balance.

Acid-Base Imbalances: Metabolic or Respiratory?

Let's consider our opening Case 2—the young woman practicing bulimia. Vomiting causes us to lose stomach acid, but only temporarily. Once we have finished vomiting, the parietal cells of the stomach immediately begin production of new stomach acid. The H^+ ions needed to make the acid are taken from blood, causing blood to become more alkaline, that is, the pH increases and HCO_3^- concentrations rise. To an extent, the alkalinization of blood is compensated for by the respiratory system (**FIGURE 9.14**). Remember that breathing is driven by H^+ ion concentration, so as that is reduced, breathing rate will decline, more CO_2 will be retained, and new H^+ ions generated. So, a metabolic loss of H^+ ions is compensated for by an increase in respiratory CO_2, readily converted to H^+ via the bicarbonate equation. The compensations serve to normalize pH, which is the important parameter for cellular function.

In Case 3, we have a different systemic problem that also will cause alkalosis. Hyperventilation because of a panic attack will cause excessive loss of CO_2, which is functionally

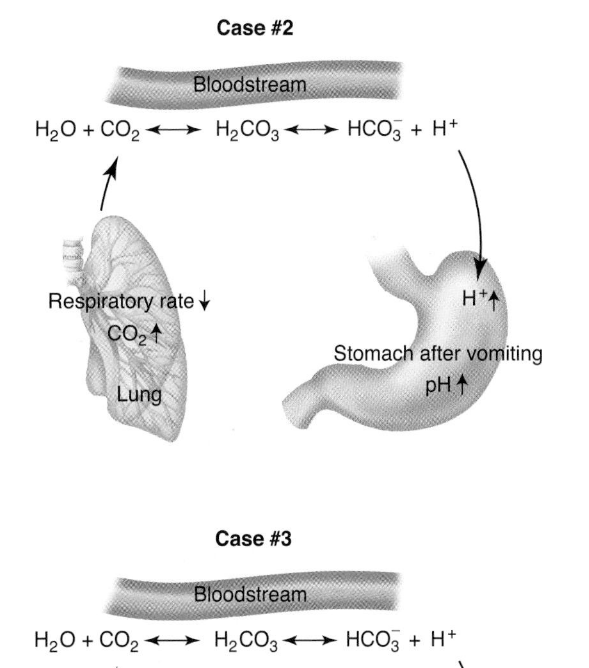

FIGURE 9.14 The lung and kidney work together to regulate blood pH.

the same as loss of H^+, and you will suffer from alkalosis, but this time from respiratory alkalosis. Compensation at the kidney will be to excrete less H^+, but this is a slow process compared to the rapidity with which a hyperventilating person can lose CO_2. If left to your own devices, you would lose consciousness, begin to breathe normally, and the problem would be solved. However, the intervention of breathing into a bag allows you to rebreathe your own CO_2 and, thereby, reestablish acid-base balance. Acid-base disorders can be multiple and complex, but the general principle is that the kidney will compensate for a respiratory disorder and the respiratory system will compensate for a kidney disorder. When the compensations are incomplete and cannot control pH, disease ensues.

Summary

The kidney tubules do a remarkable job of continuously filtering our blood to maintain ion and nutrient composition. Because of their transport of H^+ and HCO_3^- ions, the kidney also is an important contributor to acid-base balance. In addition, the kidney is an important hormonal organ, producing erythropoietin, which promotes red blood cell development, and vitamin D, which contributes to bone growth. The kidney is sensitive to parathyroid hormone, which controls Ca^{2+} balance, and to the fluid volume hormones, angiotensin II, aldosterone, and antidiuretic hormone. Given the variety of functions that the kidney must perform, it is understandable why kidney failure affects so many other systems.

Key Concepts

Filtration in the Bowman's capsule
Ion transport in the tubule
Ion transport drives osmotic reabsorption of water
Roles of each section of the tubule
Hormonal regulation of urine formation
Neural regulation of micturition
Acid-base balance

Key Terms

Bowman's capsule
Glomerular capillaries
Proximal convoluted tubule
Loop of Henle
Distal convoluted tubule
Collecting duct
Calyx
Ureter
Bladder
Juxtaglomerular apparatus
Renin
Antidiuretic hormone
Aldosterone
pH
Bicarbonate equation

Application: Pharmacology

1. Loop diuretics are commonly prescribed for patients with heart failure. How do they work? What is the result? How do they affect urine output? Total blood volume? Why would this help someone with heart failure?
2. Alcohol inhibits antidiuretic hormone (ADH). If you unwisely indulged in four beers one evening, how would you expect this to affect your urine output? Your total blood volume? What might happen to your fluid balance by morning? What hormones might be stimulated? Why?
3. Another common medication given to heart failure patients is an aldosterone inhibitor. What effect would that have on the kidney? What would be the consequences for total blood volume?

Renal Clinical Case Study

Heart Failure

BACKGROUND

The renin-angiotensin-aldosterone system is important for maintaining blood volume and blood pressure. Angiotensin II stimulates the release of ADH and aldosterone. Both of these hormones act at the kidney to increase blood volume and blood pressure. We also know that body compartments—intracelluar fluid and extracellular fluid—respond to increases in Na^+ and water. Let's put all this knowledge together to solve the case.

THE CASE

Your grandfather has always loved anchovy pizza, with extra anchovies. There is something about the saltiness of them that captures his heart and taste buds. You stop by his house one day and find him in respiratory distress, having a truly difficult time breathing. The pizza box is sitting there empty, next to a half-finished glass of water. Thirsty after the pizza, your grandfather drinks the water as he tells you how poorly he feels. As you feared, he is not taking some of the medications that have been prescribed for his heart condition, including the diuretics and the angiotensin-converting enzyme (ACE) inhibitor. As you drive him to the emergency room, you think about the physiology of what is happening.

THE QUESTIONS

1. If your grandfather is not taking the ACE inhibitor and the diuretic, what is the consequence for his total blood volume?
2. What happens to ECF when your grandfather eats a quantity of salty anchovies?
3. Why is he now short of breath?
4. What do you predict the physicians in the hospital will do for your grandfather when you arrive?

Acid-Base Clinical Case Study

Type 2 Diabetes Mellitus

BACKGROUND

Glucose is transported out of the filtrate and into the blood at the proximal convoluted tubule. Because this process occurs through a transport protein, there is a limit to the amount of glucose that can be moved between compartments, a tubular maximum transport. When blood glucose concentrations are high, the transporters fail to reabsorb all the glucose from the filtrate, and glucose is voided in urine. Remember, glucose is osmotically active, so it will increase the volume of water in urine, preventing its reabsorption.

THE CASE

Your diabetic aunt has been lax about her diet lately, as well as her medications. You come home to find her confused and lethargic. You rush her to the hospital for treatment. As you wait, you think about the specific stresses diabetes places on volume regulation.

THE QUESTIONS

1. The doctor tells you that your aunt is suffering from hyperosmotic, hyperglycemic syndrome. What does this mean?
2. What might have caused this condition?
3. What might the physicians in the hospital do to treat your aunt? Why would this help?

10

Exercise Physiology: Integration of Physiology

Case 1

For months now you have been training for a marathon, faithfully rising every morning to run, increasing your distances with each successive week. The marathon is now only a few days away, so you begin to use your knowledge of physiology to plan a strategy for the race. You ask yourself: should I run fast to get out in front and then slow down to maintain my lead? Should I hang back and wait for others to tire and then move up? Should I take all the water breaks, or save time by skipping them? What will be the physiological consequences of not drinking water? What will be the physiological consequences of sprinting at the beginning? Your physiology course ended just last week, so during this final training run you think about all the physical changes that have occurred in the cardiovascular, pulmonary, musculoskeletal, and endocrine systems during your training, and what will happen to them during the marathon. As you contemplate your own physiology and the upcoming challenge, you realize how well humans are adapted to sustained exercise.

Introduction

In physiology, we study each organ system, carefully observing how they are integrated with cellular, autonomic nervous system, and endocrine system functioning. Our organ systems work together all of the time—while we are at rest, asleep, or at play—but in everyday good health this cooperation is largely invisible. During exercise we have an opportunity to observe how major organ systems interact, as our breathing and heart rates increase, we begin to sweat, and our muscles contract, coordinating our movement. During exercise we are able to see some obvious changes, but we need to use our knowledge of physiology to understand the full scope of system integration that is actually occurring as we run this marathon.

How Do Muscles Cause Body Motion?

It is important for the physiologist to understand the cellular mechanisms of muscle contraction, but for the moment our focus will be on muscle as a tissue that generates force, propelling us through space. How can cellular muscle contraction allow us to run, jump, and swim?

Muscle cells are organized into bundles, each bundle being surrounded by a connective tissue sheath ending in a tendon. Discrete bundles of muscle cells connected to bone are identified as specific muscles—biceps, quadriceps, triceps, or gastrocnemius, for example. Skeletal muscle uses the attachment to rigid bone to generate force, which in turn moves the bone at its joint. The attachment of muscles to bone forms a lever system that allows movement.

Let's consider how your muscles move to lift a cup of coffee to your mouth. The various muscles of your fingers and hands contract to cause your fingers to close around the cup and remain contracted to hold onto it. Then the biceps, attached at the top of the humerus (upper arm) and below the elbow, on the radius (lower arm), will flex the lower arm upward when it contracts (**FIGURE 10.1**). This causes your cup of coffee to move with your hand from the table toward your mouth, as the muscle cells of the biceps muscle shorten. This is called an isotonic contraction because the load (your coffee cup) stays the same while the muscle shortens. Now imagine you tried to lift a 100-pound block of lead from the table, using that same biceps muscle—although unless you have been

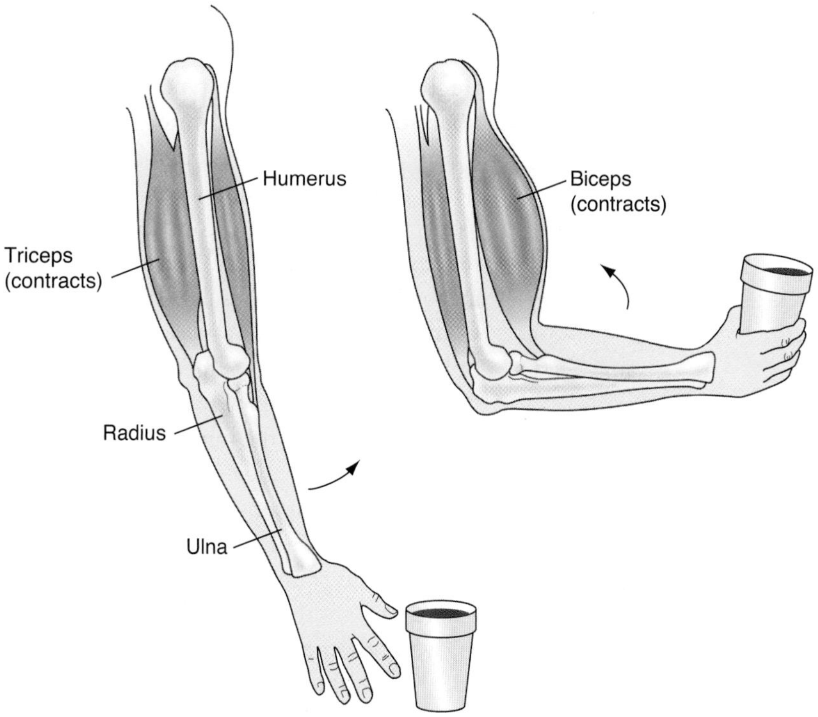

Triceps
(contracts)
Humerus
Radius
Ulna
Biceps
(contracts)

FIGURE 10.1 As muscles contract, they move the bones they are connected to.

working hard lifting weights, you may fail to actually move it from the table! Even if you can't lift it, the biceps muscle would still contract, but if it is unable to generate enough force to overcome the load (that heavy block of lead), the contraction is called isometric, meaning "same length." There is still muscle work, that is, adenosine triphosphate (ATP) usage, but no muscle shortening.

At the same time that your biceps muscle is shortening, the opposing muscle, the triceps muscle, must relax and extend. We saw this type of simultaneous and opposite innervation of muscles when we studied reflexes. As one muscle group is activated, the opposite muscle is inhibited, allowing the contracting muscle to shorten unopposed. Thus, during your run, the hamstring muscles of your right thigh will contract, flexing your leg, while the quadriceps muscles extend (**FIGURE 10.2**). Then as your right foot extends toward the ground, the quadriceps muscles contract and the hamstring muscles relax and extend. The opposite series of contraction and relaxation occurs on the left side, and you run rhythmically.

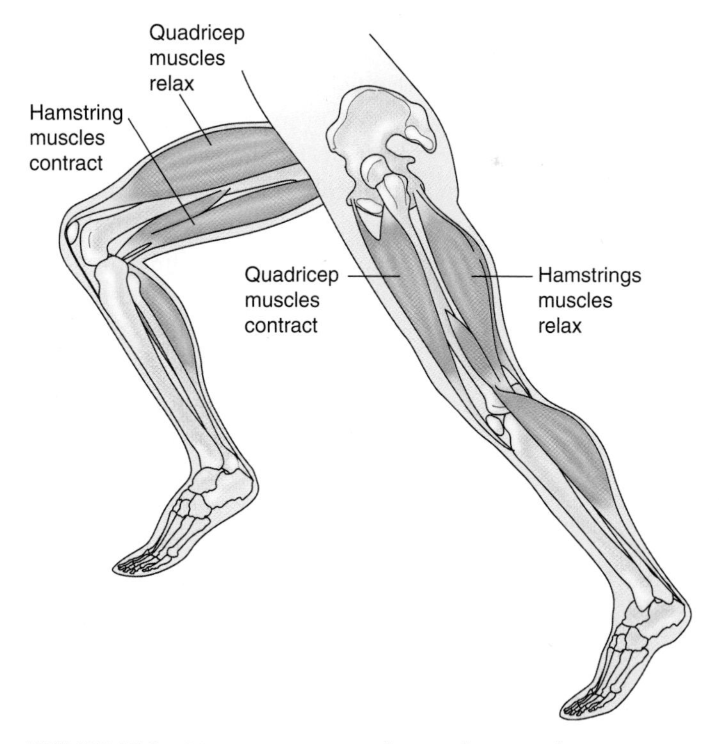

FIGURE 10.2 Movement requires the coordination of opposing muscle groups. As one muscle contracts, the opposing muscle must relax.

How Have Your Muscles Changed During Your Training Regimen?

Muscle fibers (cells) are not all identical. This is easily seen in birds and fishes, two species that separate red and white muscle cells, which we see as dark or white meat when they

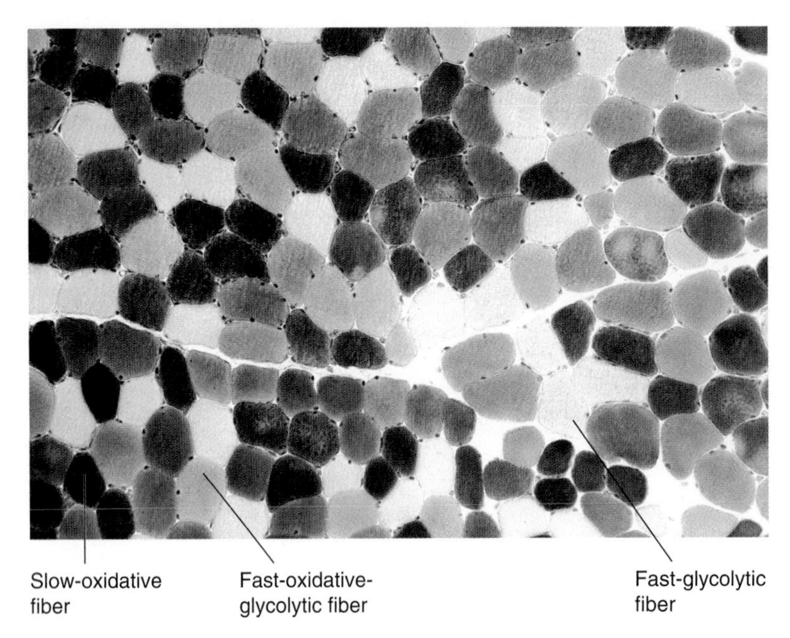

FIGURE 10.3 Oxidative (red) and glycolytic (white) fibers are mixed in humans. Intermediate types of fibers also exist. (© Dr. Gladden Willis/ Visuals Unlimited.)

come to table. The differences in color signify a difference in metabolism, which distinguishes these muscle types. Red muscle fibers contain plentiful mitochondria and a store of myoglobin. Myoglobin contains heme proteins that bind oxygen (O_2), very similar to hemoglobin, contributing to the red color of these fibers. Red muscle fibers are highly oxidative muscles. These fibers use O_2 from circulation, or released from intracellular myoglobin to metabolize glucose or fatty acids completely, via the electron transport chain of the numerous mitochondria. These muscle fibers are well adapted to chronic use. Our postural muscles are red oxidative fibers, for example (**FIGURE 10.3**). Because your training regime has been for a marathon, another sustained activity, during your months of preparation, you have been "training" these fibers. Mitochondria in red muscle fibers have proliferated, increased their concentration of oxidative enzymes, increased the number of myoglobin proteins that can bind and release O_2, and have become more efficient at ATP production that will be used to power your run.

Had you trained for a 100-meter dash, the muscle development would have been quite different. High power output comes from white, glycolytic muscle fibers. These fibers rely on glycolysis for fast ATP production. This cannot be sustained for long periods but can produce enormous power in short bursts. Glycolytic fibers lack the myoglobin of oxidative fibers and have fewer mitochondria, so the color is pale, that is, white. Training for sprints will cause an increase in all the enzymes of glycolysis, improving ATP production by this pathway. Glycolytic fibers are thicker, so when bundled into a muscle, they are more massive. This is easiest to see in weight lifters, whose muscles become larger as they increase their power output on rapid lifts of heavy weights. Unlike birds or fishes, humans do not separate white and red muscle fibers into specific muscle bundles. Instead, our biceps and quadriceps are a mixture of red oxidative and white glycolytic fibers. However, our training routines target certain fibers, causing tissue adaptation. The proportion of white glycolytic to red oxidative muscle cells that we possess is thought to be genetically determined, although there is some evidence that our activities may also influence muscle cell type.

Your training routine has also caused an increase in muscle protein production. Oxidative enzymes, glycolytic enzymes, and myoglobin have all increased, along with the obvious increase in myofibrillar proteins like myosin, actin, tropomyosin, troponin, and all of the other proteins associated with sarcomeres. Skeletal muscle cells are terminally differentiated cells that do not divide, but they can and do increase or decrease in physical size. Training increases the number of cross-bridge cycles possible by increasing muscle proteins. So, the energy you use during your workout is not simply the result of cross-bridge cycling, but also because of the ATP necessary to create new proteins. This is one reason why a workout causes an increase in metabolic rate for a sustained period, not simply during the workout.

All the months of training have increased your muscle size, which increases the amount of glycogen that can be stored within the muscle. Simultaneously, you have increased mitochondrial enzymes and, therefore, the oxidative ability of mitochondria. The skeletal muscles that expand your chest wall—the diaphragm, external intercostals, sternocleidomastoid muscles, and scalenes—have also increased in size and strength. Because you can expand your thoracic cavity more than you could before you trained, you have effectively increased your lung volume. Increased O_2 uptake by the lung and increased use by mitochondria have increased your VO_2 max, which is the maximal volume of O_2 that you can take in and use in liters per minute. During the marathon, an elevated VO_2 max will improve the aerobic use of glycogen, reducing lactic acid production and increasing ATP yield.

Increased chest wall muscle strength has improved your pulmonary vital capacity. During the race, or during your training bouts, sympathetic stimulation will relax smooth muscle in the airways, increasing airway diameter and decreasing airway resistance, further improving your effective vital capacity. Slightly elevated blood pressure will improve blood flow to the upper portions of the lung, reducing the normal, gravity induced ventilation-perfusion ratio (V/Q) mismatch. These mechanisms will further improve VO_2 max.

The additional O_2 taken in per breath because of a greater vital capacity will reduce your breathing rate, thus limiting the number of breaths per minute that you need to take. While this decreases the number of breaths taken, the work of breathing will increase during the race, as you expand your chest volume using accessory muscles for both inhalation and exhalation (**FIGURE 10.4**). This will increase the O_2 required by respiratory muscles themselves by 10 to 20 times! You will take fewer breaths, which increases alveolar ventilation, but respiratory work increases more than linearly. Nonetheless, you take in a much greater volume of O_2, which will provide ATP for all the muscles of the body.

One more mechanism will increase O_2 availability. Lactic acid released from muscles into circulation will lower your body pH slightly. This will have the beneficial effect of right-shifting the O_2-hemoglobin dissociation curve. As you recall, this means that O_2 will bind to hemoglobin less tightly, so O_2 molecules will dissociate from hemoglobin more quickly. This change will be especially pronounced within the capillary beds of working skeletal muscle. The net effect will be that more O_2 will be delivered to acidic muscles, and hemoglobin returning to the lung will be less saturated with O_2. Your deep breathing of

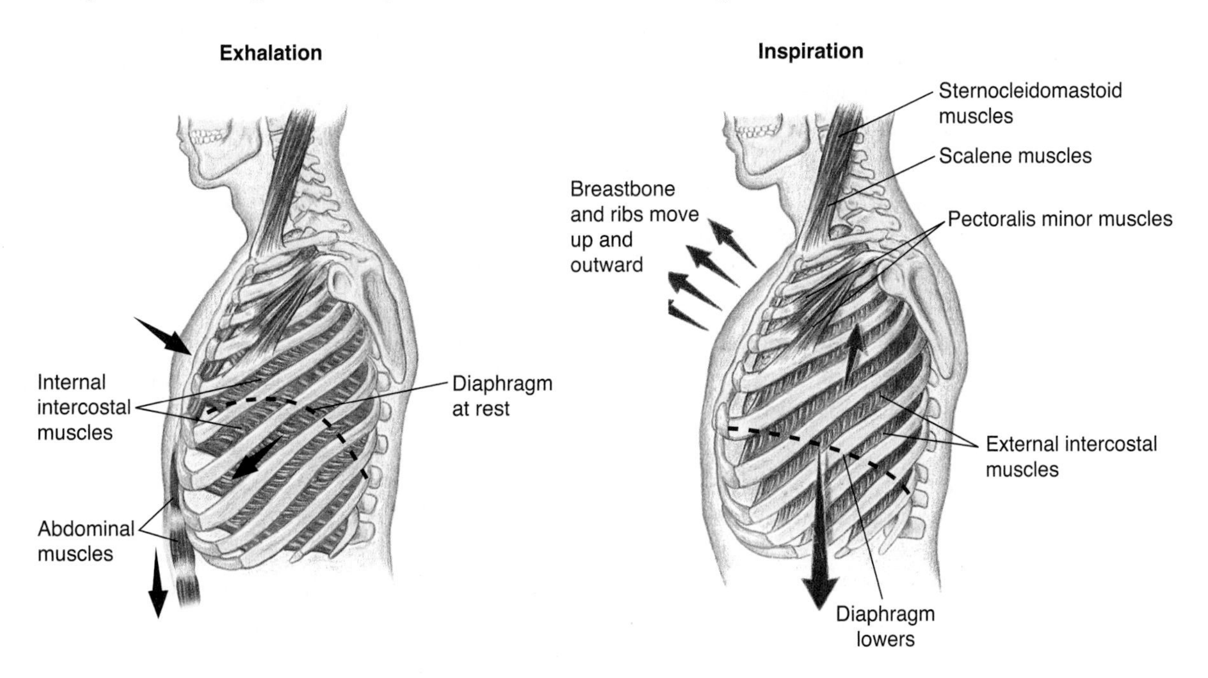

FIGURE 10.4 Primary and accessory respiratory muscles.

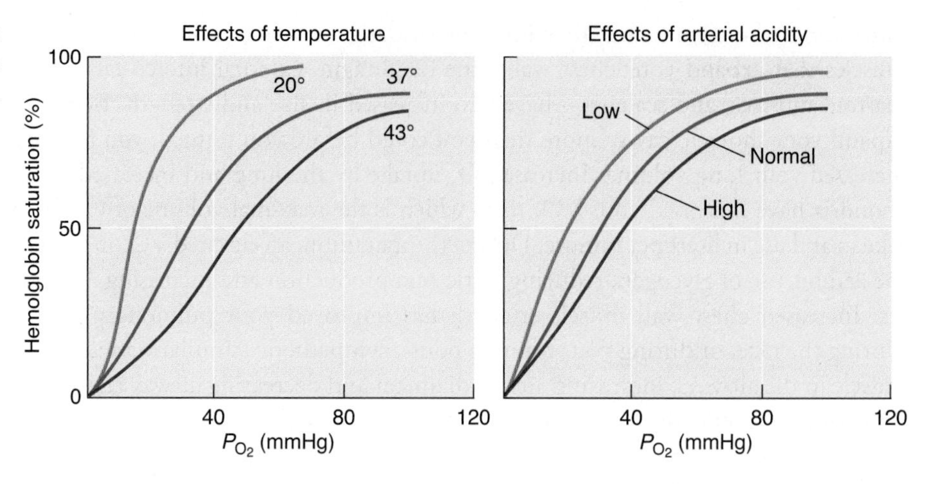

FIGURE 10.5 Even slight changes in muscle acidity will cause a right shift in the oxygen-hemoglobin dissociation curve.

cooler air will lower the temperature within the lung very slightly, which will also increase O_2 binding to hemoglobin within the lung (**FIGURE 10.5**). Thus, you will maximize both O_2 uptake and delivery.

In addition to taking in more O_2, the increase in pulmonary ventilation will also serve to regulate your whole body acid-base balance, as you exhale carbon dioxide. This is an important contribution by the lungs to the maintenance of acid-base balance, especially because the kidneys are contributing much less to H^+ ion removal from the blood, as we will see shortly.

Which Hormones Are Dominant During Your Workouts and the Marathon?

As always, during exercise, the sympathetic nervous system is active, releasing norepinephrine at nerve terminals and epinephrine from the adrenal glands into circulation. Remember that epinephrine bound to adrenergic receptors on skeletal muscle will begin a signaling cascade that ends in the activation of glycogen phosphorylase by protein kinase A (**FIGURE 10.6**). This allows intramuscular glycogen to be broken down to glucose and used by working muscle for ATP production. Remember that intramuscular glucose cannot be released into circulation, because muscle lacks glucose-6 phosphatase, which would permit glucose-6 to be converted back to glucose and potentially allow it to be transported across the plasma membrane. So, intramuscular glycogen becomes a private supply of glucose for those muscles.

Sympathetic stimulation also suppresses insulin release when epinephrine binds to an α_2-adrenergic receptor on pancreatic β cells. This receptor decreases the amount of cyclic adenosine monophosphate (cAMP), thus reducing vesicular release of insulin from the pancreas (**FIGURE 10.7**). In the absence of insulin, circulating glucose or glucose released from liver glycogen stores is not taken up by adipose tissue to be stored as fat, but rather stays in circulation for use by brain, nerves, and skeletal muscle.

How does skeletal muscle maintain its glucose transporters given the suppression of insulin? Exercise of skeletal muscle, meaning the physical stretching and contraction of the muscle itself, stimulates the upregulation of glucose transporters in the membrane, allowing skeletal muscle to take up glucose in the absence of insulin signaling.

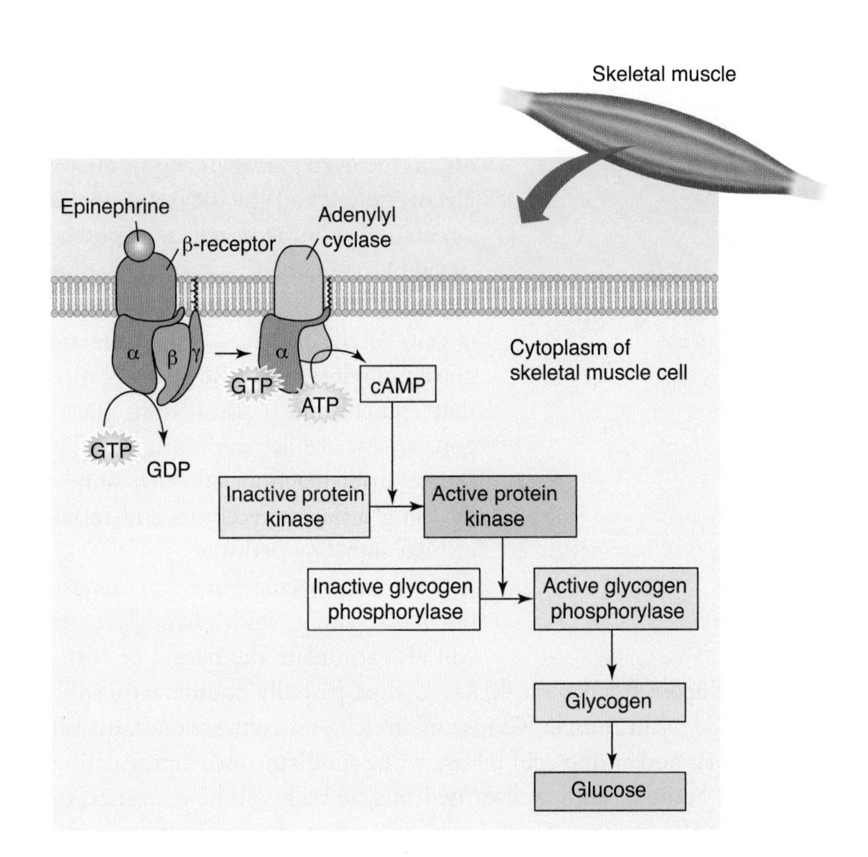

FIGURE 10.6 Epinephrine binding to β-receptors on skeletal muscle will cause the activation of glycogen phosphorylase and the hydrolysis of glycogen. Glucose-6 phosphate remains trapped within skeletal muscle, ensuring its use within the muscle cell.

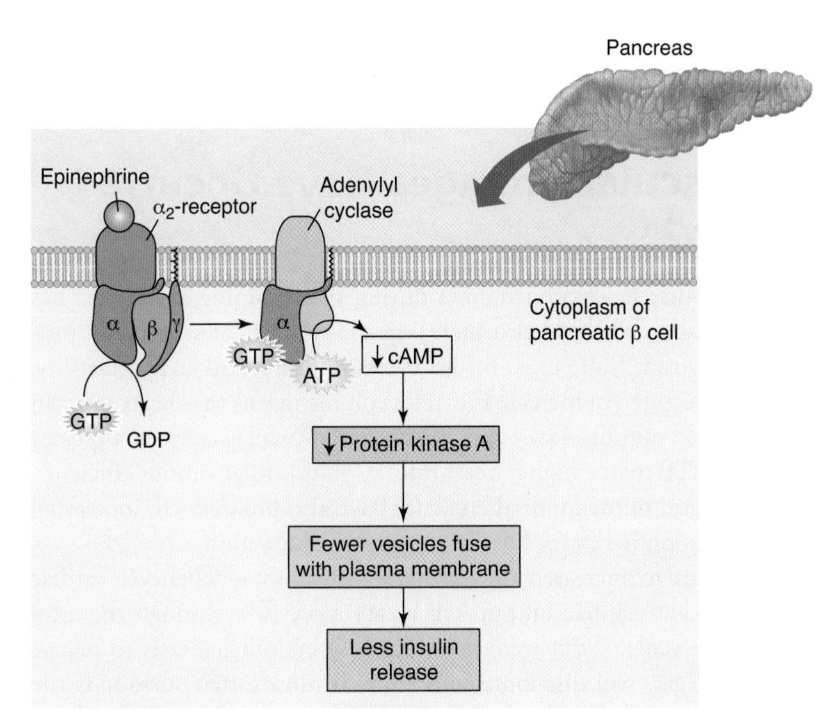

FIGURE 10.7 Epinephrine inhibits insulin release from pancreatic beta cells by binding to alpha receptors.

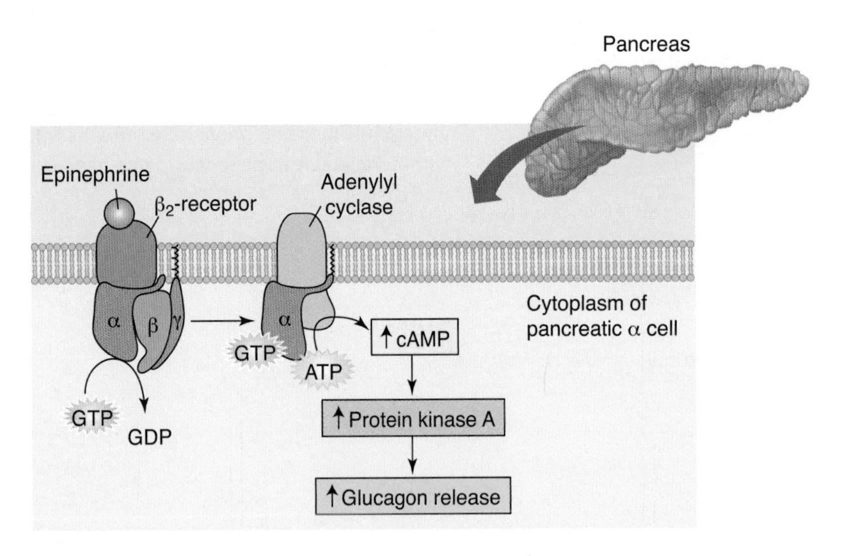

FIGURE 10.8 Epinephrine stimulates glucagon release.

Reduced blood glucose levels also stimulate glucagon release from the α-cells of the pancreas. At the target organ, the liver, glucagon will stimulate glycogenolysis and glucose release to the circulation. Glucagon release is potentiated by sympathetic stimulation. Epinephrine, binding to β2 receptors on α cells of the pancreas, will stimulate glucagon release (**FIGURE 10.8**). Notice that epinephrine is stimulating glucagon release while inhibiting insulin release from neighboring cells, simply by using different receptors and intracellular signaling pathways.

The neurogenic stress of exercise (anxiety) along with hypoglycemia will also stimulate the release of cortisol. Cortisol will decrease glucose uptake via GLUT4, thus partially counteracting the effect of upregulation of GLUT4 in muscle because of stretch and contraction. Cortisol will increase muscle proteolysis and amino acid release while simultaneously upregulating gluconeogenesis in the liver. Some of your well-earned muscle bulk will be converted to fuel during the race. Cortisol also potentiates epinephrine's action of promoting lipolysis, providing more circulating fatty acids for uptake by skeletal muscle and heart. Thus, your hormones coordinate to provide nutrients for heart, brain, and skeletal muscle during your long race. During the shorter workouts, cortisol would have a more minor influence over fuel management.

What Cardiovascular Changes Have Occurred During Training?

It isn't only skeletal muscle that has hypertrophied during your training session; so has cardiac muscle. Cardiac muscle cells have also increased the number of sarcomeric proteins including actin and myosin. More cross-bridge cycles improve cardiac contractility, stroke volume, and cardiac output. An increase in stroke volume means that heart rate can be lower for any given cardiac output. As we saw in the cardiovascular chapter, a greater stroke volume requires less ATP than a higher heart rate, so your cardiac output efficiency is improved. Mitochondria and mitochondrial enzymes have also proliferated, increasing the efficiency of ATP production necessary for all this muscle contraction.

Vasculature also responds to increased flow demands that occur whenever cardiac output increases. First, a greater cardiac output will create more flow through the arterioles. Shear stress along the walls of the arterioles will cause endothelial cells to release nitric oxide. Nitric oxide, a gas, will distribute into smooth muscle that surrounds the arteriole and cause smooth muscle relaxation by increasing calcium sequestration into sarcoplasmic reticulum. Thus, the arteriole will dilate and precapillary sphincters will open (**FIGURE 10.9**). Epinephrine, through β-receptors, will cause vasodilation within skeletal muscle beds, thus adding to the vasodilation caused by shear stress. These mechanisms will improve blood flow within skeletal muscle tissue, providing greater O_2 and nutrient supply and faster metabolic waste removal.

Second, there is increased arterial pressure that will open collateral pathways in capillary beds. Under resting conditions, many of these capillaries remain underutilized, but under exercise conditions, even slight rises in blood pressure or flow will open more routes within each capillary bed. This will decrease diffusion distance between cells (like skeletal muscle cells) and capillaries.

A long course of training can actually increase the number of capillaries that we possess through a process known as angiogenesis. Blood flow through the capillaries, nitric oxide, and a host of growth factors normally increase the number of capillaries that serve working tissue. Thus, your workouts have extended your vascular tree and reduced the diffusion distance over which O_2 and nutrients must travel.

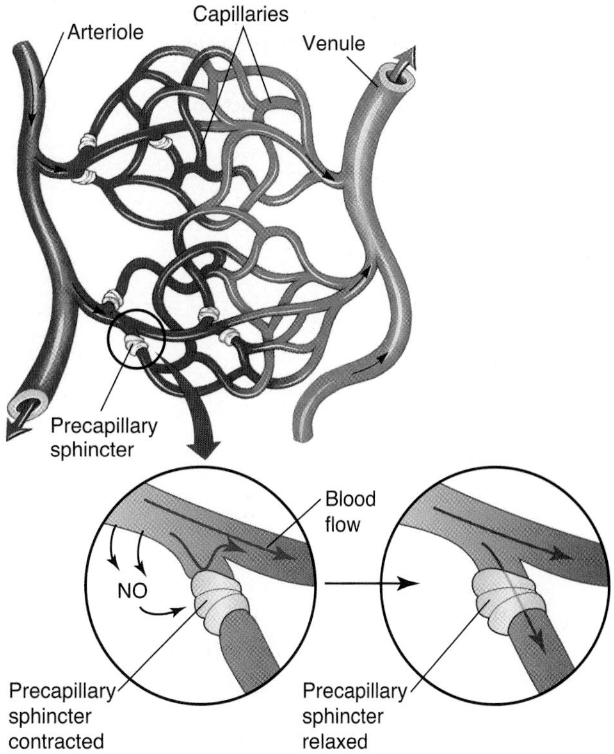

FIGURE 10.9 Nitric oxide will cause local vasodilation of precapillary sphincters and increase capillary blood flow.

What Happens in the Cardiovascular System During the Race?

Even before you begin to run, anticipation of exercise will stimulate the sympathetic nervous system. Norepinephrine released from nerve terminals at the sinoatrial node will increase heart rate. Simultaneously, phosphorylation of L-type Ca^{2+} channels in ventricular cells increases their conductivity and the amount of extracellular calcium that enters. This will increase contractility and stroke volume. Cardiac output increases with tissue demand. The marathon is causing significant tissue demand for O_2 and ATP, and, therefore, cardiac output will be maximized. Because your heart has increased in strength, with more cross-bridge cycles available, stroke volume will increase more than heart rate. This is an important adaptation. Cardiac cells within the mid-myocardial wall are at risk for hypoxia because they are not well perfused during contraction. When cardiac output is achieved by a slower heart rate and greater contractility, there is a longer period of diastole during which these cells can be perfused. Your cardiac workouts, all these months, have specifically trained your heart to perform at its peak!

Finally, we know that cardiac output is dependent on venous return. Sympathetic stimulation causes venoconstriction, constriction of the veins that are our capacitance storage vessels for blood. Constriction forces blood within the veins into the right atrium, thus increasing venous return and cardiac output. Remember that sympathetic stimulation constricts blood vessels in the gut, shunting that blood supply to the rest of the body. This also increases the blood supply available for muscles and contributes to venous return.

Lactic acid, produced by skeletal muscle and released into circulation, can be utilized by the heart as a precursor to the citric acid cycle. In fact, lactic acid is as readily used by the heart as glucose (**FIGURE 10.10**). So, the "waste" product of the skeletal muscle is not wasted at all, but once returned to the heart it provides ATP for cardiac contraction.

Doesn't all this reorganization of blood distribution cause an enormous increase in arterial pressure? This would seem logical, especially given that sympathetic stimulation increases vasoconstriction in many vessels outside of the skeletal muscles. However, increased pressure upstream opens so many parallel vessels within large working muscles

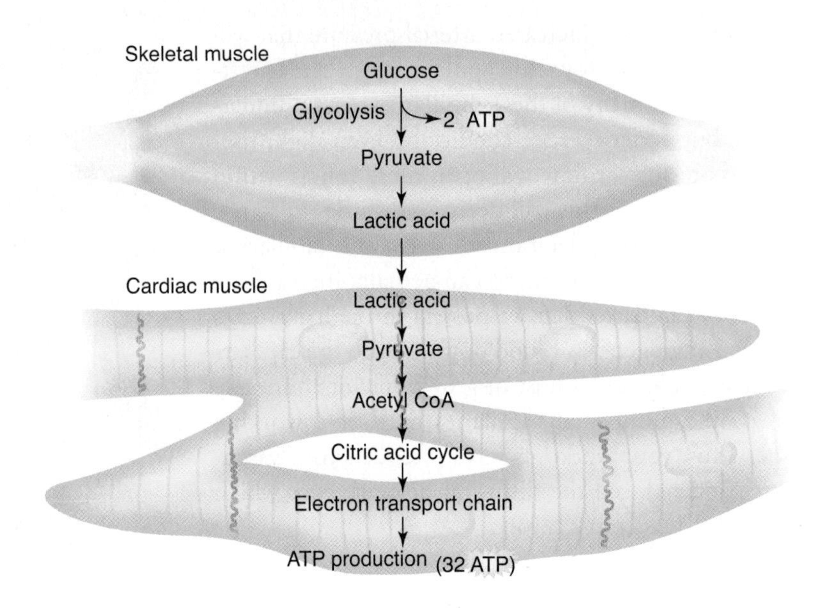

FIGURE 10.10 Lactate in circulation is taken up by cardiac myocytes and reconverted to pyruvate so it can enter the citric acid cycle and the electron transport chain for ATP production.

that the resistance to flow and, therefore, total peripheral resistance is very low. Thus, arterial pressure during the marathon will rise only slightly and will not greatly increase the afterload that the heart must overcome.

Hydration and Fluid Balance During the Marathon

You have decided to stop at the first drink station, and as you drink a cup of water, you begin to think about the physiological mechanisms of water preservation. Sympathetic stimulation has almost immediately constricted the afferent arteriole leading into the Bowman's capsule of the kidney. This reduces glomerular filtration rate (GFR). The volume of blood being filtered is reduced for the duration of your run, less urine is produced, and blood volume is maintained. The reduction in GFR also reduces the ability of the kidney to remove H^+ ions from blood, thus making the role of the lung even more important in maintaining acid-base balance.

A decrease in blood pressure in the afferent arteriole, caused by vasoconstriction, triggers renin release from the juxtaglomerular cells. This begins the cascade of the renin-angiotensin-aldosterone system and all its hormonal mechanisms for water conservation. Angiotensin II will increase vasoconstriction systemically, which will help maintain blood pressure during vasodilation of the skeletal muscle beds. At the kidney, angiotensin II constricts both the afferent and the efferent arteriole, reducing both GFR and renal blood flow (**FIGURE 10.11**). Reduced renal blood flow limits perfusion of the kidney tubular cells themselves. Long periods of reduced renal blood flow in the kidney can cause hypoxia in the tubular cells. This is unlikely to occur during your marathon, but it is a limiting factor for intense exercise in humans.

Angiotensin II increases the release of aldosterone into circulation. At the kidney, aldosterone upregulates the protein synthesis of Na^+ channels and Na^+-K^+ ATPases in the distal convoluted tubule. The increase in Na^+ transport to blood from urine by these two proteins will increase the osmotic gradient and improve water reabsorption.

Angiotensin II will also facilitate the release of antidiuretic hormone (ADH), also known as vasopressin, from the posterior pituitary. As the name vasopressin implies, this

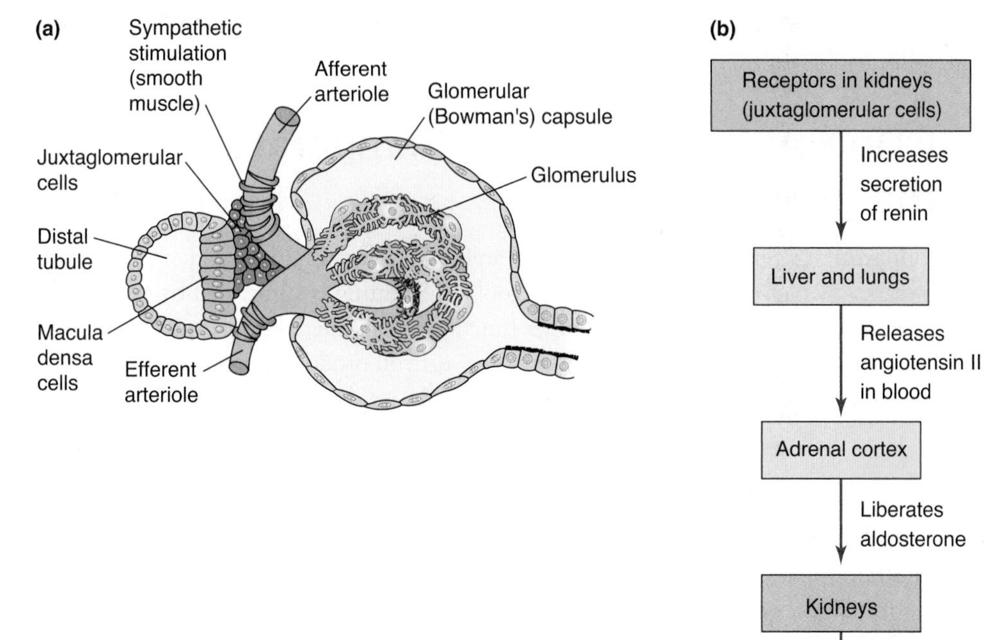

FIGURE 10.11 (a) Constriction of both the afferent and efferent arterioles by angiotensin II will reduce GFR and renal blood flow. (b) The release of renin by the kidney stimulates the formation of angtiotensin II and aldosterone, which will increase water reabsorption and maintain blood pressure.

hormone will also increase vasoconstriction systemically. At the kidney collecting duct, this same hormone will cause the fusion of vesicles containing aquaporins with the plasma membrane of the collecting duct and facilitate water reabsorption. Remember that the neurons of the hypothalamus make ADH and release it into the posterior pituitary. The hypothalamus monitors blood osmolarity. During your run, you have produced a hypotonic sweat. This means the sweat you produced contains more water than salt and is hypotonic relative to plasma. Water loss by sweating causes evaporative cooling, allowing you to maintain normal body temperature. However, this has been done at the expense of water loss. As you run, your blood osmolarity increases, which triggers release of ADH from the hypothalamic neurons that project into the posterior pituitary. While ADH will increase water reabsorption from urine, your reduced GFR will limit the effectiveness of this hormone during the marathon.

The increase in extracellular osmolarity will cause the movement of water from the intracellular space (cell cytoplasm) into the extracellular space. This will increase blood volume and restore a more normal osmolarity, at the expense of cellular water content. Because the intracellular space is so large (we have billions of cells, all containing water), the loss from any specific cell is small. However, as your dehydration continues during the race, this becomes an increased burden on all of your cells, as they lose water via osmosis to the extracellular space.

Given this information, what should your hydration strategy be during the race? Drinking small quantities of pure water at stations along the race will reduce the hyperosmolarity of the plasma and prevent cellular water loss. However, drinking large quantities of pure water during or after the race can have two detrimental effects: (1) reduce blood osmolarity below normal, causing dilution of both extracellular and intracellular compartments, and (2) inhibit release of ADH, which will increase urine volume and loss of water.

A better rehydration strategy is to drink fluids containing some Na^+, which will increase blood volume without increasing urinary water loss.

How Do You Regulate Body Temperature During the Race?

One of our greatest strengths as a species is our ability to thermoregulate in the heat. We do this by sweating, which promotes evaporative cooling. Sweating is stimulated by the sympathetic nervous system, but sweating is another function that is altered by training. The speed with which you begin to sweat and the quantity of sweat you produce increases with training. Thus, your ability to maintain a normal body temperature during the run improves with training.

Of course, sweating causes water loss, contributing to loss of blood volume. Given the option of an increased body temperature or loss of fluid volume, body temperature is maintained at the expense of total body water. Maintenance of body temperature also causes another apparent physiological contradiction: vasodilation of skin vessels. Sympathetic stimulation generally reduces blood supply to the skin, shunting blood to the muscles. During strenuous exercise, vasodilation of the surface blood vessels, under the epidermis, allows heat loss to the surface. So, blood supply to the skin actually increases during strenuous exercise, as a means of promoting heat loss. This reduces the amount of blood flow available to the muscle, but maintains body temperature within normal limits.

What Is Happening Within the Digestive System During the Race?

Sympathetic stimulation has constricted blood flow to the digestive tract and inhibited motility, so little digestion is occurring during the race. However, the pacemaker cells of the intestine, the cells that set the normal contractile rhythm within the gut, are susceptible to mechanism stimulation as well as neural stimulation. The physical jostling caused by your running motions can stimulate intestinal contractions and even mass movements within the large intestine. This may cause a need to defecate, even in the middle of the race, when it is least convenient.

Inflammatory Responses Following the Race Aid in Muscle Repair

Strenuous exercise stimulates the release of neutrophils from bone marrow into circulation. Neutrophils circulating in the blood are available to leave circulation and enter damaged tissue. Muscle tissue suffers a certain amount of "wear and tear" during the course of your marathon. Muscle cells are damaged, sarcomeres are torn, myofibrillar proteins are disorganized, and organelles may be injured as well. Some of the repair, if the cell is still intact, will occur through autophagy, literally self-eating, which is the intracellular mechanism for organelle destruction. When the cell membrane has been damaged, neutrophils, monocytes, and resident tissue macrophages will engulf the injured cells and digest them. Never fear, though, exercise also induces muscle rebuilding, and, within days, your skeletal muscle will be repaired and stronger than it was before the great race.

Summary

Our body systems function together all of the time, but during exercise we have the opportunity to see them change their function together. Our muscles contract regularly and forcefully, our hearts beat more often and more forcefully, our breathing rate and depth increases, sympathetic stimulation increases, water balance hormones are active, we regulate our body temperature, and even the immune system prepares for potential danger. As physiologists, exercise becomes an intellectual activity as well as a physical one!

Key Concepts

Isotonic contraction
Isometric contraction
Opposing muscle groups
White glycolytic fibers
Red oxidative fibers
Renin-angiotensin-aldosterone system

Key Terms

Myoglobin
VO_2 max
Glycogen phosphorylase
Glucose-6 phosphatase
Shear stress
Angiogenesis
Venoconstriction

Application: Pharmacology

1. The day after the race, you are sore and you find yourself reaching for an ibuprofen to relieve the pain. You recall that ibuprofen inhibits prostaglandin production, which is how it reduces fever, but you are less certain of the mechanism for pain reduction. Inflammatory cells increase prostaglandin synthesis during exercise, and prostaglandins sensitize several receptors in the pain pathways. If you took aspirin, would the effect be the same?
2. You are looking over the array of sports drinks on the shelf, trying to decide which one is the best for hydration. What should be your criteria for deciding about an appropriate drink? What aspect of sports drinks makes them superior to plain water for hydration? What might be the effect of a drink that contained a great deal of caffeine? What might be the effect of a drink that was hyperosmotic?
3. Some athletes take creatine phosphate (CP) supplements during training. Creatine phosphate (CP) is another source of high-energy phosphates that can be used by skeletal muscle. CP donates its high-energy phosphate to adenosine diphosphate (ADP), converting it to ATP. In this way, CP forms a separate pool of energy within the muscle. Unlike ATP, CP can be stored within muscle, and this pool can be increased by the ingestion of CP. The intracellular stores of CP have two effects: (1) they will provide a reserve of high-energy phosphate

that can fuel muscle contraction for 20–30 seconds, and (2) CP is osmotically active, so it will increase intracellular water in muscle, adding weight to skeletal muscle. Will taking CP supplements before the marathon improve your chances of success? If you were a weight lifter, would it be useful then?

Clinical Case Study

Heart Failure

BACKGROUND

Physicians have long debated about whether patients with heart failure should engage in aerobic exercise. A large-scale study completed in 2009, *Heart Failure: A Controlled Trial Investigating Outcomes of Exercise (HF-ACTION)*, established that aerobic exercise was safe for heart failure patients. To date, there is little evidence that regular exercise improves cardiac output or stroke volume in a failing heart. However, there is generally improvement in VO_2 Max, skeletal muscle strength, and general well-being. Aerobic exercise has been linked to increased skeletal muscle size as well as strength, autophagy (cellular destruction of damaged organelles and proliferation of mitochondria), and improvement of metabolic efficiency.

THE CASE

You have finally convinced your grandfather, who has heart failure, to take all of his medications. He has been faithfully taking them all for a month now. He feels better and is even beginning to take the daily walks suggested by his physician. You join him a couple of times a week, and he asks you the obvious question: will this mend his failing heart? You dislike telling him no in fear that he will stop exercising.

THE QUESTIONS

1. Given what you know about the physiological changes induced by exercise, what can you tell your grandfather about the positive benefits of exercise?
2. Regarding autophagy, why would the replacement of mitochondria in tissues be helpful?
3. Well-being is not a physiological process, but a sense of contentment and feeling healthy. A feeling of well-being is fundamental to happiness. How will this promote health in your grandfather?

Clinical Case Study: Type 2 Diabetes Mellitus

BACKGROUND

Exercise is usually prescribed for patients with type 2 diabetes. As you already know, exercise is a noninsulin dependent mechanism for moving glucose transporters into the muscle cell membrane, so exercise can lower blood sugar. Exercise also builds muscle and utilizes ATP for muscle contraction. If we figure total body energy expenditure, exercise increases energy demand and, thus, requires more fuel substrate, that is, sugar, amino acids, or fatty acids to make this ATP. The heart is a muscle and is conditioned by exercise, just as skeletal muscle is, so the heart may benefit from exercise. An increase in cardiac output causes an increase in vascular flow and shear stress within blood vessels. Shear stress is the primary signal for nitric oxide release from the endothelial cells, resulting in vasodilation. As you saw in this chapter, respiratory function is also improved by exercise, predominantly through increases in respiratory muscle strength. Exercise also increases autophagy, the destruction of damaged organelles. Obesity and diabetes both contribute to an increase in O_2-free radicals produced by the mitochondria. Those same reactive O_2 molecules can damage the mitochondria that produce them. Damaged mitochondria are inefficient producers of ATP but still accept acetyl-CoA into the citric acid cycle and the electron transport chain, regardless of the outcome. Destruction of damaged mitochondria by autophagy stimulates fission and proliferation of new mitochondria. Exercise-induced autophagy is important in many tissues besides skeletal muscle, including the brain. Exercise is, therefore, beneficial in a variety of ways.

THE CASE

Your aunt, who had trouble accepting her diabetes diagnosis, is home from the hospital and is now taking her medical condition seriously. She is checking her blood sugar regularly and following the recommended food list. She has even purchased a diabetic cookbook so she can satisfy her food cravings and still stay within a prescribed diet. You are thrilled to see that her whole attitude toward diabetes has changed from rebellion to acceptance of a new lifestyle. You are even more stunned to discover that she has begun to take long walks three times a week. On one of your visits, the two of you talk while walking. Of course, she wants to know what you think of her exercise regimen.

THE QUESTIONS

1. How will exercise affect your aunt's blood sugar levels?
2. Your aunt is overweight. Can you describe some ways in which exercise may help her lose weight? Energetically, what additional energy demands are placed on the body during and after exercise. Be specific.
3. It would be difficult to describe autophagy to a nonscientist, but what might be the benefit of autophagy in your aunt's case?

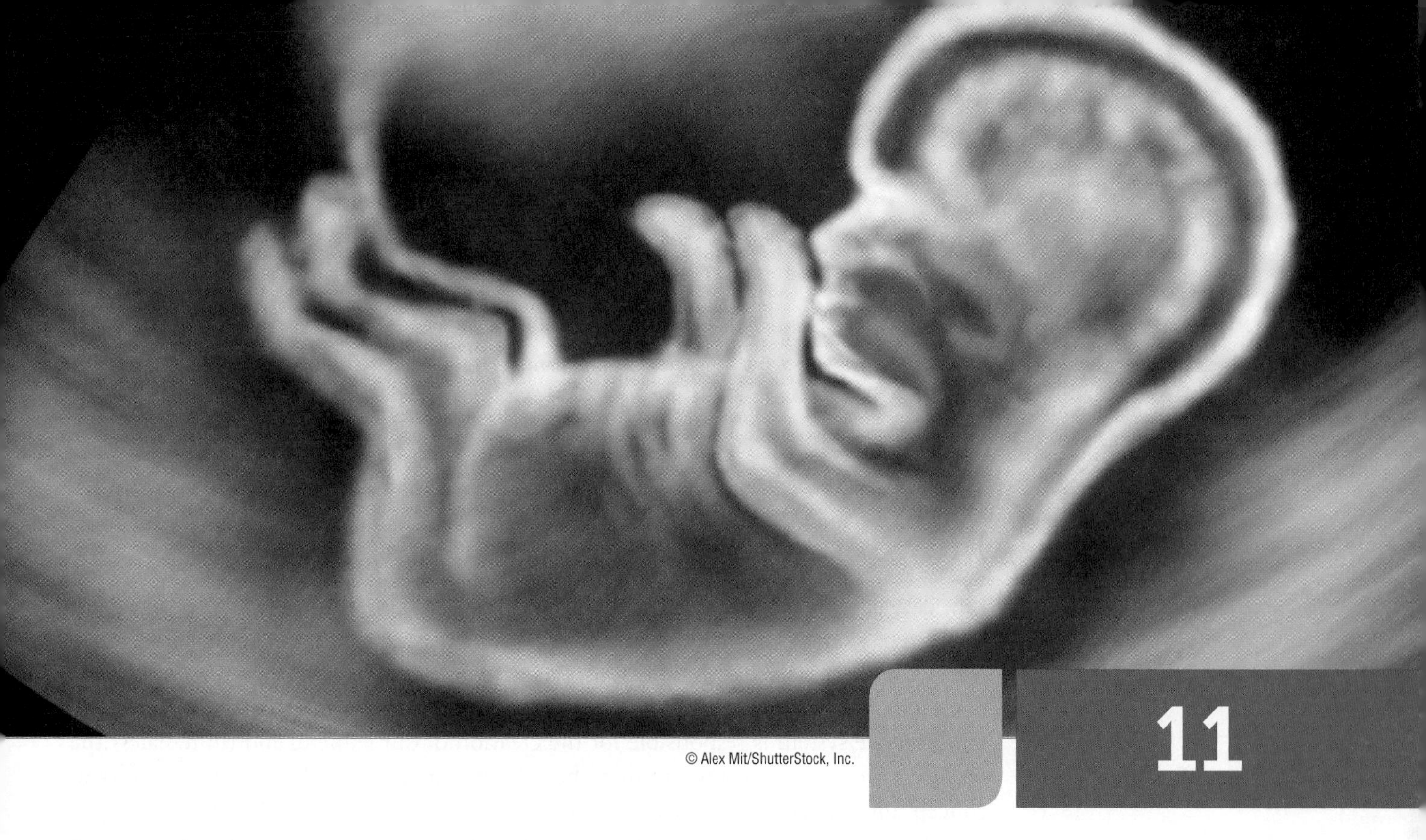
© Alex Mit/ShutterStock, Inc.

11

Reproduction and Fetal Development

Case 1

After finishing your college education and becoming settled in your new profession, you get married. You and your spouse have decided to have a baby and begin a family. Sexual intercourse, which has always been an enjoyable part of your union, has now taken on a new meaning. Several months after making your decision, no pregnancy has ensued, and you decide to monitor body temperature to predict ovulation cycles. Three months later, the early pregnancy test comes back positive! A baby is on the way. How do the male and female reproductive systems work? How does conception occur? How will pregnancy affect the mother's physiology as the child develops? How does the baby grow and change each month? What will happen during childbirth?

Introduction

Like eating, reproduction is an essential part of human physiology—and like digestion, it is well regulated by the autonomic nervous system and hormones. Unlike most mammals, human females have an estrous cycle every month, meaning that it is possible to become pregnant any month of the year. Human males are fertile at all times. The reproductive system is responsible for the creation of our gametes and (in females) the monthly preparation of the uterus to house a potential fetus. If intercourse results in conception, the fetus will develop in a predictable way for nine months before it is born in an immature form. That child will grow and develop into a fertile adult, and the cycle begins anew. Of course, every part of the reproductive system requires a delicate balance of hormones and neural input, which is what we will explore in this discussion.

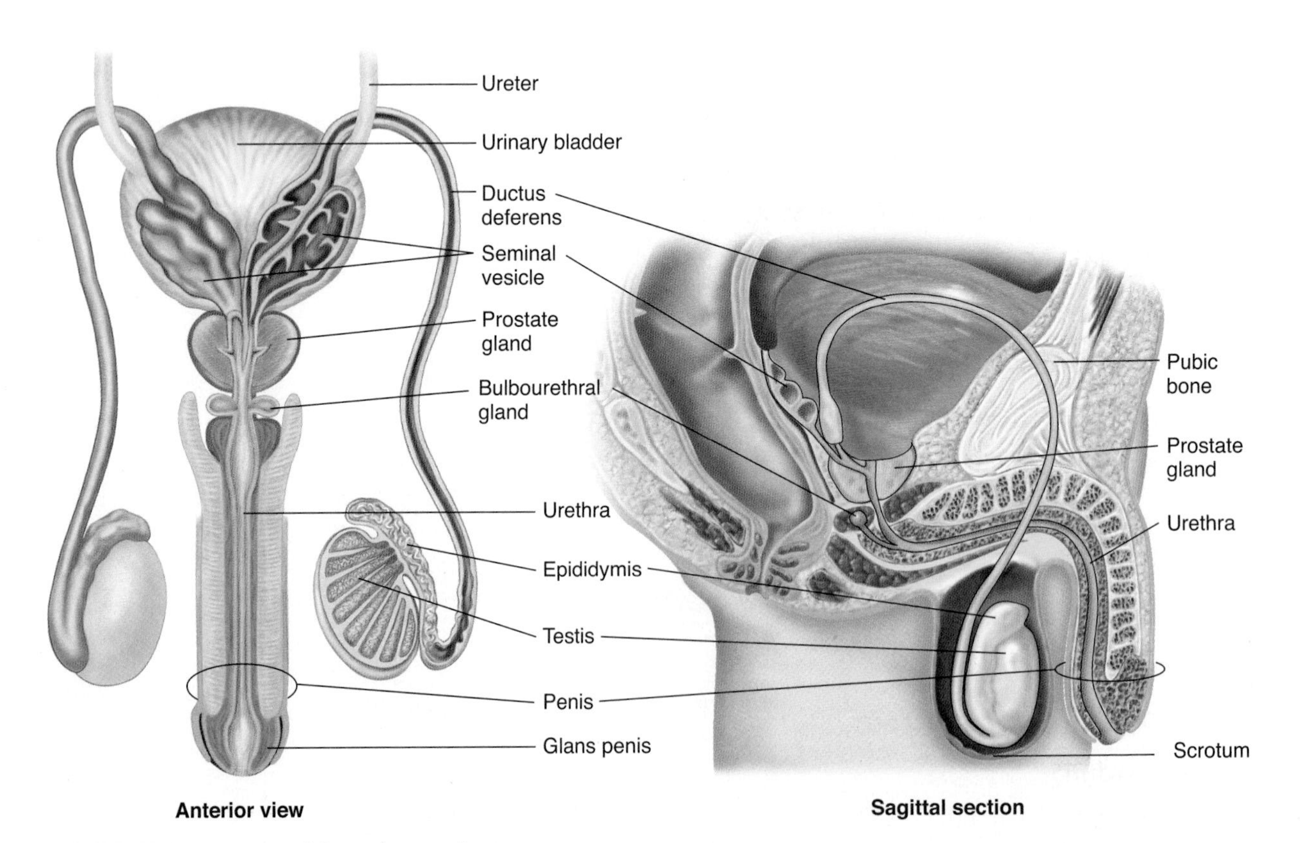

FIGURE 11.1 Anatomy of the male reproductive system.

Gross Anatomy of the Male Reproductive System

In males, gametogenesis begins in the testes. The testes contain a mass of seminiferous tubules with the tubular cells arranged concentrically within the tubule. These cells are engaged in the creation and maturation of sperm. Sperm then travel from the seminiferous tubules to the rete testis and continue on to the epididymis. The epididymis empties into a long tubule, the ductus deferens, which begins at the end of the epididymis, loops over the ureters on the bladder, and terminates in the ejaculatory duct. Sperm mature as they travel along this path (**FIGURE 11.1**). The seminal glands, bulbourethral gland, and prostate gland all contribute fluid to the semen, each emptying into the ejaculatory duct. The ejaculatory duct joins with the urethra in the penis, so semen is ejaculated through the urethra. The penis is composed of erectile tissue—a maze of vasculature and blood spaces surrounded by smooth muscle (Figure 11.1)

Gross Anatomy of the Female Reproductive System

Much of the female reproductive system is internal. Externally, the labia surround the vaginal opening. At the superior union of the labia lies the clitoris, important during arousal, but serving no other known function (**FIGURE 11.2**). The narrow vaginal canal ends in the cervix, which forms the entrance to the uterus. The uterus is a pear-shaped organ about 7.5 cm long and 5 cm wide, lying between the bladder and the colon (Figure 11.2). At the superior surface it opens into two uterine tubes, also called fallopian tubes. The uterine tubes each end in fingerlike fimbria that surround the ovaries (Figure 11.2).

In females, gametogenesis occurs in the ovaries. Unlike males, who produce gametes throughout life, females produce all of their gametes in utero, that is, before birth. These pre-formed eggs are released monthly.

Sexual Intercourse and Nervous Systems

Although our reproductive systems function all of the time, they are most noticeable during sexual intercourse. As you may recall from the sensory homunculus of the somatic nervous system, the genitals are richly endowed with touch receptors. So, the afferent nerve impulses stimulated by genital touch ascend to the sensory cortex via the dorsal column or anterolateral pathways. However, the response is not by the voluntary somatic nervous system but is triggered by the parasympathetic nervous system.

In the male, blood vessels in the penis are controlled by parasympathetic nerves that originate in the sacral section of the cord. Note that this is different from the blood vessel innervation we see in the rest of the body, where sympathetic stimulation is responsible for both vasoconstriction and vasodilation. During arousal, parasympathetic nerves release nitric oxide (NO) at their terminals on the vascular smooth muscle of the penis. NO, a gas, diffuses into smooth muscle and causes the formation of cyclic guanosine monophosphate (cGMP). In turn, cGMP reduces intracellular Ca^{2+} by increasing its uptake into the sarcoplasmic reticulum, which causes smooth muscle relaxation (**FIGURE 11.3**). Dilated blood vessels fill the blood spaces of the penis for engorgement. The distension of the penis is caused by an increase in fluid, analogous to a hydrostatic skeleton. The increased blood flow via parasympathetic nerves means erection is a parasympathetic function. The lifespan of cGMP in smooth muscle is limited by phosphodiesterase, an intracellular enzyme that breaks down cGMP. The erectile dysfunction drug, Viagra (generic name sildenafil),

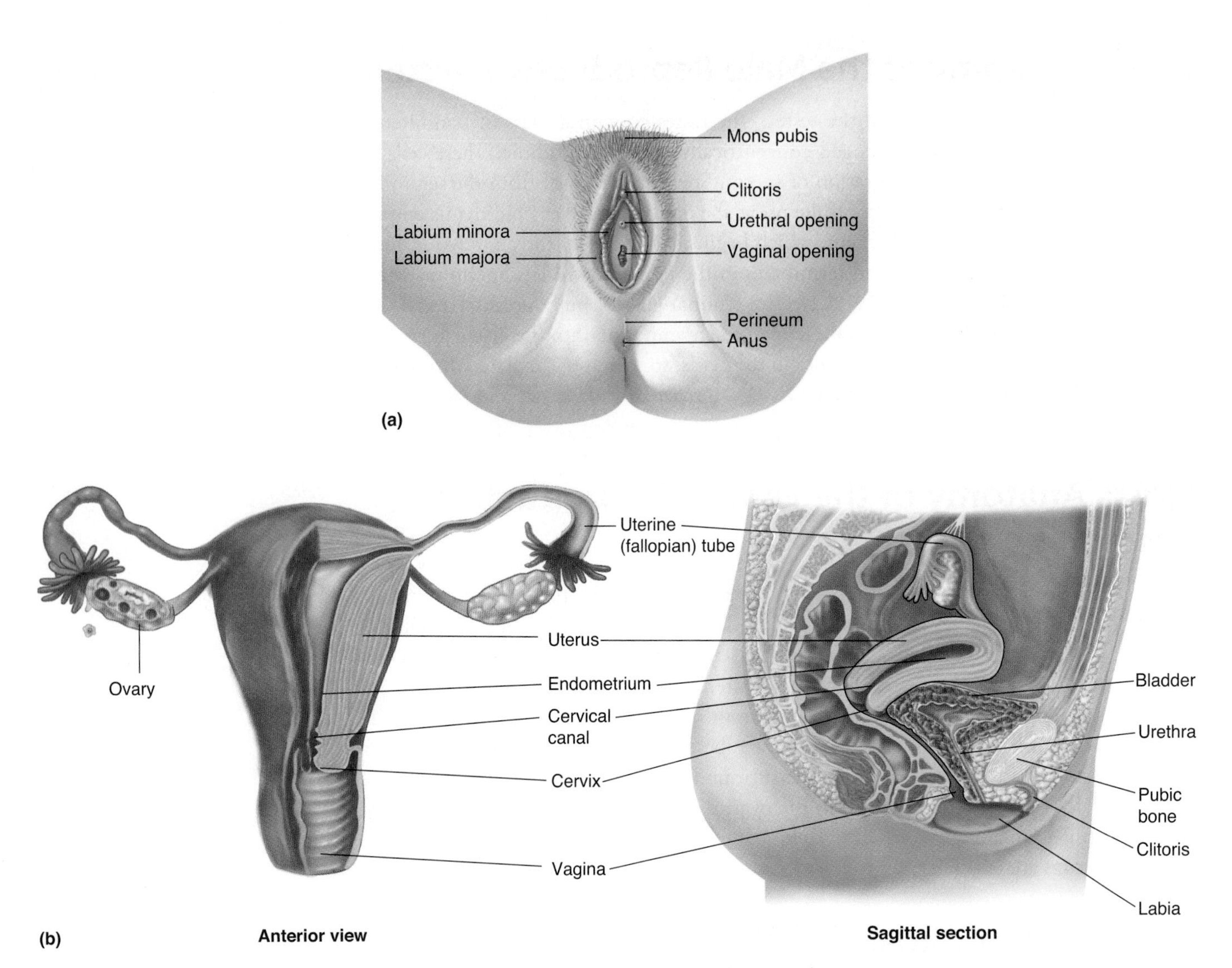

FIGURE 11.2 Anatomy of the female reproductive system. (a) External genitalia. (b) Internal anatomy.

inhibits this phosphodiesterase, thus resulting in more cGMP. This improves blood flow and prolongs the erection. Clitoral enlargement and engorgement during arousal in women are similar processes because developmentally the clitoris comes from the same parent tissue as the penis. Parasympathetic stimulation in females also affects vaginal glands, releasing mucus and fluid during arousal.

While parasympathetic stimulation is required for arousal, sympathetic stimulation is essential for ejaculation, or the release of semen. Sympathetic motor fibers from the lumbar section of the spinal cord stimulate smooth muscle contraction in the glands and tubules of the penis. In females, sympathetic fibers stimulate contraction of uterine smooth muscle. As we will see later, this facilitates sperm transport to the ovaries. Orgasm generally accompanies ejaculation in males, but not necessarily in females. While orgasm is still poorly understood, it is known that orgasm causes the release of oxytocin from the posterior pituitary in both sexes. Oxytocin targets areas of the brain, stimulating feelings of bonding and trust. In females, it also facilitates uterine contractions.

While coitus is our most obvious display of sexuality, there is a complex physiology regulating the production of gametes, ovarian, and uterine cycles in the female and secondary sexual characteristics.

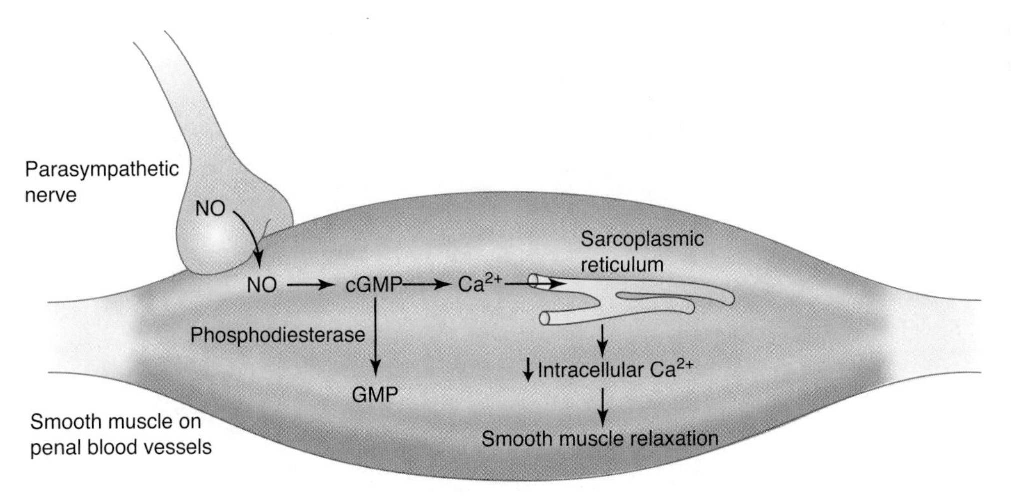

FIGURE 11.3 Nitric oxide regulates vascular diameter in the male reproductive system.

How Are Male Gametes Made?

Male gametogenesis occurs in the seminiferous tubules of the testes. The cellular organization of the seminiferous tubules is important for understanding gamete production and hormonal regulation. The outer layer of cells within the tubules is composed of spermatogonium, or stem cells that will give rise to a continual supply of primary spermatocytes by simple cell division or mitosis (**FIGURE 11.4**). Primary spermatocytes are diploid, containing 46 chromosomes and, once formed, move closer to the center of the tubule, within the blood–testis barrier. The blood-testis barrier is maintained by Sertoli cells that surround the developing spermatocytes and create a unique hormonal environment that is essential to their maturation into spermatids.

Primary spermatocytes then undergo chromosomal duplication just as for mitosis. However, during meiosis, the matching maternal and paternal chromosomes come

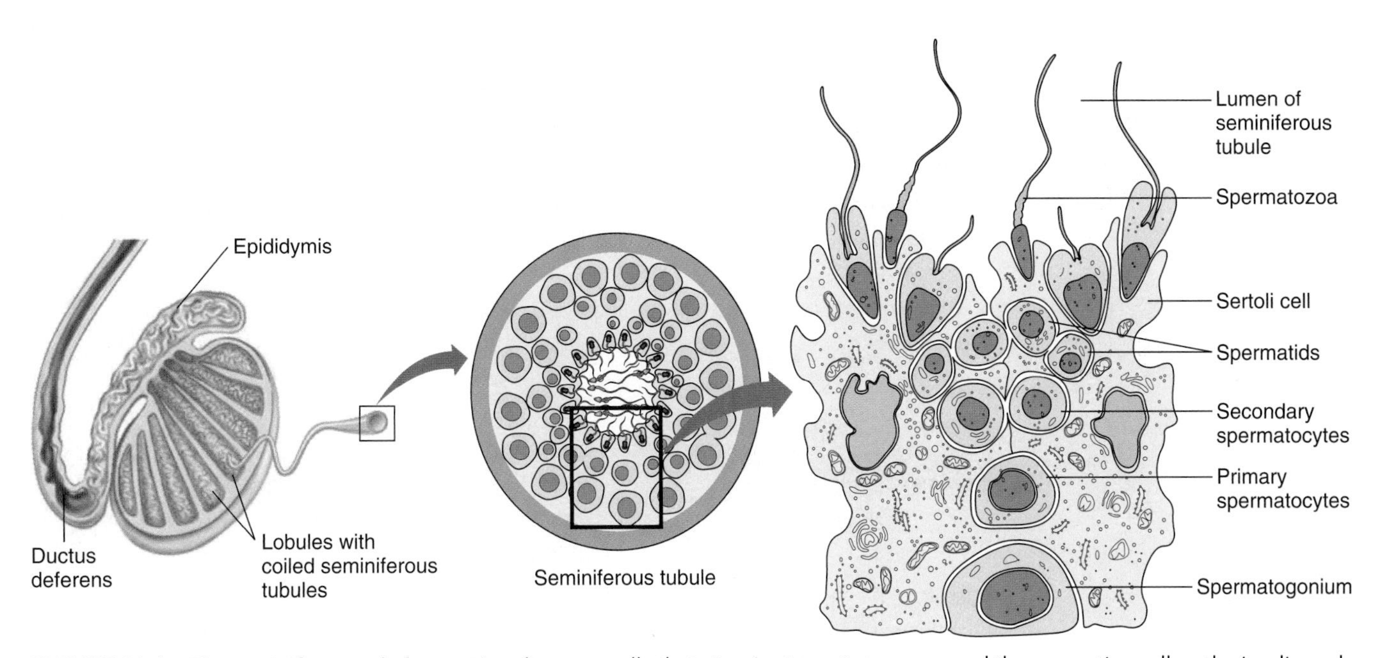

FIGURE 11.4 The seminiferous tubule contains the germ cells that give rise to mature sperm and the supporting cells—the Leydig and Sertoli cells.

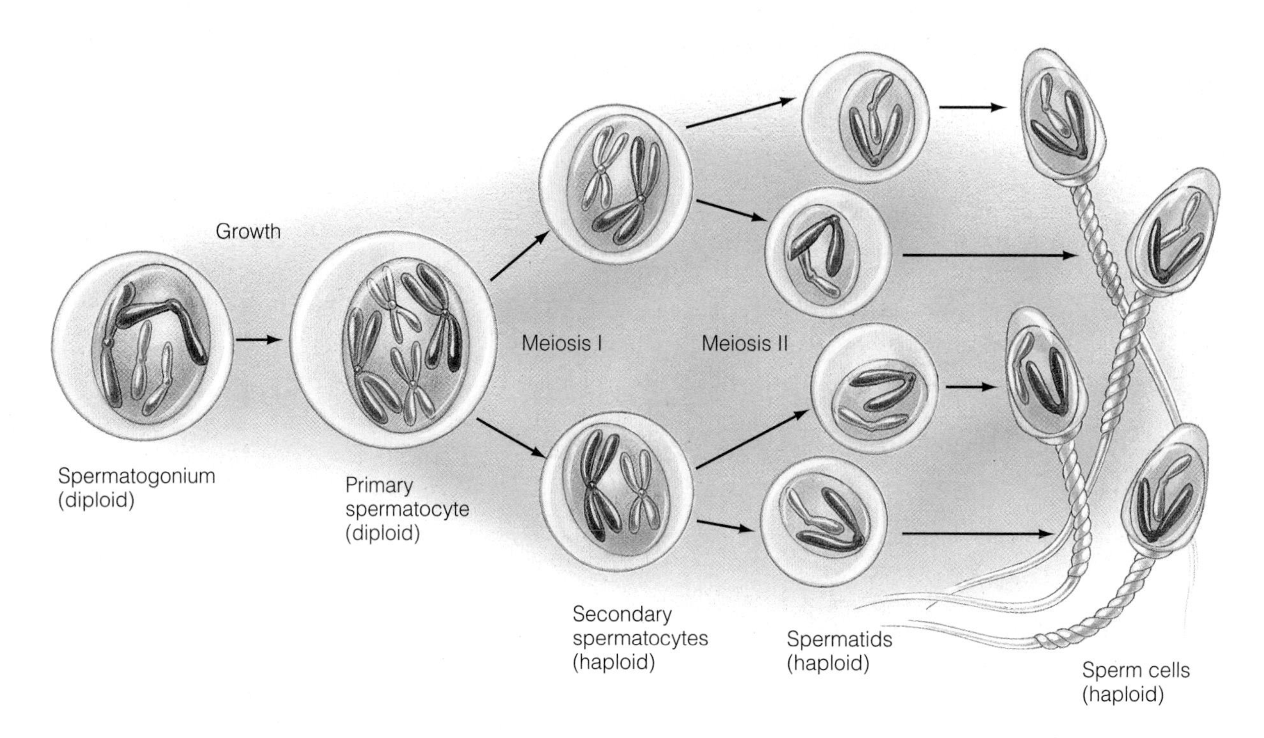

FIGURE 11.5 Meiosis in the male produces four haploid gametes from a single diploid spermatogonium.

together, during synapsis. By the end of meiosis I, both copies of a maternal chromosome are in one daughter cell, and both copies of a paternal chromosome are in another daughter cell. These daughter cells are now known as secondary spermatocytes (**FIGURE 11.5**).

Secondary spermatocytes now undergo meiosis II division, creating four spermatids, each with one copy of a maternal chromosome or one copy of a paternal chromosome (Figure 11.5). These are now haploid cells, but are not yet spermatozoa. Spermatids are still nonmotile round-shaped cells. Profound changes must occur for these spermatids to become mature, viable spermatozoa.

Spermatids are found near the lumen of the seminiferous tubule. In this space, the spermatids will transform from a round, stationary cell into a highly motile, streamlined cell before leaving the environment of the Sertoli cells and entering the epididymis. This transformation involves the loss of cytoplasm, the rearrangement of the Golgi bodies into an acrosomal vesicle located at the leading edge of the nucleus, and a localization of mitochondria around the base of the flagellar tail of the sperm (**FIGURE 11.6**). In the end, the

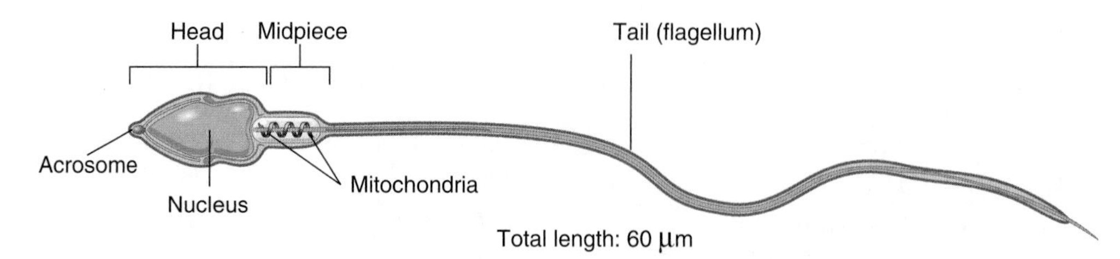

FIGURE 11.6 The mature sperm has lost most of its cytoplasm and organelles. The Golgi apparatus has become an acrosomal cap and the mitochondria are clustered near the tail to provide energy for swimming.

spermatozoan is a sleek, hydrodynamic cell stripped of all but the nucleus and acrosome at its head and possessing a flagellar tail powered by an array of adenosine triphosphate (ATP)-producing mitochondria. How is testosterone important for this process?

The Hypothalamic-Pituitary Axis Stimulates the Production of Male Hormones

Beginning at puberty, the hypothalamic gonadotropin-releasing hormone (GnRH) begins to be released more frequently and in greater quantity. It is unclear why GnRH, which is released in a pulsatile way by the hypothalamus, increases at this time, but it does. Like other hypothalmic-releasing hormones we have studied, GnRH has its target cells within the anterior pituitary. GnRH, produced by neurons of the hypothalamus, exocytoses GnRH into the portal circulation, where cells possessing GnRH receptors bind this hormone. Some of these cells within the anterior pituitary will produce luteinizing hormone (LH), and others will produce follicle-stimulating hormone (FSH) (**FIGURE 11.7**). Both LH and FSH enter the general systemic circulation and travel to their target tissue, the seminiferous tubules. There, each hormone has a specific cellular target.

LH binds to Leydig cells. Leydig cells lie outside of the testes–blood barrier, within the connective tissue layer of the seminiferous tubules. Once bound to a G-protein coupled receptor on the Leydig cell, LH stimulates the cyclic adenosine monophosphate (cAMP) and protein kinase A (PKA) pathway, causing synthesis of enzymes that will convert cholesterol into testosterone within the Leydig cells. Testosterone is not made centrally, but is made exclusively by the Leydig cells of the seminiferous tubules (**FIGURE 11.8**). Testosterone is a cholesterol-based hormone and, as such, is membrane permeant. It diffuses into the bloodstream, carrying testosterone throughout the body. Within the seminiferous tubules, it passes into the Sertoli cells. Here, testosterone can bind to androgen-binding protein (ABP) within Sertoli cells and create the high-testosterone environment necessary for sperm development. Some of the testosterone that enters Sertoli cells is converted to estradiol via the enzyme aromatase. Estradiol, also membrane permeant can diffuse back to the Leydig cell and further increase testosterone synthetic enzymes (Figure 11.8). Thus, testosterone production has a positive feedback loop that involves both the Leydig cells and the Sertoli cells.

Testosterone released into circulation has receptors in many bodily systems and is responsible for the secondary male characteristics of body shape, pattern of fat deposition, facial and body hair, and increased bone and muscle growth. Testosterone also has cellular targets within the brain. One of these targets is the hypothalamus, where testosterone inhibits the release of GnRH, thus forming a negative feedback loop.

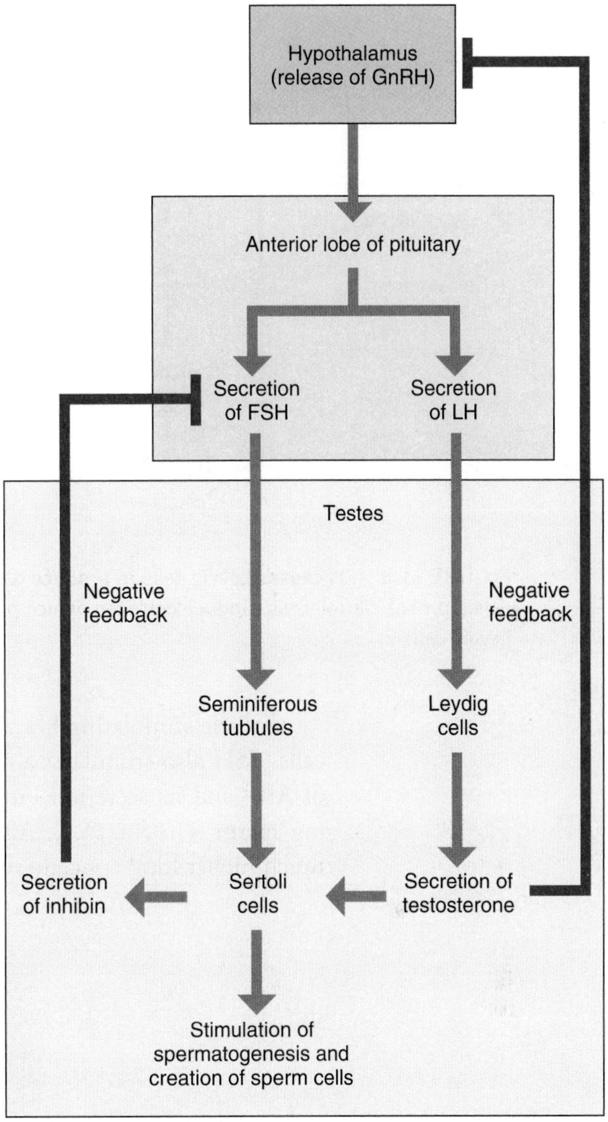

Hypothalamus (release of GnRH)
Anterior lobe of pituitary
Secretion of FSH
Secretion of LH
Testes
Negative feedback
Negative feedback
Seminiferous tublules
Leydig cells
Secretion of inhibin
Sertoli cells
Secretion of testosterone
Stimulation of spermatogenesis and creation of sperm cells

FIGURE 11.7 GnRH is released from the hypothalamus and targets secretory cells in the anterior pituitary. These cells release FSH or LH into circulation where they target cells within the seminiferous tubules.

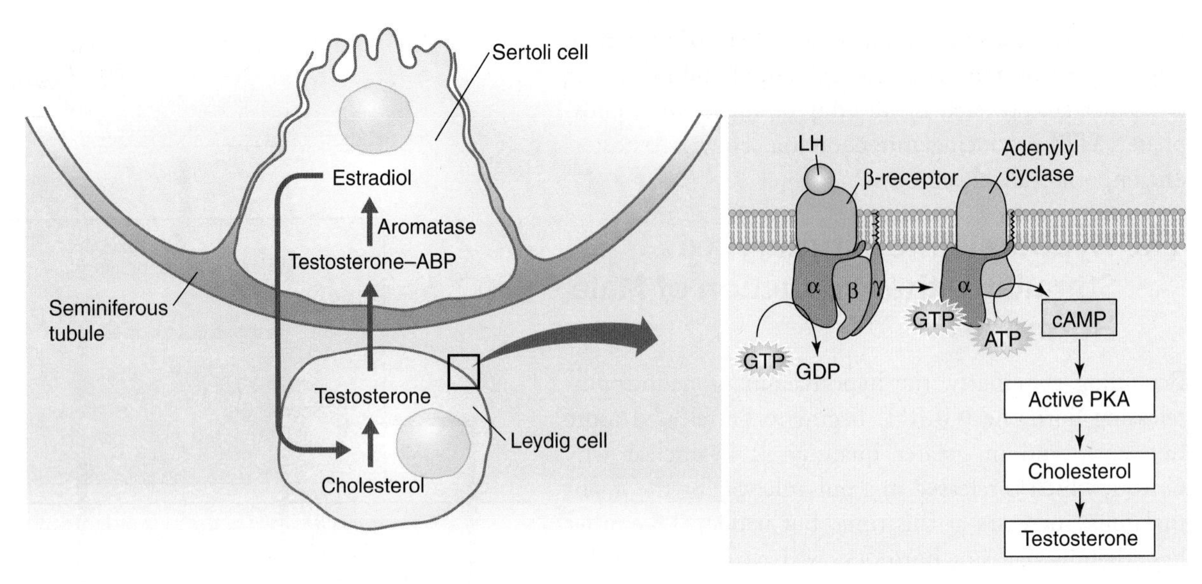

FIGURE 11.8 LH causes Leydig cells to produce testosterone, the hormone required by Sertoli cells to facilitate sperm development. Sertoli cells produce estrogen, which produces a positive feedback loop for testosterone production at the Leydig cells.

Follicle stimulating hormone, FSH, binds to G-protein coupled receptors on Sertoli cells. FSH also stimulates a cAMP–PKA signaling pathway to increase protein synthesis of ABP and its secretion into the extracellular space between Sertoli cells and developing sperm (**FIGURE 11.9**). ABP is necessary for testosterone binding, which will allow a much higher local concentration of testosterone than exists in the bloodstream. FSH also

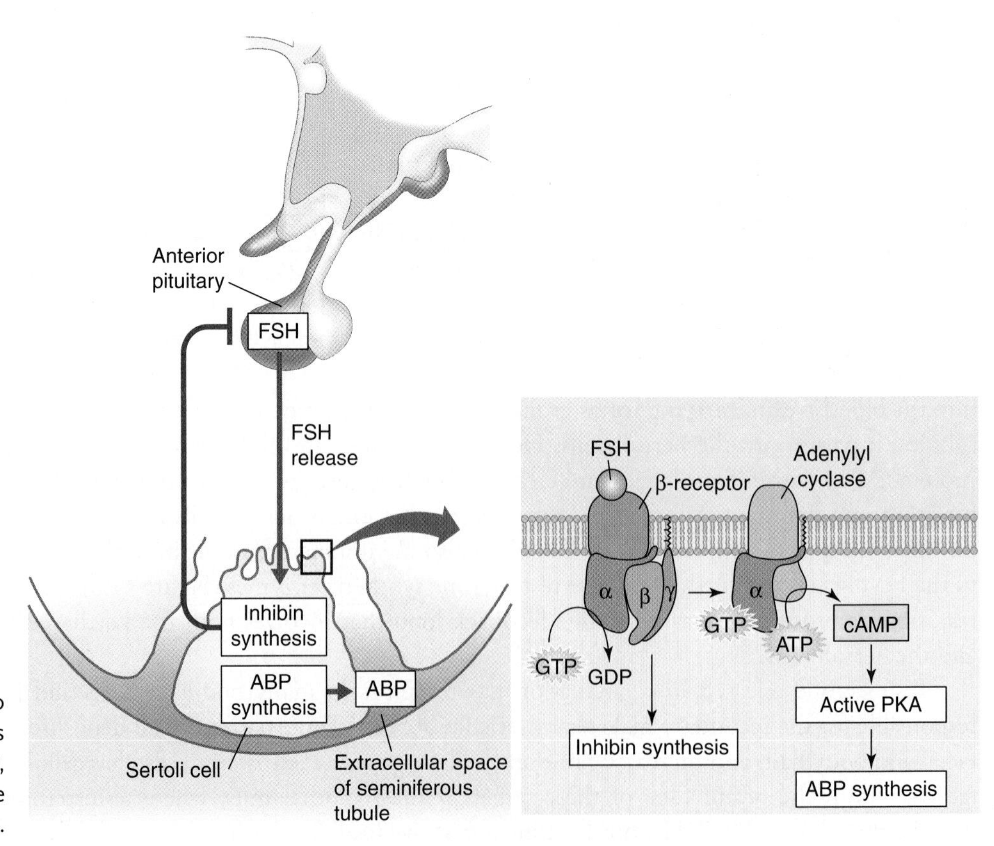

FIGURE 11.9 FSH binds to Sertoli cells causing several proteins to be produced, including ABP, which increases local testosterone concentrations.

stimulates the production of inhibin, which, as its name suggests, inhibits the production of FSH at the anterior pituitary. This functions as a negative feedback loop to prevent the overproduction of ABP and inhibin.

Please notice that the sperm cells themselves, at any level of development, do not have receptors for testosterone. The development of sperm requires testosterone, but only indirectly, because testosterone acts on the hypothalamic–pituitary axis and on the Leydig and Sertoli cells. If the Sertoli cells were absent or could not produce ABP, the concentrations of testosterone would not be high enough within the seminiferous tubules to allow sperm development. Testosterone would be made, secondary sex characteristics would be normal, and all would seem fine, but no gametes would be produced. If Leydig cells were absent, no testosterone would be produced. This would not only prevent gamete production but would have a feminizing effect on this XY male as well.

What Is the Composition of Semen?

Once a spermatid has been successful formed, it moves from the seminiferous tubule into the epididymis where it matures and is stored. As sperm moves to the end of the epididymis, it becomes more mobile, with stronger flagellar movements. During ejaculation, sympathetic stimulation causes the contraction of the smooth muscle within the vas deferens, seminal vesicles, and prostate and bulbourethral glands. Each of these glands produces a fluid that contributes to the composition of semen. The seminal vesicles contribute fructose, an important energy source for the sperm that will be used to provide ATP for the tails as they swim up the uterus. The prostate gland contributes an alkaline solution, which will neutralize the acidic uterine environment, and the bulbourethral glands contribute mucus to lubricate and facilitate semen movement through the urethra.

Presuming all is normal and sperm is being made properly and in sufficient quantity, why doesn't pregnancy occur following every act of intercourse? How is female gametogenesis controlled?

How Are Female Gametes Made?

While males make new gametes throughout their lives, females make all of their oocytes in utero. While a female fetus is herself developing, she is creating the gametes that will produce her own children! Just as in males, the stem cells undergo mitosis, producing a collection of primary diploid oocytes. These oocytes then enter meiosis I and get to the tetrad stage in prophase, when they stop (**FIGURE 11.10**). The primary oocytes will be stored in the ovaries in this suspended meiotic state until puberty. At puberty, when the ovarian cycles begin, one primary oocyte is selected each month to continue with meiosis, creating a secondary, diploid oocyte. This oocyte continues into meiosis II but remains suspended in metaphase of meiosis II until fertilization. Thus, female gametogenesis is completed during fertilization.

Female gametes face a different challenge than male gametes. Male gametes are designed compactly for speed and motility. Female gametes are not motile, but are large, saving as much cytoplasm as possible by unequal meiotic divisions, creating small polar bodies and a larger oocyte. The extra cytoplasm serves as nutrition for the embryo during the early phases of development. The oocyte will also be the only source of mitochondria for the developing embryo. Mitochondria in the male sperm are not retained. The small polar bodies dissolve, so that nuclear material is wasted, while the cytoplasm is saved in the oocyte. One mature, haploid ovum will result from a single stem cell, instead of the four gametes we saw in the male.

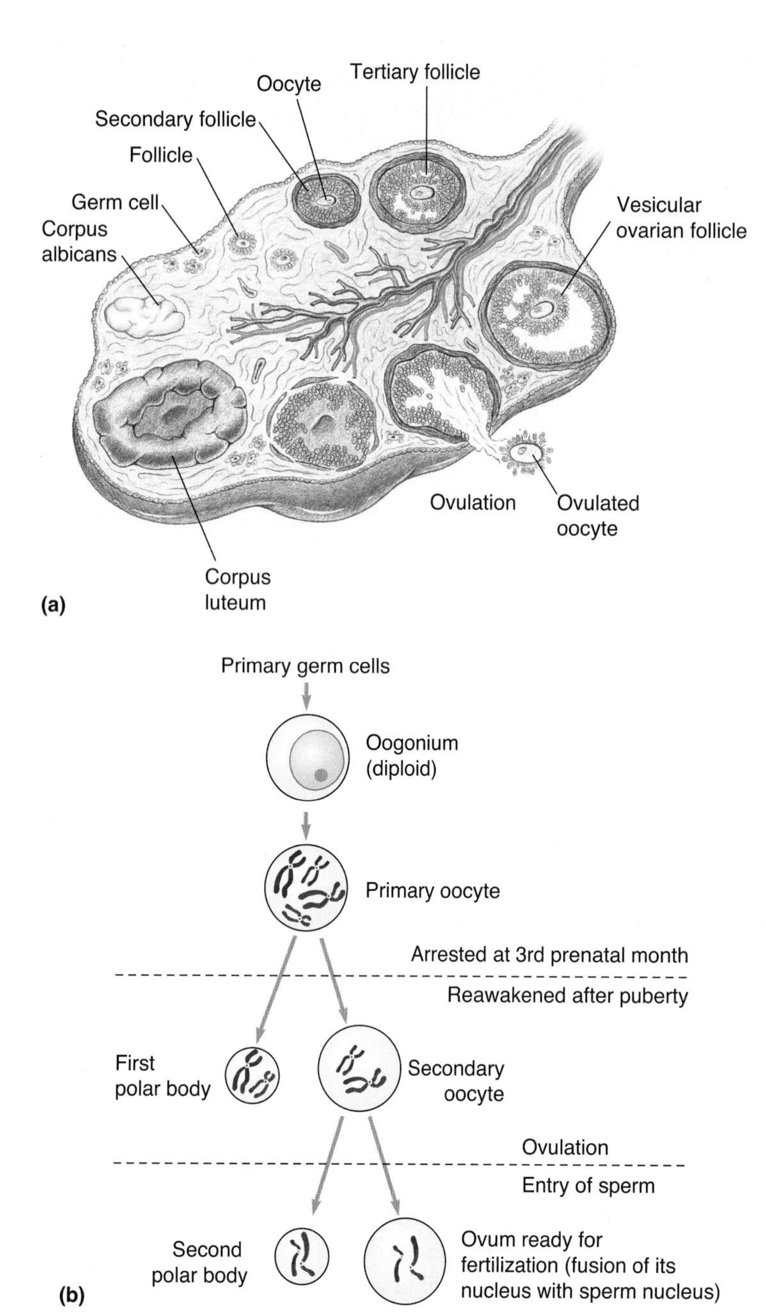

FIGURE 11.10 (a) Eggs are stored in the ovaries, where one will mature each month. (b) Gametogenesis in females produces one haploid gamete from an egg germ cell. The haploid gamete is formed only after the secondary oocyte fuses with a sperm.

Where Are the Eggs Stored and How Is Their Release Regulated?

Eggs are stored within the ovaries surrounded by a layer of follicle cells. Together, the eggs and the associated follicle cells are called the egg "nests" or primordial follicles. By a process as yet unknown, each month eggs from the nest begin to develop into primary follicles, and ultimately one will develop and be released from the ovary. Just as the male germ cell, the primary spermatozoan, is surrounded by supporting cells, so is the developing oocyte. The follicular cells around the oocyte differentiate into granulosa cells, which express receptors for FSH, and thecal cells, which have receptors for LH. These two cell types are analogous to Sertoli and Leydig cells (**FIGURE 11.11**). The aspect of female gametogenesis that is very different from male gametogenesis is that it is cyclical. Hormones are produced by the granulosa cells and the thecal cells, but the hormones and their quantity depend upon the time of the monthly cycle. Let's follow the events of the monthly ovarian cycle for clarification.

We will designate the day that primordial follicles begin to develop as day 1. At this time, the GnRH pulses are low and less frequent, so LH and FSH levels are also low. Granulosa and thecal cells begin to develop from the surrounding supporting cells, but at this early stage of development they possess few receptors, so are not greatly influenced by LH or FSH. At this point, the oocyte itself is producing factors that increase the growth of the supporting cells. Some of the primordial follicles will develop into secondary follicles, and the granulosa and thecal cells start to mature. This occurs about the third to the seventh day of the cycle (Figure 11.10). Not all of the secondary follicles will survive. By the eighth to the tenth day, only one remains, and this follicle develops into the tertiary follicle. The nonsurviving secondary follicles are reabsorbed, and those gametes are forever lost.

As the granulosa cells mature between days 3 and 10, they increase the number of FSH receptors on their surface and, thus, become more sensitive to FSH. As in the Sertoli cells, FSH binds to G-protein coupled receptors on the granulosa cells, where PKA causes the synthesis of aromatase, the enzyme that will convert testosterone into estrogen. Granulosa cells also produce inhibin, which feeds back to the anterior pituitary cells to inhibit further FSH release, a perfect negative feedback, which will maintain a steady production of estrogen.

As the thecal cells mature, they will, like the Leydig cells, produce testosterone and androstenedione. Being cholesterol-based hormones and membrane permeative, testosterone and androstenedione will pass into the granulosa cells and be converted to estrogen. So by day 8 or 9, we will begin to see a rise in estrogen. This increase will be local, around the gamete, but will also be systemic as estrogen enters the blood supply of the ovary and then the entire systemic circulation.

Now we must pause in our chronology of events to discuss a truly new situation in our understanding of physiology. At low concentrations, estrogen acts as an inhibitor of FSH and LH production. It is also thought to act the same way on the hypothalamic neurons that release GnRH. Thus, through all of the cycle thus far, estrogen has acted as a classic negative feedback regulator. However, once the granulosa cells of the secondary and tertiary follicles grow to sufficient mass and produce a threshold level of estrogen, the opposite occurs. Estrogen begins to act on the hypothalamic neurons and the anterior pituitary to *increase* GnRH pulses, GnRH release, and LH and FSH release. At higher blood concentrations, estrogen is a *positive* feedback regulator. While this may seem bizarre, it is a well-known pharmacological phenomenon. High concentrations of a drug may have completely opposite effects of low concentrations. We are seeing this phenomenon with an endogenous hormone—estrogen. The dominance of estrogen at this point of the cycle will also cause body temperature to be slightly lower by approximately 0.3°C (perhaps 0.6°F). So, ovulation can be estimated by body temperature (**FIGURE 11.12**).

By days 10 to12, GnRH pulse frequency has increased significantly, and at day 14 there is a surge of LH, with a lesser surge of FSH. It is thought that the LH surge is responsible for the most important event of day 14: the rupture of the follicle and release of the egg—ovulation. The female gamete is now free of its supporting cells and has been released from the ovary itself, suspended in metaphase of meiosis II.

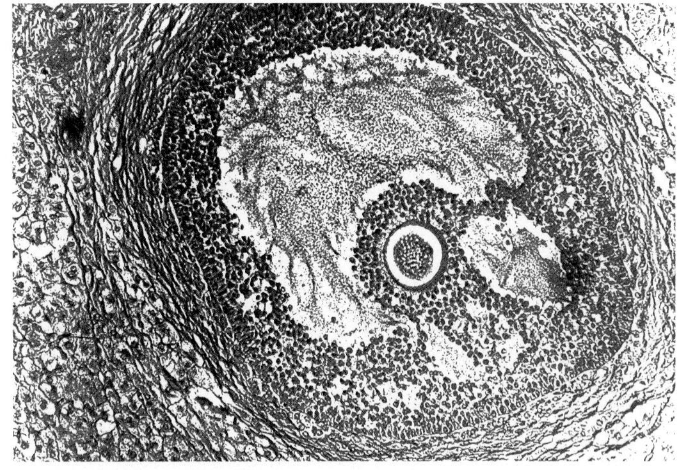

(a)

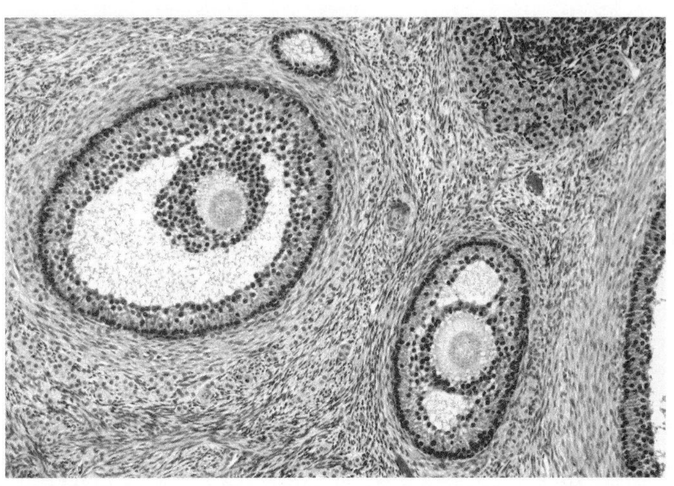

(b)

FIGURE 11.11 Granulosa cells and thecal cells are analogous to Sertoli cells and Leydig cells in the male. (a) Light micrograph of a section through a mature Graafian follicle from a mammalian ovary. This contains the secondary oocyte (round object at center), which becomes an ovum (egg cell) when it is released into the oviduct (fallopian tube). The secondary oocyte is surrounded by cells (corona radiata) within the cavity of the follicle, which is filled with follicular fluid. The follicle wall (outer area) consists of an inner layer of granulosa cells and an outer fibrous layer. (× 100) (© Dr. Keith Wheeler/Science Source.) (b) Light micrograph of a section through secondary follicles in an ovary. Developing eggs (oocytes) are orange. They are surrounded by a fluid-filled cavity (follicular antrum, light pink) and granulosa cells (dark pink). Thecal cells are those in the orange circular band just outside of the granulosa cells. (× 80) (© Steve Gschmeissner/Science Photo Library/Corbis.)

Now a strange journey occurs. The egg is outside of the ovary but is captured by the "fingers" of the ovarian tubes, the fimbrae. There the egg will be moved by ciliary action

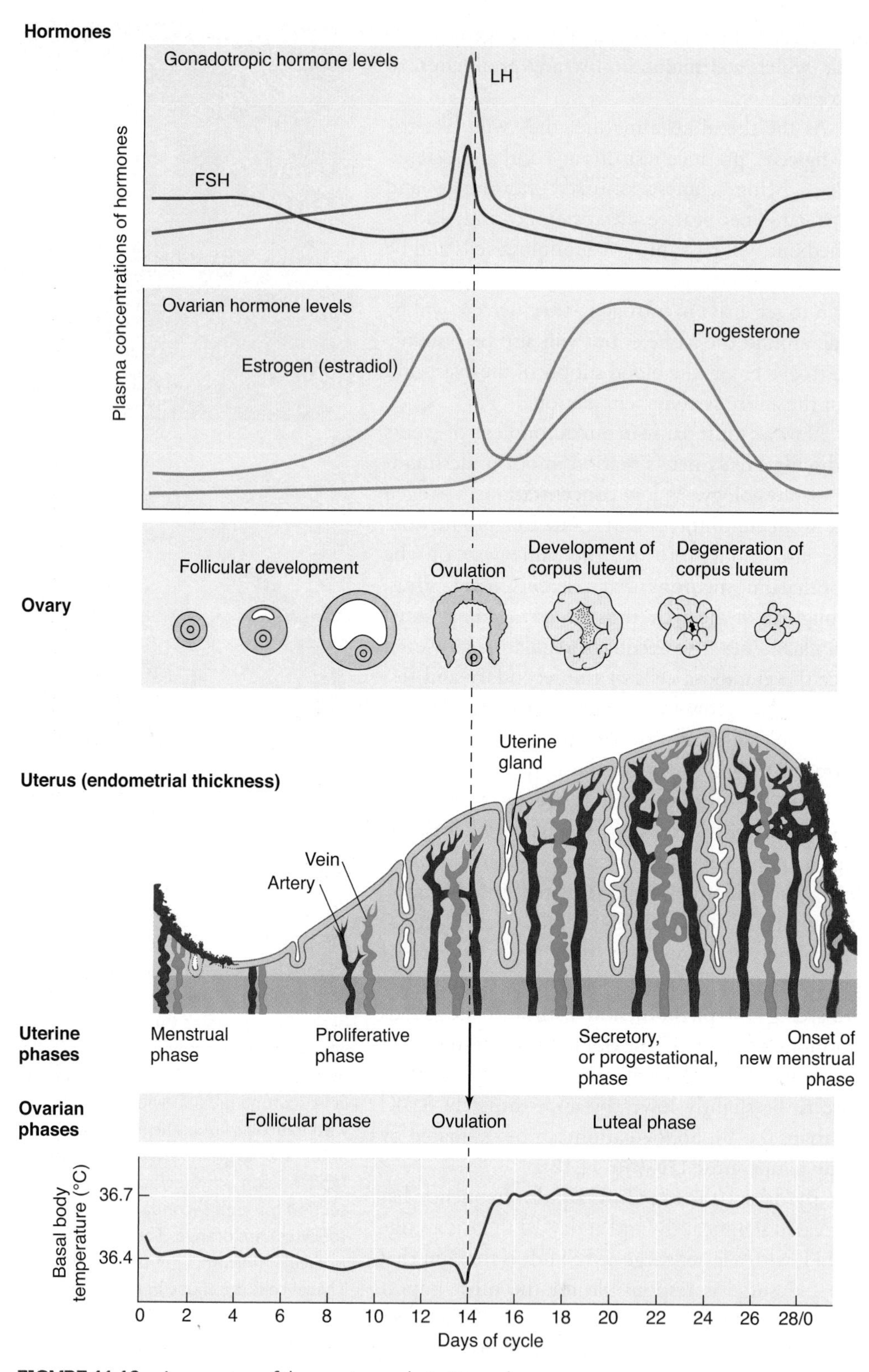

FIGURE 11.12 A summary of the ovarian and uterine cycles.

toward the main portion of the uterine tubes. Fertilization, should it occur, generally does so within the uterine tube, within a day of ovulation.

Preparing for Baby

Now, at day 14 or 15, with the egg within the uterine tube being moved gradually toward the uterus, we confront another problem. Is the uterus ready to accept an embryo? Is it available to nurture a child? The estrogen produced by the combined work of the granulosa and thecal cells had another target besides the hypothalamus and anterior pituitary cells. The uterus is also greatly affected by estrogen. At the uterus, the effects of estrogen are anabolic, causing the proliferation of the uterine endometrium, the innermost tissue layer of the uterus. It also causes angiogenesis, increasing the blood vessels of the endometrium, and it increases the size and number of uterine glands (Figure 11.12). In short, the entire endometrium is thicker, better vascularized, and laden with uterine glands that possess a nutrient solution that will provide energy for the developing embryo in its early days. This profound change in endometrial composition occurs every month, so that the female is prepared to accept a fertilized egg. If no fertilization occurs, this endometrial layer is lost. But what is the signal for that?

The Post-Ovulatory Ovarian Cycle

After ovulation, the supporting cells remain in the ovary while the egg travels along the uterine tubes. But the supporting cells do not die. Blood vessels to the mature follicle are ruptured during ovulation, and blood fills the cavity left by the egg. The supporting cells, now in a structure called the corpus luteum, transform as their space is invaded by fibroblasts and blood into slightly different forms, producing primarily progesterone instead of estrogen (Figure 11.10). Now, estrogen levels are lower, and progesterone levels soar (Figure 11.12).

The target of progesterone is twofold. Progesterone inhibits GnRH release, thus reducing the amount of LH and FSH released from the anterior pituitary. At the same time, progesterone increases secretions by the uterine glands and enhances blood vessel proliferation of the uterus. This will continue from the time of ovulation at day 14 until day 21. After day 21, unless the corpus luteum is "rescued" by a developing embryo, the cells of the corpus luteum die by day 28.

As the corpus luteum loses its function, there is less progesterone and less estrogen. As progesterone levels decline, there is less support for the developing endometrium. Prostaglandins, generally inhibited by progesterone, are now free to constrict the newly created blood vessels. Uterine glands and endometrial tissue, deprived of nutrition, slough off, resulting in menses. Thus, menstrual fluid is a combination of blood, tissue, and fluid from uterine glands. During menses, the primordial follicles begin their development, and the monthly cycle begins anew.

This brings us to the alternate path. Several months after you and your spouse decided you wanted a pregnancy and a child, the female partner's menses occur as usual. The two of you wonder if your timing is askew, and the female partner begins the adventure of taking a daily temperature. The change in temperature is slight, occurring in tenths of degrees, so you need a fine-scale thermometer that registers only temperatures between 95° to 100°F or 35° to 37.8°C. The female partner takes her temperature each morning. At the time of ovulation, her temperature will drop slightly and then rise by several tenths of a degree Celsius thereafter. The two of you spend two months replicating the temperature profile (you are a scientist, after all) and begin to recognize the days of peak fertility, based on body temperature. During the third month, menses fails to appear and you look forward triumphantly to the next nine months.

Fertilization—All the Events That Occurred While You Were Waiting

The two weeks between ovulation and the date when menses would have normally occurred was a busy time. Egg and sperm were united, fertilization was completed, and the developing embryo embedded in the uterine wall, all before you even knew you had achieved a pregnancy! Of course, there were significant regulatory mechanisms in place, which we will now investigate.

During ejaculation, sperm along with their nutrient semen were released into the vaginal canal. Viable sperm are vigorous swimmers, but their trip to the uterine tubes is facilitated by uterine contractions that sweep the sperm along in waves of smooth muscle contraction. This helps the sperm make the trip to the uterine tubes within about 24 hours. Sperm reaching the egg bump into the corona radiata, a circle of cells that surrounds the egg itself. The egg is still suspended in metaphase of meiosis II. As you recall, the head of each sperm contains an acrosomal cap. The acrosomal cap contains two proteases: hyaluronidase and acrosin. These enzymes are vital now for dissolving the extracellular matrix proteins that hold the cells of the corona radiata together (**FIGURE 11.13**). Without the acrosomal enzymes, the sperm would have no access to the surface of the egg. One sperm is not sufficient for this process, but many sperm must participate to weaken the corona radiata enough for a single sperm to break through.

When the corona radiata has been sufficiently dispersed, one sperm will make contact with the surface of the egg, the zona pellucida. Why one sperm? The first sperm that makes contact with receptors on the zona pellucida will cause the membrane of the egg to

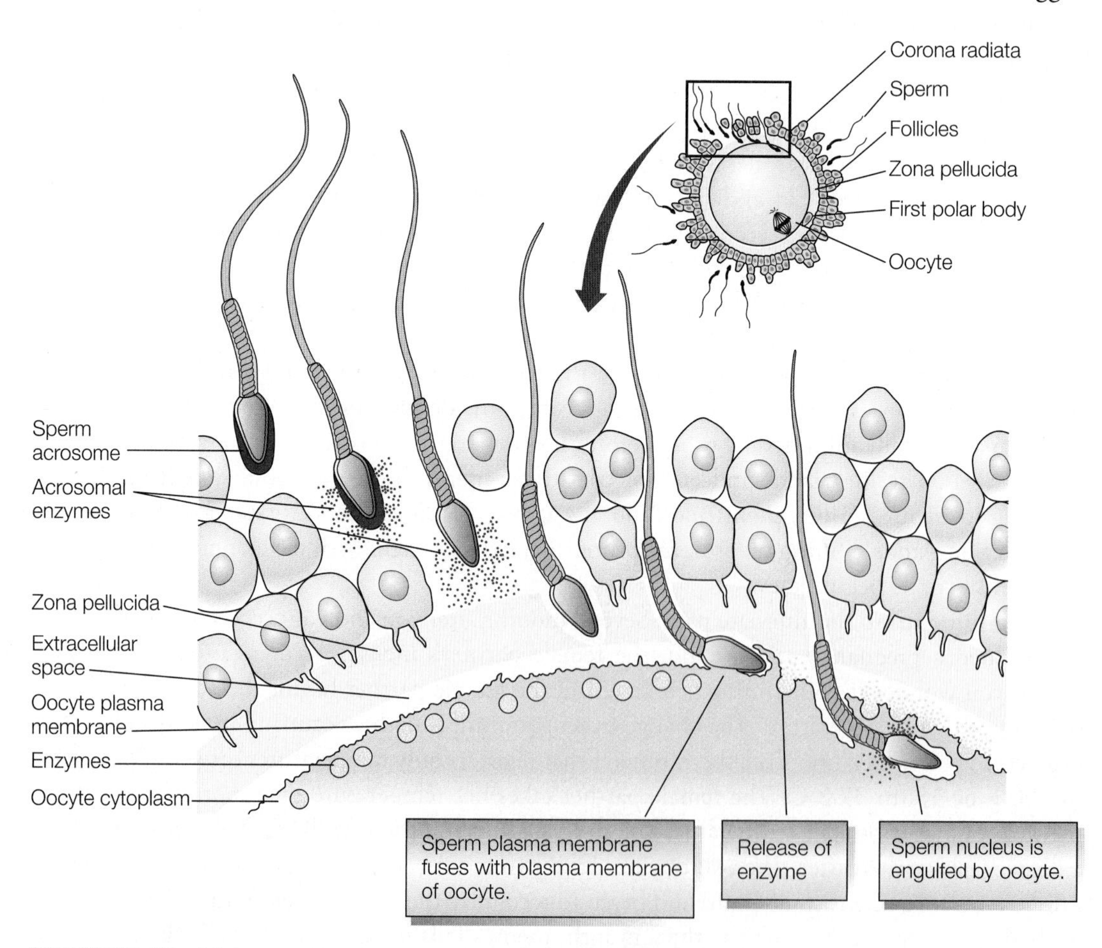

FIGURE 11.13 It takes many sperm to get through the corona radiata.

depolarize. This causes an increase in intracellular Ca^{2+} in the egg and causes exocytosis of enzyme-filled vesicles that inactivate sperm receptors and make the zona pellucida impermeant to sperm. This prevents polyspermy, or multiple sperm present within the egg. The entire sperm gamete enters the egg. The flagellar tail and mitochondria are dissolved by proteolysis. Therefore, only maternal mitochondria usually remain.

The increase in intracellular Ca^{2+} also has another effect: it triggers the completion of meiosis. The final polar body is created, and the egg is now, finally, haploid. The last event of fertilization is the union of the male and female nuclei, a process called amphimixis. When the two nuclei fuse, the first division of the new embryo begins, 24 to 48 hours after ovulation. So, two days after intercourse, conception has occurred and the embryo began developing.

Remember that the uterus is now early in the secretory phase, under the influence of progesterone from the corpus luteum. The embryo is still in the uterine tube, being moved toward the uterus by cilia as it continues to divide. The only nutrition available for the embryo at this stage is the cytoplasm of the egg itself and fluid within the uterine tube. By day 6 following fertilization (day 7 following ovulation), the embryo has formed a blastocyst with a trophoblast shell that will form part of the placenta and an inner cell mass that will become the child. By this point of the secretory phase, the uterus is well

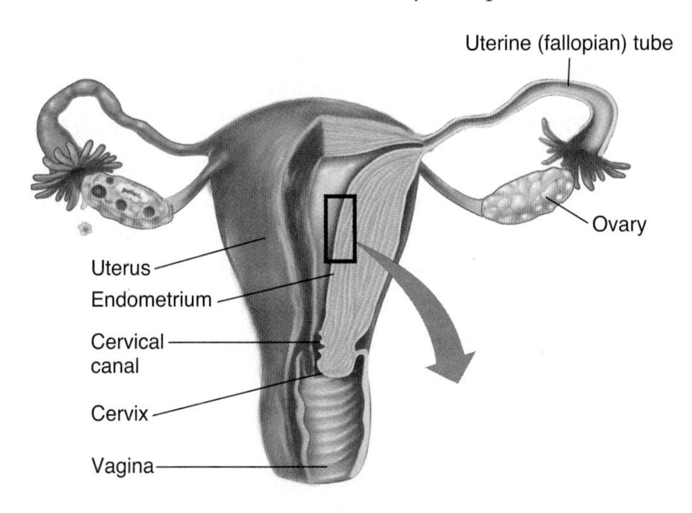

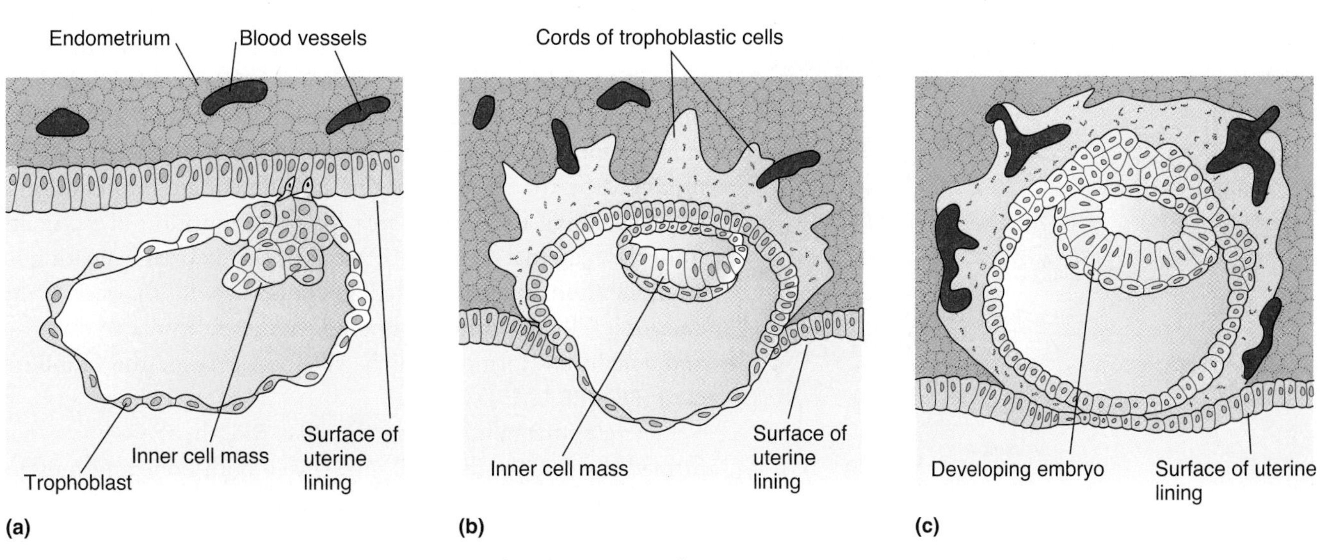

FIGURE 11.14 By day 10, the embryo is implanted in the uterine wall.

supplied with both uterine glands and spiral arteries. The blastocyst settles onto the uterine wall, where the trophoblast cells (now differentiated into syncytial trophoblasts) begin to secrete hyaluronidase. This dissolves the extracellular matrix between cells of the uterine glands and some of the arterial walls (**FIGURE 11.14**). Fluid from the uterine glands and blood from the arteries form into lakes, or lacunae, within the trophoblast layer and now can provide nutrition to the blastocyst by diffusion from the fluid to the embryo. This is the initial implantation of the embryo, which is complete by day 10.

The First Trimester of Fetal Development

Now the embryo begins to participate in its own survival. The syncytial trophoblast makes the hormone human chorionic gonadotrophin (hCG), a hormone similar to LH. Because the maternal blood vessels have been breached, this hormone enters the maternal blood supply, where its target is the corpus luteum of the ovary. Without hCG, the corpus luteum would age and become the corpus albans; it would stop producing progesterone, the spiral arteries would constrict, and the uterine lining would slough off. With hCG, the corpus luteum remains a viable endocrine organ that continues to produce progesterone throughout the pregnancy. Thus, the embryo "rescues" the corpus luteum and ensures the maintenance of a favorable uterine environment. It is hCG that the early pregnancy test detects. By day 14, when you would have expected menses to occur, the embryo is already altering maternal hormonal balance.

The blastocyst has a different genetic makeup from the mother, and yet it is invading maternal tissue. Why isn't the early embryo attacked by the maternal immune system? While the blastocyst is in such intimate contact with maternal blood, it secretes a variety of immunosuppressive factors, one of which is hCG itself. Other immune-inhibiting factors include interleukins 1α and 6, interferon α, immunosuppressive factor, and prostaglandin E2. The result is that the early embryo can "hide" from the maternal immune system. Later, as the placenta forms, there is a greater separation of fetal and maternal blood supply, and the immunosuppression is reduced.

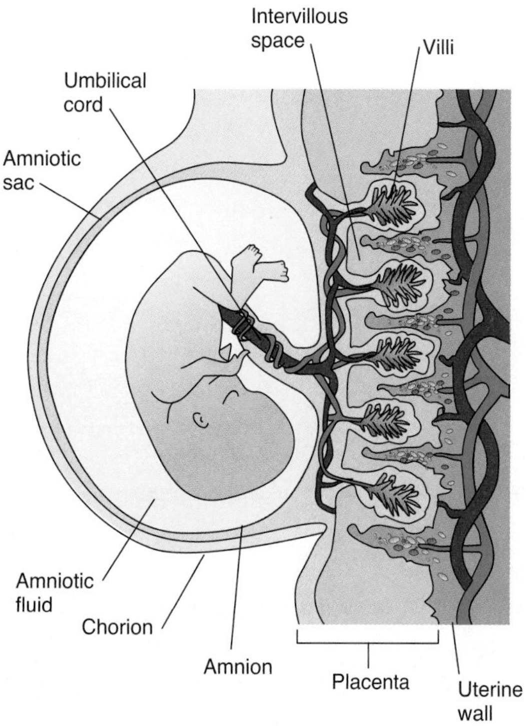

FIGURE 11.15 Chorionic villi early in the pregnancy quickly become a mature placenta.

Fetal Development: First Month

We have already made it to day 10, but there are profound changes that occur in the fetus and the placental structure in the next 20 days. As the embryo continues to divide and the tissues differentiate, the need for additional nutrition increases as well. Simple diffusion through the syncytial trophoblast cannot support the embryo for long. The trophoblast layer, along with mesoderm produced by the embryo, begins to differentiate into chorionic villi, increasing the surface area for the acquisition of nutrients from the maternal lacunae. By week 3, the embryo is floating in amniotic fluid and surrounded by chorionic villi. By week 4, the chorionic villi are limited to one side of the encapsulated embryo, and a stalk has formed, which will soon become the umbilical cord (**FIGURE 11.15**).

Developmentally, during this first month, the embryo has formed the somites that will give rise to both bone and muscle. The neural tube is formed, beginning the creation of the brain. The heart has begun to beat, and the lungs, trachea, and digestive tract are beginning to develop.

Fetal Development: Second Month

During the second month, the placenta and the umbilical cord mature. The chorionic villi still draw nutrition from maternal lacunae, essentially formed from broken vessels within the uterine wall. These open blood spaces surround the embryonic vessels of the chorionic villi allowing oxygen (O_2) and carbon dioxide (CO_2) to exchange, along with nutrients, without mixing embryonic and maternal blood. The umbilical cord serves as a conduit for these vessels to reach the embryo. By week 10, this structure is complete. The placenta not only serves for nutritional transport, it is also an important endocrine organ. The placenta itself begins to produce progesterone, because the corpus luteum is not capable of making a sufficient quantity to maintain the pregnancy. The placenta will also make GnRH, thyrotropin-releasing hormone and thyroid-stimulating hormone, corticotropin-releasing hormone, and growth hormone-releasing hormone to facilitate growth of the placenta and the embryo.

How are gases exchanged across the placenta? Maternal blood in the lacunae bathes the vessels of the chorionic villi (**FIGURE 11.16**). Because O_2 and CO_2 are membrane permeant, they can cross the endothelial cells that make up the chorionic blood vessels. Oxygen and

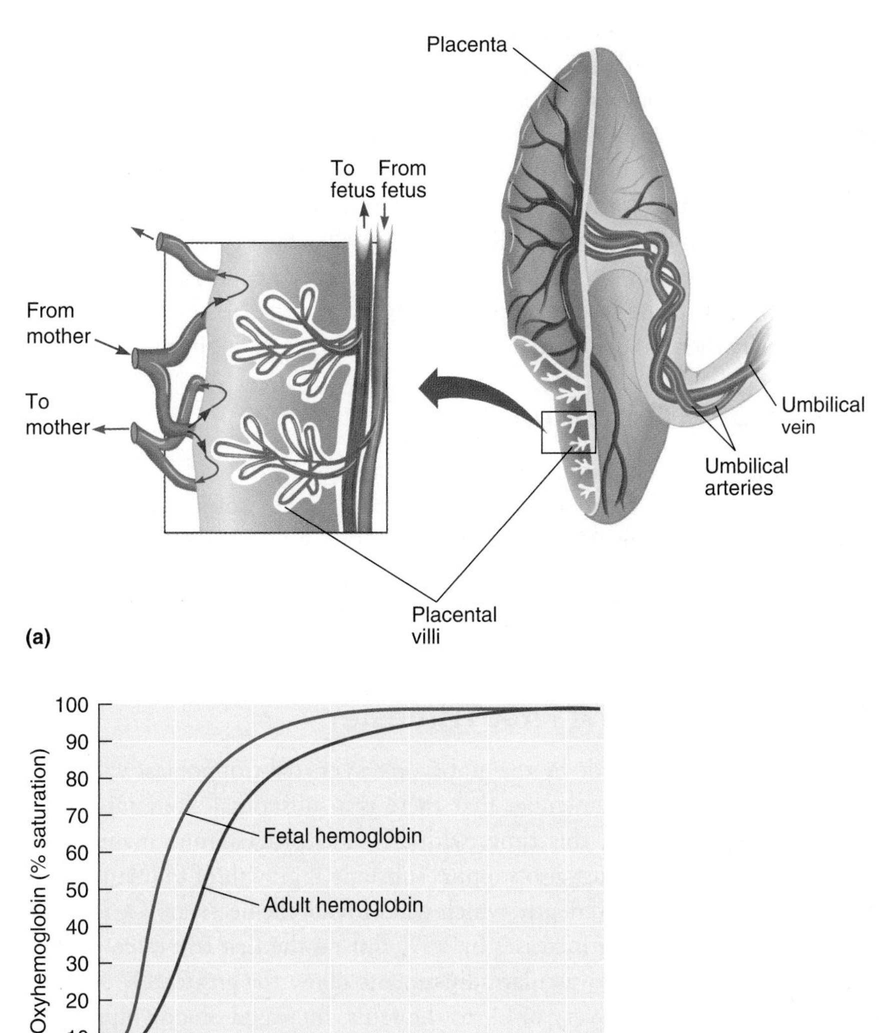

FIGURE 11.16 (a) Maternal blood bathes the blood vessels of the infant in their placental villi. Oxygen, carbon dioxide, and nutrients can be exchanged without mixing the blood of mother and child. (b) Fetal hemoglobin has a greater affinity for oxygen than adult hemoglobin. This allows the fetus to load oxygen from maternal blood.

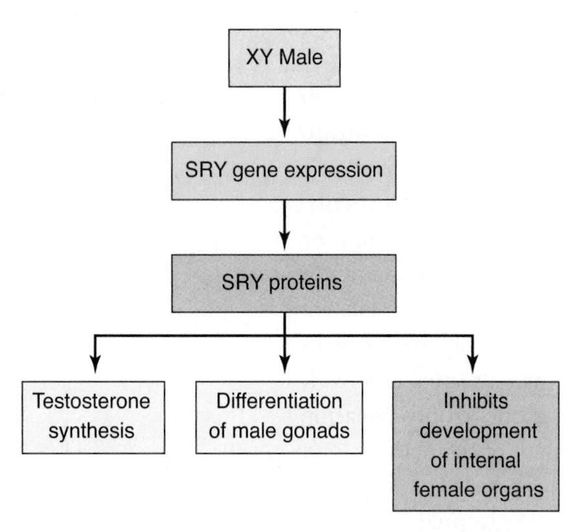

FIGURE 11.17 In the absence of SRY protein, the primordial reproductive organs develop into females.

carbon dioxide are transported by simple partial pressure gradients. However, the fetus works with one advantage, a unique isoform of hemoglobin that has a higher binding affinity for O_2 than maternal hemoglobin. So, at any given partial pressure of O_2, fetal hemoglobin is more likely to bind O_2 than maternal hemoglobin. The result is that fetal hemoglobin loads O_2 from maternal plasma. On the delivery end, fetal tissues are slightly more acidic, retaining a bit more CO_2 than they will as an adult. This causes a right shift in the hemoglobin dissociation curve and release of O_2. Thus, the placental blood can remove O_2 from maternal blood and deliver it to the fetus.

During the second month, the embryo develops sweat glands, hair, limbs, the beginnings of muscle, central and peripheral nervous systems, thymus gland, pituitary and adrenal glands, heart chambers and blood vessels, more lung structure, a more mature digestive system, and kidneys. Junior, still an embryo, is quickly coming to resemble a human being!

In the discussion of male and female gamete formation, we noted the similarity of Sertoli and Leydig cells to granulosa and thecal cells. In the embryo at this stage, the reproductive organs are undifferentiated and could be either male or female. Tubules that will become either ductus deferens or uterine tubes connect to gonads that will become either testes or ovaries. If the embryo has an XY genotype, then a gene on the Y chromosome, the sex-determining region Y (SRY) will be expressed (**FIGURE 11.17**). The products of the SRY gene inhibit female development and initiate differentiation of the male gonads and the production of testosterone. In the absence of SRY, the fetus will default to female. Therefore, any interruption in the expression of the SRY protein will also cause female development. Knowing that sexual organs come from a common progenitor gonadal structure makes it easier to understand why there is such homology between the gamete-supporting cells.

Fetal Development: Third Month

At the beginning of the third month, the embryo is now a fetus. By the end of this month, all major organ systems will be formed. In the third month, skin covers the body and ossification centers appear in bone. Muscle development in the limbs begins. The spinal cord and brain are in place, although not fully developed. Blood vessels have invaded bone, facilitating growth. The tonsils and thymus glands are formed and gondal differentiation occurs. By the end of this trimester, the fetus is recognizable as human.

Maternal Physiology in the First Trimester

Early in the first trimester, hCG made by the embryonic syncytial trophoblast will cause morning sickness in the mother. Remember that there is a substantial alteration in the maternal hormonal environment at this time, with LH and progesterone maintaining high concentrations. The mother must also support substantial growth of the fetus, causing an increase in respiratory rate and depth, which will provide the necessary O_2 for ATP production. Maternal blood volume increases by 45% during the first trimester, as does cardiac output. Both of these cardiovascular adjustments allow for greater O_2 carrying capacity of the blood and better delivery of O_2 to the fetus. Increased blood volume will also require greater glomerular filtration rate, because the maternal kidney is responsible for clearing both fetal and maternal waste from the blood. Neither blood volume nor cardiac output will increase much more during the remainder of the pregnancy.

Fetal and Maternal Changes in the Second Trimester

During the second trimester, the fetal organ systems set up during the first trimester develop more complexity and mature. The spleen, liver, and bone marrow take shape, and alveoli of the lungs are formed by the end of this trimester. The fetus grows from 26 to 640 grams, increasing its size by more than 24-fold!

Maternal energy requirements increase as the female supports fetal growth as well as her own physiological needs. Placental hormones, along with maternal thyroid hormone, growth hormone, estrogen, and progesterone, will initiate growth and proliferation of mammary glands during this period, preparing to feed the newborn at the end of the next trimester.

Fetal and Maternal Changes in the Third Trimester

During the third trimester, the fetus grows in size from 0.64 kg to 3.2 kg, or from 1.4 to 7 lb, another enormous increase in size. Maternal size also increases, with the fetus now occupying a significant portion of the abdominal cavity. Obviously, the uterus, once the size of a pear, has now expanded to house the full-grown fetus. One of the final stages of fetal development is the formation of the septal cells within the lung that make pulmonary surfactant, the molecule that reduces the air-water interface tension in the alveoli and reduces the work of breathing. During the final month, development is complete and the infant only increases in size. We are now ready for childbirth.

Childbirth

For nine months the fetus has lived and grown within the mother. Childbirth requires a change in this homeostasis, which as you could predict by now, means we must have a positive regulatory feedback loop to create change. What initiates this loop and starts the process of childbirth? As you would expect, it is not a single stimulus, but several factors that potentiate the effect and begin childbirth.

Placental progesterone has suppressed uterine contractions during the pregnancy. Late in pregnancy, maternal estrogen production increases, while progesterone declines, changing the ratio of these two hormones. Maternal estrogen increases oxytocin release. As you recall from the earlier section on orgasm, oxytocin promotes uterine contraction. Together, estrogen and oxytocin increase prostaglandin production in the uterus, which also increases contractions (**FIGURE 11.18**). So, three maternal hormones are contributing to increased smooth muscle contractility of the uterus.

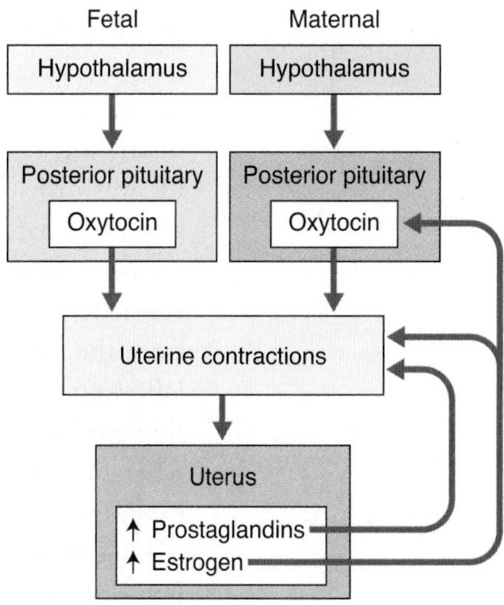

FIGURE 11.18 Several hormones create positive feedback loops that initiate parturition.

Uterine stretching stimulates reflexive contraction, which is facilitated by estrogen. Thus, the large fetus, moving within the uterus and stretching it, contributes to uterine contractions and the beginning of its own delivery. At the same time, the fetus begins to produce oxytocin in the hypothalamus, releasing it in response to estrogens. More oxytocin causes more uterine contractions; so again, the fetus is doing his part to initiate childbirth.

As the contractions become stronger and move the fetus within the uterus, the contractions and movement of the fetus become yet another

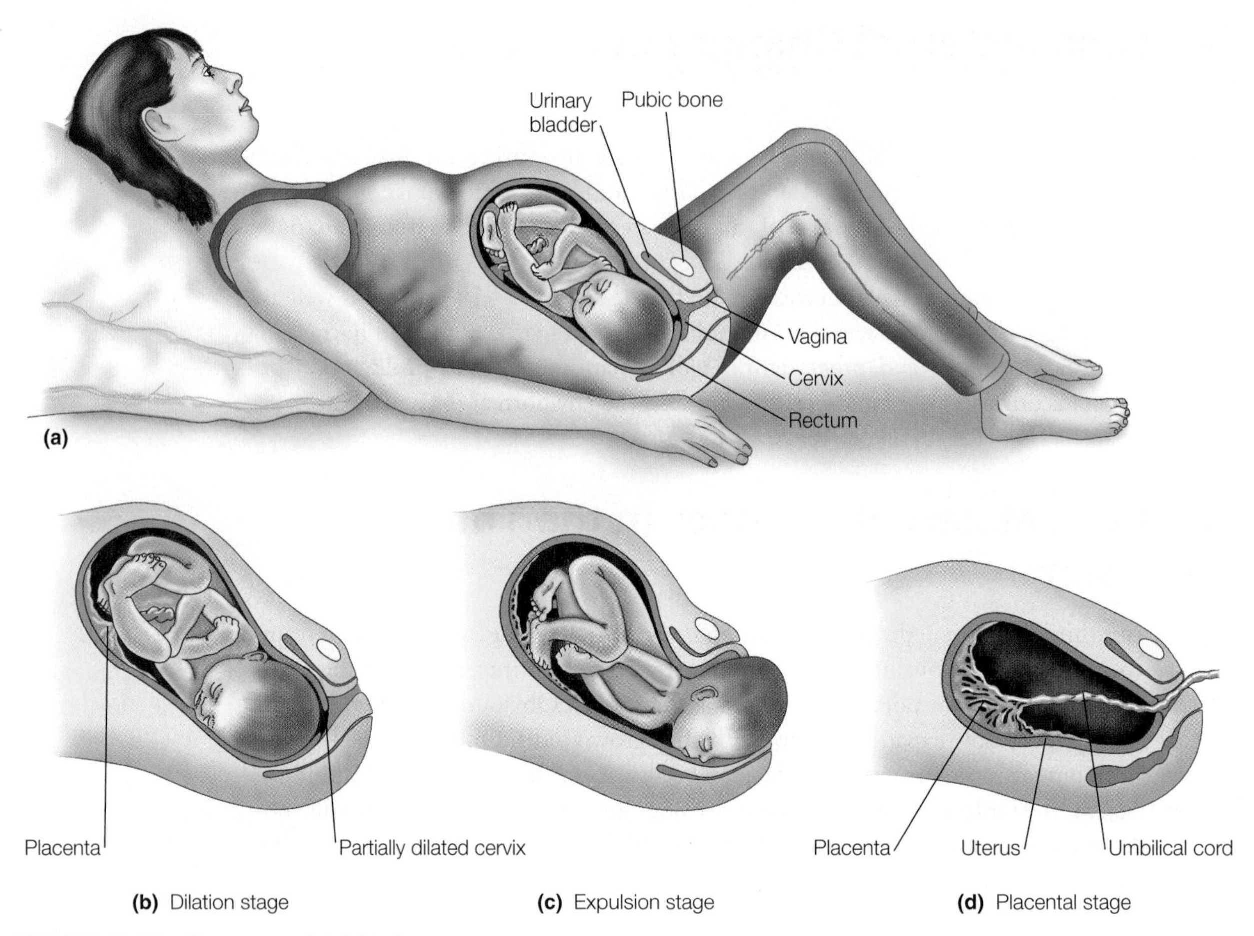

FIGURE 11.19 The stages of childbirth.

positive feedback loop. Thus, childbirth, or parturition, is begun (**FIGURE 11.19**). Early stages in parturition involve dilation of the cervix, to allow passage of the fetus. The head and shoulders of the fetus present the largest profile, so it is important that they pass through first, when the cervix is most dilated. Thus, the best position for the infant before delivery is head downwards. The fetus, now an infant, is expelled first, followed by the placenta.

A remarkable change now occurs in the infant's physiology. For the first time the infant must breathe air. The first breath causes several seminal changes. Before birth, O_2 was provided by the placenta. The umbilical vein, returning to the fetal heart, carried oxygenated blood and entered the inferior vena cava through a vessel known as the ductus venosus (**FIGURE 11.20**). Oxygenated blood entered the right atrium and the right ventricle, just as it would in an adult. Two structures make fetal circulation unique. Blood from the right atrium can flow into the left atrium through the foramen ovale, enter the left ventricle, and be pumped into the systemic circulation, bypassing the lung. Blood that enters the right ventricle from the right atrium is pumped into the pulmonary trunk, where it faces high resistance because the fetal lung is collapsed. Instead of traveling to the lung, the blood from the pulmonary trunk is shunted into the aorta via a duct known as the ductus arteriosus and goes into systemic circulation. At birth, when the infant takes its first breath, the expansion of the lung opens the extra-alveolar blood vessels as well as the alveoli. This immediately lowers the pressure in the left atrium, closing the foramen ovale by pressure. Thus blood, still coming through the ductus venosus, is directed from the right atrium to the right ventricle and into the pulmonary trunk. High O_2 concentrations of blood following the first breath cause constriction of the ductus arteriosus, so blood within the pulmonary trunk is no longer shunted into the aorta. So, within a few seconds, the fetal heart circulation pattern changes to that of an adult. The ductus venosus remains viable for several hours, but the lack of blood supply from the placenta will cause its closure and dissolution.

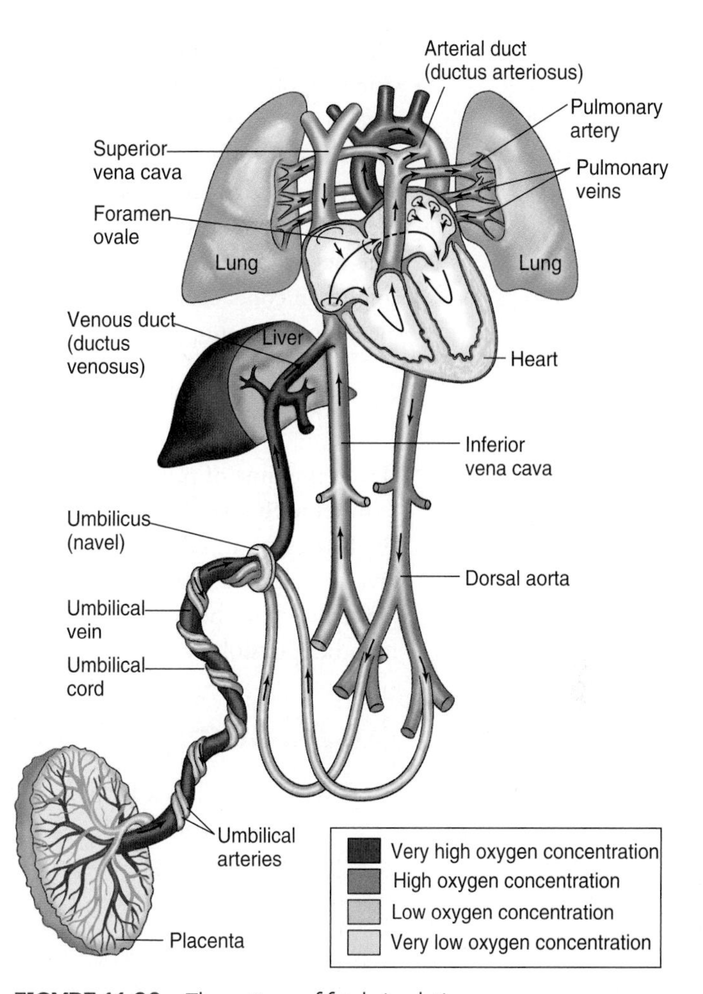

FIGURE 11.20 The pattern of fetal circulation.

Lactation

The mammary glands have proliferated, preparing milk to feed the infant. However, the release of milk from the mammary glands is a separate process from the production of milk. Once again, the hormone oxytocin is important. Oxytocin is released by two mechanisms: physical touch to the nipples, caused by suckling of the infant, or emotional stimulation, often caused by the sound of an infant crying. Remember that the neurons that produce oxytocin are located in the hypothalamus, a center for our emotions as well as hormone release. Oxytocin causes contraction of smooth muscle within the mammary glands and their ducts, thus releasing milk from the nipple. Milk production by the mother requires energy (**FIGURE 11.21**).

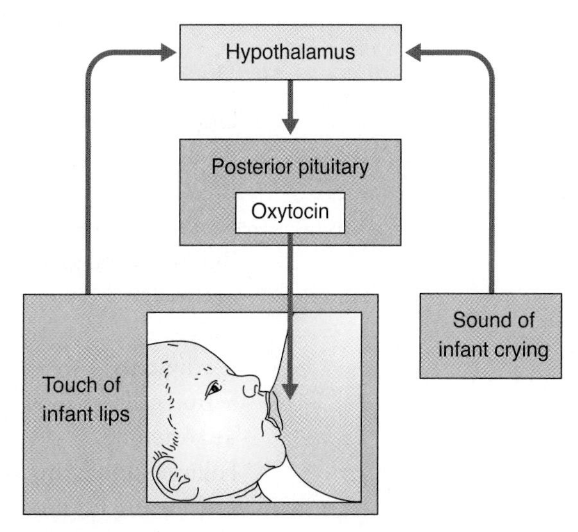

FIGURE 11.21 The hormonal mechanism of milk let-down.

Early Development

Human beings are born immature. They are not simply small adults that grow larger, but humans that are not fully formed. The nervous system is still developing, as we can easily see by an infant's inability to walk, feed itself, or control urination or defecation. They lack the coordination they will develop as older children and adults. In fact, our brains are not fully mature until we are in our late teens or early twenties.

The immune system is also vestigial. Some immunity is passed from mother to child in breast milk, but the child will also form its own antibodies, unique to its environment and the immunological challenges it faces.

Menopause

Your adventures in childbirth have made you think about your mother, who has recently entered menopause. As you know, the supply of maternal eggs in the ovary is limited. Several are wasted each month as they become secondary follicles that never mature. Other eggs in the nest simply undergo atresia, or deterioration, so they are unavailable as potential oocytes. Remember that the oocytes and their supporting cells, the granulosa and thecal cells, are the sole source of estrogen. Estrogen targets many tissues in the body, signaling protein production, so the loss of estrogen has systemic effects.

At approximately 50 years of age, women begin menopause, characterized by the cessation of menstrual periods. GnRH is still released, as are FSH and LH. However, the negative feedback control exerted by estrogen is reduced, so FSH and LH tend to be higher in concentration than normal. This may contribute to the hot flashes that so often accompany menopause.

Summary

From the time of puberty (11 to 15 years of age), humans are remarkable in their fertility and reproductive ability. In men, this fertility begins at puberty and is lifelong. In women, it also begins at puberty but lasts only until menopause (approximately 50 years of age on average). Unlike many animals, we have a long lifespan beyond our reproductive life. This contributes to the cluster of ailments that populate our "old age." But, that is the topic for another physiology text.

Key Concepts

Negative feedback loop
Positive feedback loop
Fertilization
Gametogenesis
Ovulation
Fetal circulation
Lactation
Menopause

Key Terms

Ductus deferens
Epididymus
Erectile tissue
Seminal glands
Bulbourethral glands
Oxytocin
Coitus
Leydig cells
Sertoli cells
Testosterone
Follicle-stimulating hormone
Luteinizing hormone
Diploid
Haploid
Primordial follicles
Granulosa cells
Thecal cells
Estrogen

GnRH
Corpus luteum
Progesterone
Estrogen
Menses
Ejaculation
Acrosomal cap
Corona radiata
Syncytial trophoblast
HCG
Chorionic villi
Placenta
Parturition

Application: Pharmacology

1. Birth control pills have been used for over 50 years to prevent pregnancy. They are generally pills that contain both estrogen and progesterone but, when taken daily, prevent ovulation. Despite the lack of ovulation, there are still menses. Why? Would you expect that a woman taking birth control pills would enter menopause at the same age as someone who did not? Why?
2. Viagra was discussed during the chapter. It causes vasodilation in the vessels serving the corpus cavernosa and corpus spongiosum, or blood spaces, of the penis. If a man were taking Viagra to improve his erectile function, what effect would this have on his fertility? Why?
3. Induced labor has been a tool for decreasing time in childbirth labor for many years. The drug pitocin, a derivative of oxytocin, is the drug generally given. How would pitocin shorten the time in labor? Be specific in your answer.

Clinical Case Study: Diabetes

BACKGROUND

Having any type of diabetes—type 1 diabetes, type 2 diabetes, or gestational diabetes—adds another risk factor to pregnancy. Diabetes, especially poorly controlled diabetes, can cause high blood sugar in both the mother and the child. This tends to increase the size of the fetus during development, resulting in a larger fetus. A large fetus will be more difficult to deliver naturally and may require surgical intervention or a cesarean section. This may be stressful on the fetus.

For the mother, uncontrolled diabetes during pregnancy is a risk factor for preeclampsia, a condition that includes high blood pressure and protein in the urine. Preeclampsia is, of course, the preliminary stage of eclampsia, a condition that can cause seizures and the separation of the placenta from the uterine wall before labor begins. This is a serious condition indeed and can be life-threatening to both mother and child.

THE CASE

Your friend is so taken with your new baby that she decides to have one of her own. It was a decision she had been trying to make for a long time, but your beautiful baby inspired her to have one of her own. Your friend has had type 2 diabetes for many years, always struggling with both her weight and her diet. When she announces to you that she wants to have a baby, you feel obligated to discuss her medical condition, so that she takes it seriously.

THE QUESTIONS

1. What will you say to your friend about the potential hazards to her health if she gets pregnant?
2. What will you say to her about the potential hazards to her child if she doesn't control her diet during pregnancy?
3. What can she do to improve her chances of having a normal and successful pregnancy?
4. If she needs to lose weight, what advantage will pregnancy have for her? What about lactation?

A

The Electrocardiogram (ECG)

The ECG is a simple, noninvasive tool that gives a clinician valuable information about the electrical activity of the heart. We know that electrical activity is required for depolarization of cardiac muscle, allowing it to contract and force blood out through the blood vessels. Remember, however, that the ECG can only directly give us electrical information. Let's look at a normal ECG (**FIGURE A.1**) to review its meaning and then posit a few electrical abnormalities and how they would appear on an ECG.

In a normal ECG, the sinoatrial (SA) nodal action potential appears as the first upward deflection of the P wave. The depolarization of the atria makes up the remainder of the P wave. The electrical action potential now reaches the atrioventricular (A-V) node, where the impulse is slowed. Because there is no net change in the depolarization occurring at this point, the P-R interval is a flat line. The P-R interval duration is the time that the action potential is delayed at the A-V node. The QRS complex is the electrical depolarization of the ventricle. The S-T segment is again flat, or isoelectric, because the ventricle remains depolarized for a short time. The T wave is the repolarization of the ventricle. Now there is an interval of time until the next SA nodal action potential, occurring while the ventricle is at its resting membrane potential, which will show on the ECG as a flat line. The regularity of each wave or wave complex and their duration can be important diagnostic tools.

The SA Node

The SA node is our normal pacemaker. If the SA node did not depolarize quickly enough, we could experience a slow heart rhythm, or sinus bradycardia (regular rhythm less than 60 bpm) (**FIGURE A.2**).

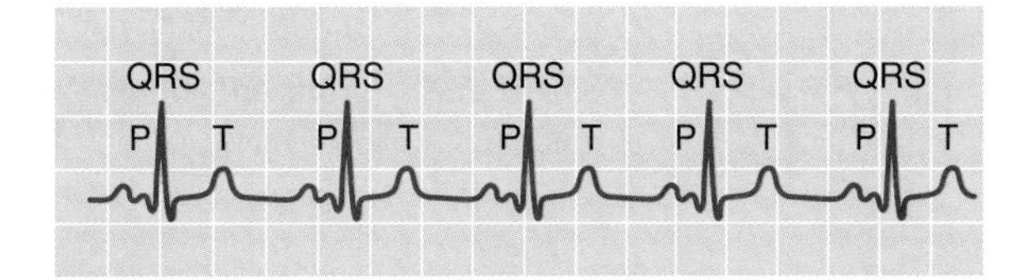

FIGURE A.1

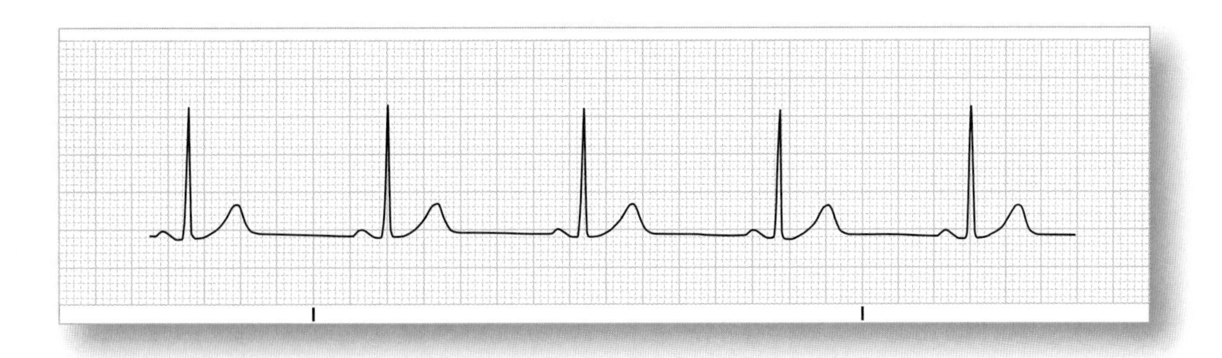

FIGURE A.2 (From *Arrhythmia Recognition: The Art of Interpretation*, courtesy of Tomas B. Garcia, MD.)

This is not generally a serious condition and is common among athletes and trained individuals. It does illustrate how the SA node regulates cardiac depolarization.

Atrial Fibrillation

What would happen if the SA node depolarizes regularly and normally but the atrial pathways are disordered? Age or disease can cause the remodeling of atrial tissue in such a way as to derange electrical conduction and promote atrial fibrillation. The problem is not with the SA node but with electrical conduction within the atria themselves. The absence of a coordinated depolarization will reduce normal atrial contraction, thus reducing the contraction-driven blood flow from the atria into the ventricle. The A-V node can still stimulate ventricular contraction, but some SA nodal "escape" depolarizations may promote irregular ventricular depolarizations (**FIGURE A.3**).

A-V Block

Sometimes the SA node and the atria function well electrically, but the action potential is not transmitted through the A-V node normally. If there is simply a delay in the action potential leaving the A-V node, then it is termed *first-degree block* (**FIGURE A.4**). On the ECG this would appear as a longer P-R interval (**FIGURE A.5a**). In second-degree block, P waves might appear without a subsequent QRS complex (**FIGURE A.5b**). That is, the SA nodal action potential doesn't always reach the ventricle. If the block is complete, then there is no electrical connection between the atria and the ventricles, and the ventricle depolarizes because of action potentials begun within the Purkinje fibers. These depolarizations are very slow and are not conducive to health.

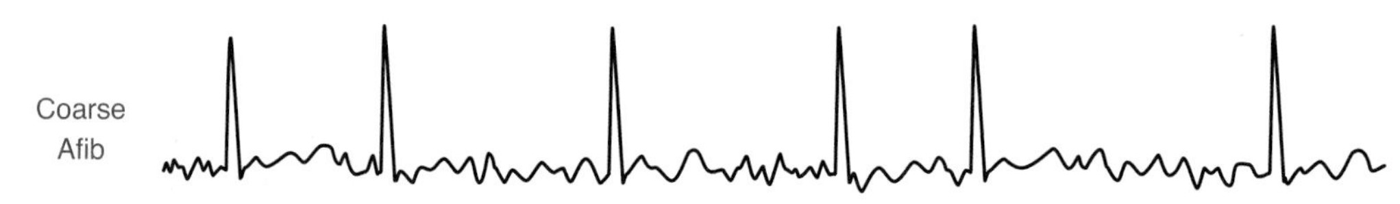

FIGURE A.3 (From *Arrhythmia Recognition: The Art of Interpretation*, courtesy of Tomas B. Garcia, MD.)

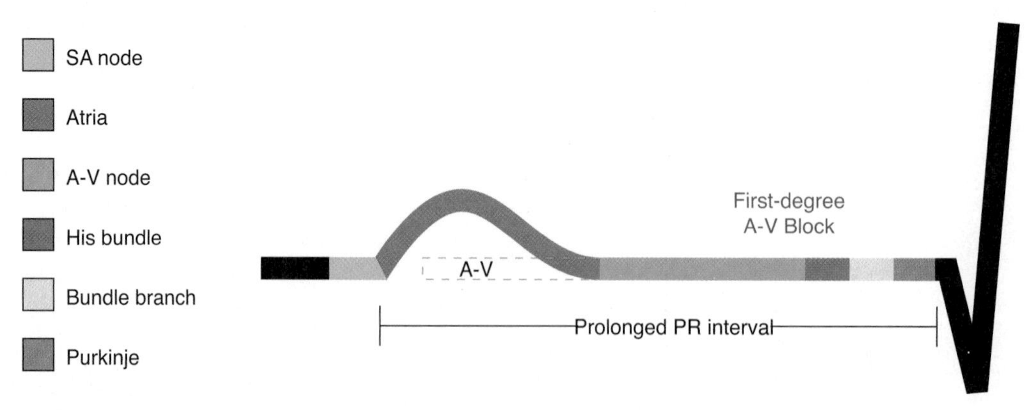

FIGURE A.4 (From *Arrhythmia Recognition: The Art of Interpretation*, courtesy of Tomas B. Garcia, MD.)

Ventricular Arrhythmias

The cardiac action potential can proceed normally from the SA node, through the atria and the A-V node to the ventricle, and become irregular, or dysrhythmic, within the ventricle. This is most commonly caused by electrically heterogeneous areas of the ventricle, usually resulting from myocardial infarction. Delays in depolarization and re-entry arrhythmias can cause ventricular fibrillation. While there is a great deal of electrical activity during ventricular fibrillation, it is not coordinated enough to allow muscle contraction and blood is not pumped out of the heart (**FIGURE A.6**). This is a most serious and life-threatening electrical disturbance.

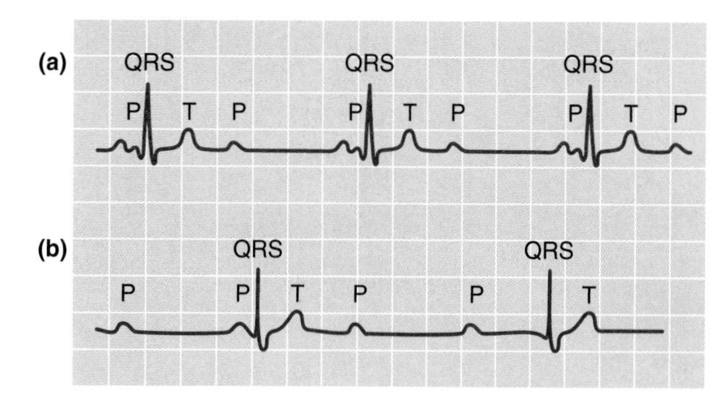

FIGURE A.5

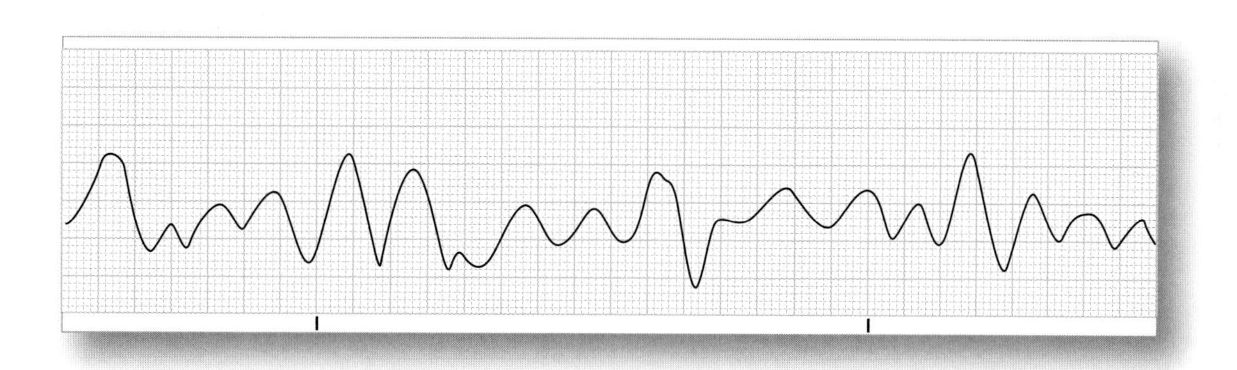

FIGURE A.6 (From *Arrhythmia Recognition: The Art of Interpretation*, courtesy of Tomas B. Garcia, MD.)

B

Measuring Pulmonary Capacities and Volumes

The spirometer pictured in **FIGURE B.1** is almost never used clinically anymore, but it is the basis for the classic illustration of pulmonary volumes and capacities. Briefly, this is how a water-filled spirometer works. The upper, inverted bell is air-filled and floats on water, while snugly fit over a water-filled bell. As the subject inhales and exhales into the tube, pressures change within the upper bell such that it rises on exhalation and lowers on inhalation. As this is happening, the upper bell is connected to a pen that records the entire cycle of breathing. Please note that when the upper bell descends, it causes the pen to *ascend* and vice versa. The graph that results from these tests is shown in **FIGURE B.2**.

Several pulmonary volumes are of importance. Tidal volume is the amount of air inhaled or exhaled during quiet breathing—usually about 500 mL. At the end of a quiet inspiration, however, you still have the ability to inspire more air; this is the *inspiratory reserve volume*. Similarly, at the end of a quiet expiration, you can expel even more air—the *expiratory reserve volume*. However, there is a volume of air in the lung you can never exhale—the *residual volume*. Spirometry cannot measure this volume, and no amount of effort will allow you to expel it. This is the air trapped in alveoli by closing pressures. Residual volume serves to keep the alveoli inflated and, as we find in our investigations of acid-base balance, it helps maintain a constant pH environment within the lung by retaining a constant level of carbon dioxide (CO_2).

Pulmonary capacities are the sum of two or more volumes. *Inspiratory capacity* is the sum of tidal volume and inspiratory reserve volume, thus equaling our total capacity to inspire. *Expiratory capacity* is the sum of tidal volume and the expiratory reserve volume. *Vital capacity* is the entire respiratory excursion: tidal volume + inspiratory reserve volume + expiratory reserve volume. *Total lung capacity* is tidal volume + inspiratory reserve volume + expiratory reserve volume + residual volume. Notice that this capacity cannot be measured by spirometry because it includes residual volume.

A clinical use of spirometry comes with the forced expiratory volume (FEV) test. Instead of a quiet vital capacity measurement, the subject is required to force the inspiration and expiration. This may take a bit of "coaching" because this is a voluntary action, not an autonomic function. After a maximum inspiration to determine the inspirational capacity (IC), the subject is asked to exhale as much as he or she can, as fast as possible. The volume of air exhaled in the first second (FEV_1) gives a clinician a great deal of information concerning the mechanics of a person's lung.

A person whose airways are constricted—from asthma or a fibrotic lung, for example—will have increased resistance to flow and will be unable to exhale the same volume of air in the first second as someone with no airway constriction. The FEV_1 will be reduced. If the

FIGURE B.1

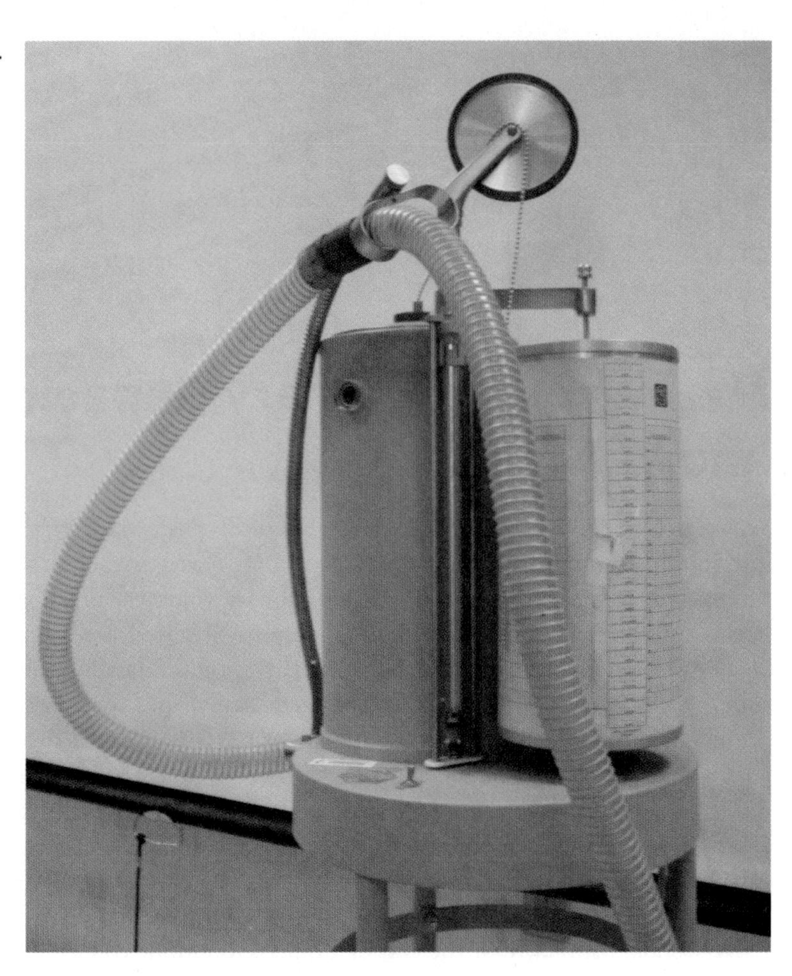

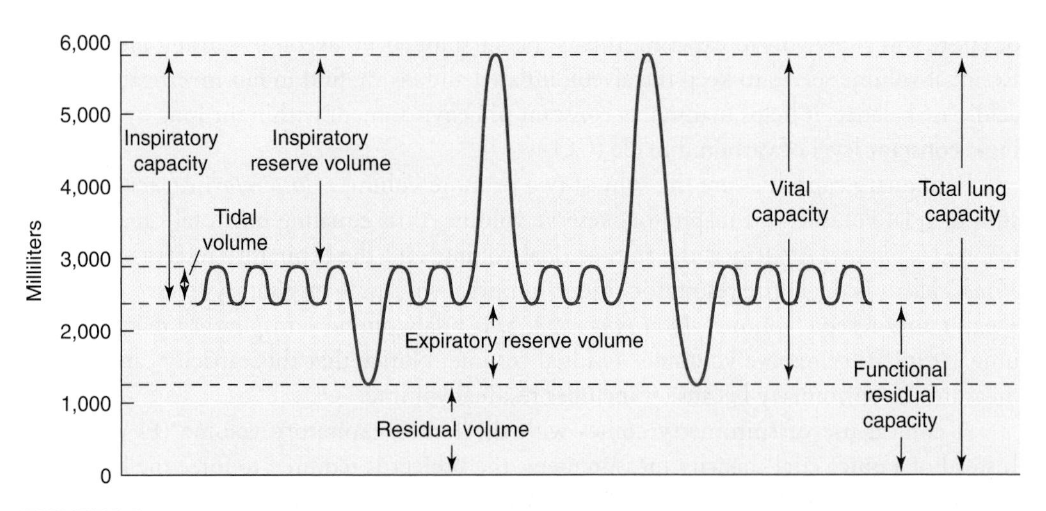

FIGURE B.2

test is extended to three seconds (FEV_3), a standardized amount will eventually be exhaled. This indicates an increased resistance to flow without airway closure.

Someone with emphysema, however, has less elasticity in the alveoli. It is the recoil of the elastic fibers that generally forces air from the alveoli into the airways. When elastic recoil is reduced, less air leaves the alveoli, and, therefore, less air leaves the lungs.

Furthermore, increasing pressure during a forced expiration, as during an FEV_1 test, will cause airway closure, trapping air within the alveoli. Therefore, FEV_1 in an emphysemic patient will be much reduced and will not improve much during FEV_3 because the airways have closed.

A simple, noninvasive breathing test can be effective in determining changes in pulmonary function.

C

Measuring Glomerular Filtration Rate and Clearance

Glomerular filtration rate (GFR) is the volume of plasma that is filtered by the nephron per unit of time. This can be an important clinical measurement of nephron function. Let's look at how it is calculated.

First, we need to find a molecule that is freely filtered at the Bowman's capsule but neither reabsorbed nor secreted by the tubule. The classic molecule used is inulin (note that this is *inulin* the polysaccharide NOT *insulin* the pancreatic hormone). Inulin is not a native component of plasma and must be injected for this measurement.

We can easily measure the inulin concentration excreted in urine. If we also measure urine volume over time, we can obtain:

$$\text{Urine inulin concentration} \times \text{urine volume/time}$$

Because the volume of plasma filtered per unit of time equals GFR, we can use the concentration of inulin in plasma to calculate the amount of plasma filtered.

$$\text{Plasma concentration of inulin} \times \text{GFR} = \text{urine inulin concentration} \times \text{urine volume/time}$$

Reorganizing that equation:

$$\text{GFR} = \frac{\text{urine concentration of inulin} \times \text{urine volume per unit time}}{\text{Plasma inulin concentration}}$$

So, by measuring urine volume over time along with urine inulin concentration and plasma inulin concentration, you can determine the rate at which plasma is filtered. If not all nephrons are working, GFR will be lower than normal—that is, you will see less inulin in the urine than would be expected in someone with healthy kidneys, because damaged kidneys are not as efficient at filtering plasma. Therefore, GFR can be a method for determining the health of nephrons and general kidney function.

Clinically, inulin is rarely used. Instead, creatinine, a naturally occurring molecule that is also freely filtered, is generally used for this calculation. Because creatinine is usually present in plasma in constant amounts, it is easy enough to measure it in urine to see if there's a discrepancy. Some creatinine is secreted, so GFR is slightly overestimated using this method. The obvious advantage of this method, however, is that it negates the need to inject inulin and measure plasma inulin. Only noninvasive urine measurements of creatinine are required.

Clearance is the amount of substance "cleared" from the plasma per unit time. In the case of inulin, because all plasma inulin ends up in urine, the clearance will equal GFR.

Every substance, however, has a clearance rate that can be calculated. Glucose is freely filtered at the Bowman's capsule, but because it is actively reabsorbed in the nephron, its clearance rate is zero. The estimate of GFR using creatinine is actually a measurement of creatinine clearance. It just happens that creatinine clearance is a good estimate of real GFR.

Glossary

α-adrenergic receptor A seven-pass, G-protein, coupled receptor that binds norepinephrine more avidly than epinephrine.

β-hydroxybutyrate A ketone body formed from fatty acid metabolism in the liver.

Acetoacetate A keto acid produced in the liver as a by-product of fatty acid breakdown.

Acetylcholine A neurotransmitter of the neuromuscular junction and the autonomic nervous system formed from acetyl CoA and choline.

Acrosomal cap The cap of the mature sperm, derived from the Golgi apparatus, containing the enzymes hyaluronidase and acrosin.

Actin A protein within muscle that interacts with myosin to form cross-bridge cycles.

Action potential A membrane depolarization that reaches threshold opens ion channels and propagates an electrical signal over a distance.

Activation Opening of activation gates, particularly of the sodium channel.

Adrenal medulla The portion of the adrenal gland that releases epinephrine.

Adrenocorticotrophic hormone (ACTH) Hormone released from the anterior pituitary in response to hypothalamic release of CRH. ACTH targets the adrenal gland, causing the release of cortisol.

Afferent arterioles The arterioles that carry blood into the glomerular capillaries of the nephron.

Airway closure Collapse of the respiratory or terminal bronchioles.

Aldosterone A steroid hormone released from the adrenal gland in response to plasma K^+ or angiotensin II stimulus. It acts at the distal convoluted tubule to increase Na^+ reabsorption into the blood.

Alveoli The primary gas-exchange surface in the lung.

Amacrine cells Communicating cells between the rod cells and the optic nerve.

Analgesic A substance that reduces pain.

Anatomical dead space The conducting airways that do not exchange gases.

Angiogenesis The proliferation of blood vessels.

Angiotensin II A hormone produced by the coordinated action of several organs. The original substrate is angiotensinogen, produced by the liver. This is cleaved by renin from the kidney to become angiotensin I, which is enzymatically converted to angiotensin II in the lung. It is a powerful vasoconstrictor, causes the release of ADH and aldosterone, and is important in maintaining blood volume.

Anterolateral pathways Sensory pathways, also known as the spinothalamic tracts, that sense pain and temperature.

Antibodies Proteins produced by B-lymphocytes that protect against infection.

Antidiuretic hormone (ADH) A hormone produced by neurons of the hypothalamus and release via the posterior pituitary in response to an increase in blood osmolarity. It targets cells of the collecting duct in the kidney to stimulate water reabsorption.

Aorta The great vessel arising from the left ventricle and serving the systemic circulation.

Apoproteins The lipid-binding protein portion of lipoproteins.

Aquaporins Water channels that are inserted into the membrane of tubular cells in response to ADH.

Arachnoid granulations Outcroppings of the arachnoid membrane where cerebrospinal fluid enters the dural sinuses.

Arachnoid membrane One of the dural layers of the brain, between the pia and the dura mater.

Arterial baroreceptors Pressure receptors located in the carotid sinus and aortic arch.

Asthma Disease caused by narrowing of the terminal bronchioles.

Astrocytes Glial cells in the brain modulating neuronal nutrient uptake. They also form part of the blood-brain barrier.

Atenolol A drug that blocks β1-adrenergic receptors.

Atherosclerosis Inflammation and remodeling of the vascular wall involving macrophages and fat deposits.

ATPase A protein that uses hydrolysis of ATP to perform its primary function.

Atria The two upper chambers of the heart that receive blood from the superior and inferior vena cavae (right) or from the pulmonary circuit (left).

Atrial naturetic peptide (ANP) A hormone released from the right atria in response to stretch. It stimulates sodium loss at the kidney and subsequently water loss.

Atrioventricular valves Valves between the atria and the ventricles.

Atrium One of the two upper chambers of the heart (*see* Atria).

Atropine A drug that blocks muscarinic cholinergic receptors.

Auditory cortex Portion of the cortex that receives input from the auditory system.

Autophagy The destruction and reabsorption of damaged cellular organelles. The word literally means "self-eating."

A-V node The atrioventricular node lies between the sinoatrial node and the bundle branch fibers.

Axoplasmic transport Transport within the neuron that goes either from the cell body to the synaptic bulb or from the synaptic bulb to the cell body.

B cells B lymphocytes.

B lymphocytes Immune cells produced in the bone marrow.

β-adrenergic receptor A seven-pass G-protein coupled receptor that binds epinephrine more avidly than norepinephrine.

Basilar membrane A portion of the organ of Corti that carries vibration.

Basophils White blood cells that store histamine.

Beta oxidation The process of fat metabolism in the mitochondria, where fatty acids are broken down, two fatty acids at a time.

Bicarbonate equation $CO_2 + H_2O \leftrightarrow H_2CO_3 \leftrightarrow HCO_3 + H^+$.

Biceps Muscle of the upper arm, contraction of which raises the lower arm.

Bile A substance formed by the liver and stored in the gallbladder that is important in fat digestion.

Biphosphoglycerate (BPG) A product of red blood cell metabolism that reduces hemoglobin's affinity for oxygen.

Bladder Muscular organ that stores urine.

Blastocyst The developing zygote after the morula stage.

Blood-brain barrier The separation of blood capillaries from the cerebrospinal fluid of the brain. This is largely managed through tight junction between capillary endothelial cells and astrocytes.

Blood-testis barrier The barrier produced by the Sertoli cells that separates the spermatocytes from the systemic blood supply.

Bone resorption Remodeling of bone, which decreases the amount of bone in some areas. This can happen during reclamation of bone calcium.

Botulinum toxin A bacterial toxin that inhibits the movement of acetylcholine vesicles in the synaptic bulb.

Bowman's capsule The capsule that holds the filtrate of the glomerular capillaries.

Boyle's law Within a closed cavity, volume and pressure are inversely related.

BPG Biphosphoglycerate.

Bradykinin A hormone that causes vasodilation.

Bronchioles Small conducting airways of the lung that are surrounded by smooth muscle.

Bronchoconstriction Smooth muscle contraction around bronchioles.

Bronchodilation Smooth muscle relaxation around bronchioles.

Bulbourethral glands Glands of the male reproductive system that provide an alkaline mucus to semen.

Bundle branch fibers Fibers in the heart that carry the ventricular action potential down the interventricular septum to the Purkinje fibers.

Calmodulin An intracellular calcium-binding protein.

Calyx Portion of the kidney that receives urine from all of the collecting ducts of the nephrons.

Carbon monoxide A neurotransmitter.

Carbonic anhydrase The enzyme that facilitates the transitions of the bicarbonate equation.

Cardiac output The amount of blood pumped out of the heart per minute.

Central chemoreceptors Chemoreceptors in the brain that regulate breathing in response to CO_2 concentrations.

Cephalic phase The first phase in digestion, which initiates digestive responses in the mouth even before eating begins.

Cerebellum The hindmost portion of the brain that coordinates movement.

Cerebrospinal fluid A specialized fluid produced by ependymal cells that circulates throughout the brain.

Cerebrum The uppermost portion of the brain, responsible for our conscious actions.

Cervix The entry to the uterus from the vaginal canal.

cGMP A cyclic form of GTP, analogous to cAMP.

Chaperone protein Proteins that assist in the folding of other proteins within the cytoplasm.

Chemokines A special form of cytokine that acts as a leukocyte attractant.

Chemosensors Cells that respond to specific chemicals.

Chief cells Cells in the gastric glands of the stomach that make pepsinogen.

Cholecystokinin (CCK) A hormone released from the duodenum that inhibits stomach motility and acts at the gallbladder and pancreas to increase their secretions.

Cholesterol An essential portion of the plasma membrane, the parent molecule of steroid hormones, and a nutrient in food.

Chorionic villi Extensions of the chorionic membrane that will form the primary exchange mechanism of the placenta.

Choroid plexus Blood vessels surrounding the ependymal cells.

Chylomicrons A collection of fatty acid, cholesterol, apoproteins, and lipoproteins made by the enterocytes of the digestive track.

Chyme Partially digested food formed within the stomach.

Citric acid cycle The rearrangement of citric acid to yield NADH, $FADH_2$, GTP, and 2 molecules of CO_2.

Classical pathway A cascade of blood proteins that recognize antibody-linked bacteria.

Coagulation Clotting of blood to stop its flow.

Cochlea Shell-shaped structure in the ear that houses the cells responsible for hearing.

Coitus Sexual intercourse.

Collecting duct The last portion of the nephron, important for the concentration of urine.

Colony-stimulating factor A factor within the bone marrow, stimulated by IL-6 and TNF-α, that increases the production of white blood cells.

Complement Circulating blood proteins that are important for targeting bacteria for destruction by the immune system.

Compliance Change in volume/change in pressure. Compliance is a greater increase in volume for any given change in pressure.

Conducting arteries Large arteries.

Conformational change Change in the shape of a protein from one semi-stable state to another.

Contractility Force of contraction of the heart.

Corona radiata A layer of cells that surround the ovum.

Corpus cavernosum Erectile tissue within the penis.

Corpus luteum The hormonally follicular active cells that remain after ovulation.

Corpus quadrigemini The superior and inferior colliculi, located in the midbrain and responsible for central reflexes to light and sound.

Corpus spongiosum Erectile tissue of the penis.

Corticobulbar tract Portion of the corticospinal tract that innervates the cranial nuclei and serves to control voluntary motion above the spinal cord.

Corticospinal tract The primary motor nerve tract for voluntary motion.

Corticotropin-releasing hormone (CRH) A hormone released from the hypothalamus to the portal circulation of the anterior pituitary. It stimulates the release of ACTH.

Cortisol An important steroid hormone for glucose homeostasis. It also has suppressant effects on the immune system.

Cranial nerves The nerves that arise above the spinal cord and serve the head and other organs.

C-reactive protein A protein made by the liver that tags microorganisms for destruction by macrophages or neutrophils.

Creatine phosphate A high-energy molecule that is stored and used in skeletal muscle.

Cribiform plate A portion of the ethmoid bone where cranial nerves exit the brain and project to the face.

Cross-bridge cycle The interaction of actin and myosin, including ATP hydrolysis, hinge movement, and release.

Crossed extensor reflex A coordinated set of responses that allow withdrawal of one limb and extension of another.

Cyclic AMP An intracellular signaling molecule formed from ATP by adenylate cyclase.

Cytokines A small signaling molecule.

Cytotoxic T cells T cells that target infected self cells.

Dantrolene sodium A drug that blocks the skeletal muscle calcium release channel of the sarcoplasmic reticulum.

Defecation Elimination of feces.

Dendrite A projection off the cell body of a neuron that receives input from other neurons.

Dendritic spine A small projection off of a dendrite.

Depolarization Change in membrane potential to a less negative voltage.

Dermatome A body section served by a single pair of spinal nerves.

Diabetes mellitus Metabolic dysregulation stemming from poor uptake of glucose from circulation. Different types of diabetes occur as a result of various causes (autoimmune, gestational, and metabolic factors). All types are frequently associated with more complex disease complications.

Diabetic neuropathy Nerve damage caused by high circulating blood sugar.

Diacylglycerol An intracellular signaling molecule produced from membrane phospholipids by phospholipase C.

Diaphragm Skeletal muscle that divides the thoracic cavity from the abdominal cavity. It is the primary muscle in quiet breathing and is innervated by the phrenic nerve.

Diarrhea Frequent, watery feces.

Diastole Period between heartbeats.

Diffusion Movement of molecules from an area of higher concentration to an area of lower concentration.

Diploid Having two sets of chromosomes, one maternal and one paternal.

Distal convoluted tubule The portion of the nephron just prior to the collecting duct. It is sensitive to aldosterone and parathyroid hormone.

Diuretic Any substance, either physiological or pharmacological, that increases urine output.

Diurnal An event that occurs daily.

DNA Deoxyribonucleic acid chains located in the nucleus of the cell that encode our genetic material.

Dorsal column pathways Sensory pathways responsible for sensing fine touch, proprioception, and vibration.

Downregulation In regards to receptors, the systematic removal of receptors from the plasma membrane.

Dual innervation Innervation of an organ by both the parasympathetic and sympathetic nervous systems.

Ductus arteriosus The connection between the pulmonary trunk and the aorta that exists before birth.

Ductus deferens Also known as the vas deferens, it connects the epididymis to the seminal vesicles.

Ductus venosus The vessel that carries oxygenated blood from the placenta to the fetal heart.

Duodenum The first part of the small intestine.

Dura The meningeal membranes covering the brain and spinal cord.

E-C coupling The events that link the electrical potentials at the sarcolemma to the cross-bridge cycling in cardiac muscle.

ECG Electrocardiogram.

Eclampsia A pregnancy complication that causes seizures in the mother and can cause separation of the placenta from the uterine wall before birth.

Efferent arterioles Arterioles that leave the glomerular capillaries and enter the peritubular capillary bed around the nephron.

Ejaculation The explosive ejection of semen from the penis.

Electrocardiogram (ECG) Measurement of the electrical changes in the heart as a whole via surface electrodes.

Electron transport chain A chain of proteins located in the inner membrane of the mitochondria that are essential for the production of ATP during oxidative phosphorylation.

Embryo A fertilized egg, during the first trimester.

Emphysema Disease characterized by the loss of elastic recoil of the lung.

Endogenous pyrogens Endogenous molecules that produce fever, including IL-1, IL-6, and TNF-α.

Endothelial cell Hormonally active cells that line blood vessel lumen.

Endothelin A substance produced by endothelial cells that causes constriction of vascular smooth muscle.

Enteric nervous system The nervous system that controls the circular and longitudinal muscle of the digestive tract.

Eosinophils A type of white blood cell that targets parasites and some infectious diseases.

Ependymal cells Cells within the ventricles of the brain and spinal cord that make cerebrospinal fluid.

Epididymus Site of maturation of sperm.

EPSP Excitatory postsynaptic potentials.

Erectile tissue Tissue of the penis containing blood spaces that serve to cause erection.

Erythrocytes Red blood cells.

Erythropoietin Hormone produced and released by the kidney that stimulates the production of red blood cells.

Estrogen Female hormone produced by the granulosa cells.

Excitation-contraction coupling The process that joins the electrical depolarization of a muscle membrane and the interaction of actin and myosin to cause muscle contraction.

Exocrine pancreas Cells of the pancreas that make digestive fluids and enzymes and release them to the duodenum through ducts.

Exocytosis Movement of molecules out of the cell through fusion of a vesicle with the plasma membrane.

External intercostal muscles Inspiratory muscles used during higher volume breathing.

Facilitated diffusion Diffusion that requires movement through a protein or the assistance of a protein. Facilitated diffusion always moves down a concentration gradient.

Factor X Part of the coagulation pathway that is common to both the intrinsic and extrinsic pathways.

$FADH_2$ A molecule that transports electrons from the citric acid cycle to the electron transport chain to allow ATP production during oxidative phosphorylation.

Fatty acids Breakdown product of fats that can be taken up by adipose tissue for storage or other tissues for metabolism.

Fertilization The unification of a haploid sperm and the ovum.

Fetal circulation The unique circulatory pattern of the fetus.

Fetus The term used for an unborn infant during the second and third trimester.

Fever Increase in body temperature that is maintained by the hypothalamus.

Fibrin Part of the coagulation pathway that helps to form the clot over a wound.

Fibrinogen A precursor to fibrin.

Filling time Amount of time for filling of the atria and ventricles.

Fimbria Finger-like structures at the ends of the uterine tubes that first accept the egg after ovulation.

Fission Cleavage of a mitochondrion into two separate mitochondria.

Follicle-stimulating hormone The hormone released from the anterior pituitary as a result of.

Follicular cells The cells that surround the developing oocyte.

Frank-Starling law The principle that the heart will increase contractility in proportion to the amount of blood that returns to the heart.

Frontal lobe The most anterior portion of the cerebrum, responsible for rational thought, moral judgment, and decision making.

Fructose A monosaccharide.

Fusion Combination of two separate mitochondria into one mitochondrion.

GABA An inhibitory neurotransmitter.

Galactose A monosaccharide.

Gametogenesis The maturation of gametes.

Gap junctional proteins Proteins that directly connect the cytoplasm of two cells.

Gastric glands Glands located in the stomach that contain chief cells, G cells, and parietal cells.

Gastric phase The second stage of digestion that begins with food entering the stomach.

Gastrin A hormone produced by the G cells that stimulates gastric motility and secretion from parietal and chief cells.

Gastrocnemius The calf muscle of the leg.

Ghrelin A hormone produced peripherally by the stomach and centrally by the brain. It stimulates hunger and growth hormone release.

Glomerular capillaries Capillaries within the Bowman's capsule that provide the filtrate that will become urine.

Glomerular filtration rate The rate at which a substance is cleared from the blood by the kidney.

Glucagon Hormone released from α-cells of the pancreas in response to several signals, including low blood sugar. The target for this tissue is the liver, where it stimulates glycogenolysis.

Gluconeogenesis The creation of new glucose from deaminated amino acids.

Glucose A monosaccharide.

Glucose-6 phosphatase Enzyme that dephosphorylates glucose-6 phosphate and allows it to be transported out of the cell. This enzyme is present in liver but not in skeletal muscle.

Glucose-dependent insulinotrophic peptide (GIP) A hormone released from the duodenum in response to nutrients that causes reduced gastric motility and reduced secretions from chief cells and parietal cells.

Glucose transporter type 4 (GLUT4) An isoform of glucose transporter. It is located in skeletal and cardiac muscle and adipose tissue.

Glycine An inhibitory neurotransmitter.

Glycogen phosphorylase Enzyme that stimulates the breakdown of glycogen into glucose.

Glycogenolysis Hydrolysis of glycogen.

Glycolysis Enzymatic rearrangement of glucose to yield pyruvate and 2 ATP molecules.

Gonadotrophin-releasing hormone (GnRH) A hormone produced within the hypothalamus that stimulates production of LH and FSH in the pituitary.

Graded potential A change in membrane potential that does not reach threshold and dies away across distance and time.

Granulosa cells Cells of the follicle that respond to FSH and produce aromatase.

Growth hormone (GH) Hormone released from the anterior pituitary in response to GHRH release. It targets adipose tissue, muscle, and liver.

Growth hormone–releasing hormone (GHRH) A hormone produced in the hypothalamus and released to the anterior pituitary via the portal circulation.

GTP A high-energy molecule like ATP that contains guanine instead of adenine.

Gustatory cortex The portion of the cerebral cortex that receives sensory information from the taste buds.

Gyri Raised folds of the brain.

Haploid Containing one half the number of chromosomes present in a somatic cell.

HCN channel Hyperpolarization activated and cyclic nucleotide gated channel; one of the channels that maintain the SA nodal rhythm in the heart.

HDL High-density lipoproteins.

Helper T cells T cells that bind to MHC-antigen presenting cells and attract cytotoxic T cells.

Hemoglobin A four-subunit protein within red blood cells that is responsible for carrying oxygen in the blood.

Hepatic portal circulation The circulation that goes from the stomach and intestines to the liver.

Hepatocytes Cells of the liver.

Hippocampus A portion of the limbic system important in the formation of memories.

Histamine A molecule released from mast cells that stimulates vascular permeability and swelling.

Homeostasis Maintenance of a constant internal environment.

Horizontal cells Communicating cells between the rods and the optic nerve.

Human chorionic gonadotrophin (HCG) A hormone produced by the embryo/placenta and acts as an LH substitute, maintaining the corpus luteum.

Hyaluronidase A proteolytic enzyme present in the acrosomal cap in the spermatozoa and in the syncytial trophoblast of the developing embryo.

Hydrolysis Breakdown of a molecule that requires water.

Hydrophilic "Water-loving" molecules that have an affinity for water and are polar.

Hydrophobic "Water-hating" molecules that do not have an affinity for water and are generally nonpolar.

Hydrostatic pressure Pressure exerted by a fluid along a column.

Hyperlipidemia Excess lipids circulating in the blood.

Hyperpolarization Change in membrane voltage to a more negative membrane potential.

Hypertension Increased systolic and/or diastolic blood pressure.

Hypothalamus An area of the diencephalon below the thalamus that regulates autonomic nervous system function, hormone release, temperature control, appetite, thirst, osmoregulation, and some unconscious muscle motions, particularly those of the face.

IL-6 A cytokine with a myriad of functions, including a role in fever generation.

Immunity Protection against disease.

Inactivation Nonconducting ion channel state that is not closed.

Inferior vena cava The vessel that returns blood to the right atrium from all vessels inferior to it.

Inflammation Tissue response to injury.

Innate immunity Nonspecific immunity.

Inositol phosphate 3 An intracellular signaling molecule formed by phospholipase C from membrane phospholipids. It binds to a receptor on the endoplasmic or sarcoplasmic reticulum and causes calcium release.

Insulin A molecule produced by β cells of the pancreas that binds to tyrosine kinase receptors on the membranes of insulin sensitive cells. It is important for the production and movement of glucose transporters into the plasma membrane.

Insulin-like growth factor (IGF-1) A hormone produced by the liver in response to growth hormone, which promotes bone and cartilage growth.

Intercalated cells Cells of the collecting duct important for acid-base balance.

Internal intercostal muscles Expiratory muscles used during forced exhalation.

Interneuron A neuron located between two other neurons. It frequently serves to distribute the action potential to several neurons.

Internodal spaces Also called the nodes of Ranvier, these are spaces of nonmyelination in a myelinated nerve.

Intestinal phase Third phase of digestion that occurs in the intestine.

IPSP Inhibitory postsynaptic potential.

Isoform One type of a protein or enzyme.

Isometric contraction Contraction of skeletal muscle, in which the length of the muscle remains the same.

Isotonic contraction Contraction of skeletal muscle against a constant load.

Juxtaglomerular apparatus Located between the distal convoluted tubule and the proximal convoluted tubule, the juxtaglomerular apparatus monitors flow in the kidney and releases renin when flow is low.

K^+ leak channel A nonvoltage dependent potassium channel that is generally open and is responsible for the majority of the resting membrane potential.

Kinase A protein that phosphorylates another protein.

Kupffer cells Resident tissue macrophages of the liver.

Labeled line The path between a sensory receptor and the sensory cortex.

Labia A part of the female external genitalia.

Lacrimal glands Glands located at the lateral portion of the eye that produce tears.

Lactation Human milk production.

Lacteal Lymphatic capillary within the villi of the intestine.

Lactose Milk sugar.

Laminar flow Flow of a fluid that is smooth and unidirectional.

LDL Low-density lipoprotein.

Lectin pathway The complement pathway activated by mannose present in bacterial walls.

Leptin Hormone released from adipose tissue and binds to receptors in the hypothalamus to promote a sense of satiety.

Leydig cells Testosterone-producing cells of the male reproductive system.

Ligand-gated ion channels An ion channel that opens in response to the binding of a ligand.

Linear acceleration Acceleration in the forward or backward direction.

Lipolysis Hydrolysis of fats.

Lipophilic Able to dissolve or bind to lipids. An example is a nonpolar protein or portion of a protein that resides within the fatty acid layer of the membrane.

Lipopolysaccharide Molecule found in the cell walls of bacteria.

Lipoprotein lipase Enzyme located in capillary walls and on the surface of adipose tissue that cleaves fatty acids from lipoproteins.

Locus coeruleus An area of the pons that synthesizes epinephrine.

Long-chain fatty acids Fatty acids with 12 or more carbons in their tails.

Loop of Henle The portion of the nephron between the proximal convoluted tubule and the distal convoluted tubule.

L-type calcium channel A voltage-gated calcium channel with a relatively long open time.

Luteinizing hormone Hormone released from the anterior pituitary in response to GnRH release.

Lymph nodes Nodes located along the lymphatic circulation with large collections of B and T lymphocytes.

Lymph vessels Vessels containing lymphatic fluid collected from capillary beds and circulated back to venous circulation.

Lysozyme An antibacterial enzyme present in tears.

Macula densa Cells located at the ascending loop of Henle that sense the Na^+ concentration of the filtrate.

Mannose-binding protein Produced by the liver, a protein that tags bacteria for destruction by the immune system.

Mass movement Rapid forward movement of chyme in the large intestine during a cessation of segmentation.

Mean arterial pressure Pressure within the arteries; calculated by MAP = pressure during diastole + (pressure during systole – pressure during diastole)/3.

Mechanoreceptors Receptors that respond to mechanical stimulation, like touch or vibration.

Medulla The portion of the brainstem that arises from the spinal cord.

Megakaryocytes A white blood cell that gives rise to platelets.

Memory B cells B cells that retain the "memory" of an antigen and can quickly produce antibodies against it when it is next encountered.

Menopause The cessation of ovulation and menses, generally occurring in the fifth decade.

Menses The loss of the endometrial layer that occurs once during a 28-day cycle.

Metformin An antidiabetes drug that antagonizes the action of glucagon and increases insulin receptor sensitivity.

MHC protein Major histocompatibility protein that serves as a cellular marker of self.

Microglial cells A glial cell that operates like a resident brain macrophage.

Microvilli Tiny fingerlike projections off of villi within the intestine.

Micturition Urination.

Midbrain A specific area of the brain that includes the corpus quadrigemini and the cerebral peduncles.

Minute ventilation The volume of air inhaled per minute.

Monocytes A form of white blood cell that can transform into macrophages.

mRNA Messenger RNA transcribes DNA and participates in translation at the ribosome.

mtDNA Mitochondrial DNA residing within the mitochondria and unique from nuclear DNA.

Muscarinic receptors Receptors located on target organs of the autonomic nervous system that bind acetylcholine.

Muscle contraction Shortening of muscle tissue to generate force.

Myelin Fatty substance produced by oligodendrocytes in the central nervous system and Schwann cells in the peripheral nervous system.

Myoglobin A single-unit protein that resembles hemoglobin in structure. Myoglobin is found in red muscle and the heart, where it serves as storage for oxygen.

Myosin A muscle protein consisting of a head region with an enzymatic ATPase, hinge region, and tail region. Myosin is important in muscle contraction and relaxation.

Myosin light chain kinase The protein that phosphorylates myosin light chain in smooth muscle.

Na^+-Ca^{2+} exchanger A protein that moves intracellular Ca^{2+} out of a cell in exchange for extracellular Na^+ coming into the cell.

Na^+/K^+ ATPase An important active transporter that moves Na^+ out of the cell and K^+ into the cell, each against their concentration gradient.

NAD Nicotinamide adenine dinucleotide, the oxidized form of NADH.

Natural killer cells Cells that bind to abnormal antigens on other cells.

Negative feedback loop Regulatory mechanism where the product inhibits an earlier step in its own production.

Neuromuscular junction The anatomical location where a nerve terminal synapses with skeletal muscle.

Neuron The primary cell of electrical communication.

Neutrophil A form of white blood cell capable of ameboid movement.

Nicotinic receptors Receptors located at the neuromuscular junction and between the preganglionic and postganglionic neurons of the autonomic nervous system.

Nitric oxide A gaseous signaling molecule formed from arginine by nitric oxide synthase.

Norephinephrine A neurotransmitter.

Novocaine A derivative of lidocaine.

Occipital lobe A posterior lobe of the brain that houses the visual cortex.

Olfactory cortex A portion of the brain that receives signals from the olfactory cells of the nose.

Oligodendrocytes Myelin-producing cells of the central nervous system.

Oocytes Gametes of the female, suspended in meiosis.

Organ of Corti Portion of the ear, within the cochlea, that transduces sound into action potentials.

Osmoreceptors Receptors that sense the osmolarity of a body fluid.

Osmosis The movement of water "down" its concentration gradient. It is also correct to say that it is the movement of water toward a higher solute concentration.

Osmotic pressure The pressure exerted by osmotically active particles.

Otoliths Calcium carbonate crystals within the vestibular system that contribute to our sense of balance.

Ovulation The release of the oocyte from the follicle.

Oxygen-hemoglobin dissociation curve A graph that describes the percentage of hemoglobin saturated with oxygen, at any given partial pressure of oxygen.

Oxytocin Hormone released from the posterior pituitary that is important in childbirth and milk let-down.

P wave The wave of the ECG that designates the atrial depolarization.

Pacemaker cells The cells of the SA node that initiate depolarization.

Pain receptors Nociceptors that respond to noxious stimuli: chemical, temperature, or mechanical.

Pancreatic amylase Enzyme released from the pancreas that contributes to carbohydrate digestion.

Pancreatic lipase Enzyme released from the pancreas that contributes to fat digestion in the intestine.

Papillary muscles Muscles connected to the A-V valves.

Paracrine Local cell-signaling mechanism that does not involve release of factors into the blood supply.

Parasympathetic nervous system A branch of the autonomic nervous system, sometimes called the system of "rest and digest."

Parathyroid hormone A hormone produced by the parathyroid glands that is important in the regulation of plasma calcium.

Parietal cells Cells of the stomach, located within the gastric glands, that produce HCl.

Parietal pleura The portion of the pleural membrane that is continuous with the body wall.

Parturition Childbirth.

Pattern generators Cells that create repetitive patterns of action potentials.

Pepsinogen A proenzyme produced by the chief cells that is converted to pepsin by HCl.

Peptidases Enzymes that cleave peptides into amino acids or smaller peptides.

Perforins A protein complex that forms pores in targeted cells causing their death.

Peristalsis Forward movement of chyme resulting from the coordinated action of circular and longitudinal muscle.

Peritubular capillaries Capillaries that surround the nephron and take up water, ions, and nutrients from the extracellular space and return it to circulation.

Perivascular space Space around the blood vessels.

pH The negative log of the H^+ concentration of a solution.

Phosphofructokinase An enzyme in the glycolytic cycle.

Phospholamban A protein that regulates the function of the Ca^{2+} ATPase of the sarcoplasmic reticulum.

Phospholipase C A bound enzyme that cleaves phospholipids into diacylglycerol and inositol phosphate 3.

Phospholipid A membrane molecule with fatty acid tails and phosphate heads.

Phrenic nerve The nerve that innervates the diaphragm.

Physiological work Any use of ATP.

PI-3 kinase Phosphatidylinositol 3 kinase; important for moving intracellular vesicles of glucose transporters to the plasma membrane.

Pia The innermost layer of dura, closest to the brain and spinal cord.

Placenta The structures of fetal origin that will allow transport of nutrients from the mother to the embryo/fetus.

Platelets Portions of shed cytoplasm from megakaryocytes.

Podocytes Cells that partially cover the glomerular capillaries.

Polar bodies Waste products of meiosis containing nuclear DNA but little cytoplasm.

Porphoryin ring The moiety in hemoglobin that contains Fe^{2+}.

Positive feedback loop Regulatory mechanism where the product facilitates an earlier step in its own production.

Postganglionic The neurons that arise from an autonomic ganglia and end at target organs.

Potentiation To augment or increase a signal.

Power stroke The part of myosin-actin interaction that yields muscle shortening.

Preganglionic The neurons that arise from the spinal cord and end at the autonomic ganglia.

Presynaptic facilitation Action on a presynaptic nerve that increases its release of neurotransmitter.

Presynaptic inhibition Action on a presynaptic nerve that decreases its release of neurotransmitter.

Primordial follicles The first stage after recruitment of eggs from the ovarian egg nest.

Principal cells Cells of the collecting duct that are important for the transport of Na^+ from the filtrate into the extracellular fluid.

Progenitor cells Parent cells, or stem cells.

Progesterone Female hormone produced by the corpus luteum and by the placenta during pregnancy.

Proprioception Sense of body position.

Prostacyclins A type of prostaglandin made by endothelial cells.

Prostaglandin Lipid compounds derived from fatty acids that have important effects in the body, including regulating contraction and relaxation of smooth muscle tissue.

Prostate gland Accessory male reproductive gland that contributes an acidic fluid to semen.

Protein kinase A A protein kinase that is activated by cAMP.

Protein kinase C A protein kinase that is activated by diacylglycerol.

Protein synthesis The formation of proteins from amino acids using ribosomes, tRNA, mRNA, and ATP.

Proteolysis Breakdown of protein. This can occur during digestion but can also be a regulatory process.

Proximal convoluted tubule The portion of the nephron that lies just past the Bowman's capsule where many ions and nutrients are absorbed along with water.

Pulmonary hypoxic vasoconstriction Constriction by pulmonary blood vessels in response to low O_2 levels.

Purkinje fibers Fibers in the heart that carry the ventricular depolarization up the ventricular walls.

QRS interval The time in an ECG between the beginning of the Q wave and the end of the S wave. This encompasses all of ventricular depolarization.

Q-T interval The time in an ECG between the beginning of the Q wave and the end of the T wave. This includes all of ventricular depolarization and repolarization.

R wave The depolarization wave of the ventricle as shown in the ECG.

Receptor field Area of skin served by a single sensory nerve.

Receptor signaling An intracellular cascade of protein interactions that begins with the binding of a ligand to a membrane receptor.

Receptor tyrosine kinases Membrane receptor that is itself a kinase and phosphorylates itself.

Red oxidative fibers Skeletal muscle fibers containing myoglobin and numerous mitochondria. These fibers rely on oxidative phosphorylation for most of their ATP production and are used for sustained work.

Refractory period A time after the opening of an ion channel when it is not available to open. Sodium channels have a refractory period while the channel is inactivated. Refractory period generally refers to a time when a cell cannot be stimulated for another depolarization.

Renal blood flow The amount of blood that circulates around the nephrons.

Renin An enzyme produced at the kidney responsible for the cleavage of angiotensinogen into angiotensin I.

Resistin A hormone produced by adipose tissue that increases insulin resistance.

Respiratory bronchioles The most terminal of the bronchioles before the alveoli. These bronchioles are capable of some gas exchange.

Resting membrane potential The membrane potential of an electrically excitable cell at rest. The actual voltage of the resting membrane potential is cell-type specific.

Reticulocytes Immature red blood cells.

Retrograde transport Transport along the neuron, from the synaptic bulb to the cell body.

Rhodopsin The pigment portion of visual mechanism, derived from vitamin A.

Ribosomal RNA An important component of the ribosome. The RNA is noncoding.

Ribosomes A combination of proteins and RNA essential for protein synthesis.

Rigor mortis Rigor after death induced by the lack of ATP, which allows myosin head detachment from actin.

Ryanodine receptor The calcium release channel of the sarcoplasmic reticulum.

SA node The sinoatrial node; it establishes cardiac rhythm.

Saccules Portion of the inner ear and vestibular system that contain otoliths.

Sacral The lowest portion of the spinal cord.

Salivary amylase Enzyme in saliva that is capable of cleaving sugars.

Salivary glands Glands in the mouth that produce saliva.

Saltatory conduction The type of conduction that occurs in myelinated neurons.

Sarcoplasmic reticulum An endoplasmic reticulum located within muscle tissue. It has the additional function of serving as Ca^{2+} storage sites within muscle.

Satiety The sensation of having eaten enough food.

Scalene muscles Accessory respiratory muscles used during forced inspiration.

Schwann cells Cells that make myelin to cover neurons in the peripheral nervous system.

Secretin A hormone released from the duodenum in response to low pH. Secretin targets the stomach and reduces motility, chief cell secretion, and parietal cell secretion.

Segmentation Movement of chyme in both the forward and backwards direction that promotes mixing.

Semen Ejaculatory fluid from males that contains sperm, nutrients, and fluids from accessory glands.

Semicircular canals The anatomical structures of the vestibular system.

Seminal vesicles Accessory male reproductive glands that contribute high fructose fluid to semen.

Sensory cortex The portion of the cortex that receives somatic sensory input.

Sensory nerve pathways Pathways that carry sensory information from the receptors to the sensory cortex.

SERCA pump Ca^{2+} ATPase of the sarcoplasmic reticulum.

Serotonin A hormone and neurotransmitter that is important as a signaling molecule. It is released from platelets and causes vasoconstriction.

Sertoli cells Also known as sustentacular cells or nurse cells, these supporting cells form the blood–testes barrier and the environment necessary for developing spermatozoa.

Shear stress The stress placed on a vessel wall in a parallel direction to the wall; the force exerted along a surface, typically by a moving fluid.

Slow waves Subthreshold contractions of smooth muscle in the intestine.

Somatostatin A hormone that antagonizes growth hormone release.

Spatial summation Summing of action potentials over space.

Spermatid A haploid spermatozoan that is not yet mature.

Spermatozoan A male gamete.

Spinocerebellar tracts Sensory pathways with the muscle spindle fibers and Golgi tendon organs as their receptors, carrying information about body position and muscle contraction to the cerebellum.

Stereocilia Hair cells of the vestibular system.

Sternocleidomastoid muscles Accessory respiratory muscles used during forced inspiration.

Steroid hormones Hormones formed from the cholesterol. They are generally membrane permeant.

Stroke volume The amount of blood pumped out of the ventricles/beat.

Substance P A neurotransmitter of the analgesic system.

Sucrose A disaccharide.

Superior vena cava Great vessel that carries blood back to the right atrium from all blood vessels superior to it.

Suprachiasmatic nucleus A cluster of neurons thought to control diurnal rhythms.

Surface tension The tension that exists at an air–water interface.

Surfactant Dipalmitoylphosphatidylcholine, produced by intraalveolar cells to reduce pulmonary surface tension.

Sympathetic nervous system A branch of the autonomic nervous system sometimes called the system of "fight or flight."

Symport Transport of molecules in the same direction.

Synaptogamin An important protein in the regulation of vesicular movement of vesicles for exocytosis.

Syncytial trophoblast The extraembryonic membrane that invades the endometrium.

Syncytium Cells within a tissue working together as a single cell.

Systole Period during the contraction of the ventricle.

T cells Immune cells that are formed in the bone marrow but mature in the thymus.

Tastant Molecules that can be tasted.

Tectorial membrane Upper membrane of the organ of Corti that bends the hair cells and allows vibration to be transduced into action potentials.

Tectum Includes the roof of the midbrain and the corpus quadrigemina.

Temporal lobe The lateral lobe of the brain.

Temporal summation Summing of action potentials over time.

Testosterone Hormone produced by the Leydig cells of the male and the thecal cells of the female.

Thalamus Brain region of the diencephalon important for sensory nerve pathways and integration of motor pathways.

Thecal cells Follicle-associated cells of the female that produce testosterone in response to luteinizing hormone.

Thermoreceptors Receptors that sense temperature, either hot or cold.

Thromboxane A2 Released by platelets and causes vasoconstriction.

Thyroglobulin Transport protein for iodine with the thyroid gland.

Thyroid hormone Hormone responsible for regulating the body's metabolic rate. It acts by increasing the Na^+-K^+ ATPase, which increases ATP hydrolysis, and by increasing ATP synthetic enzymes.

Tissue osmotic pressure Pressure exerted by proteins and osmotically active particles within the tissue.

Total peripheral resistance Resistance to blood flow within the entire circulatory system.

Trachea The main airway connected to the pharynx and dividing into the right and left bronchi.

Transcription Formation of mRNA from DNA.

Transcription factor Molecule that initiates mRNA formation.

Transient calcium channel Voltage-gated calcium channel within the SA node that has a short open time.

Translation Formation of proteins from mRNA.

Trehalose A disaccharide.

Triceps The opposing muscle to the biceps. This muscle must relax when the biceps contracts.

Trigger calcium Extracellular calcium required for calcium release from the sarcoplasmic reticulum in the heart.

Triglycerides Fatty acids with a glycerol backbone.

Tropomyosin A protein associated with actin that covers the actin-binding site.

Troponin C A regulatory protein associated with tropomyosin that regulates the position of tropomyosin over the actin-binding site.

Tumor necrosis factor-α (TNF-α) A cytokine.

Turbulent flow Flow of a fluid in several directions, increasing the resistance to flow.

Tympanic membrane The eardrum.

Upregulation In regards to receptors, the systematic increase in the number of receptors in the plasma membrane.

Urea Benign metabolite of ammonia that is important in the concentration of urine in the nephron.

Ureters Paired tubular structures that carry urine from the kidney to the bladder.

Utricles Portion of the inner ear and vestibular system that contain otoliths.

Vaccine A portion of a virus or bacteria that is injected in order to stimulate antibody production by B cells.

Vasodilation Widening of blood vessels caused by relaxation of the surrounding smooth muscle.

Vena cava Either of two great veins bringing blood back to the right atrium.

Venoconstriction Constriction of venous vessels.

Venous return The amount of blood that returns to the heart during diastole.

Ventilation Moving air into and out of the lung.

Ventilation-perfusion inequalities Ventilation of nonperfused alveoli or perfusion of nonventilated alveoli.

Ventricles The two lower chambers of the heart.

Ventricular action potential The action potential of the ventricles of the heart, distinct from the atrial or SA nodal action potentials.

Vestibular system Sensory system responsible for balance and equilibrium.

Vestibulocochlear nerve Cranial nerve that carries input from the vestibular system and the cochlea to their respective nuclei.

Viagra A drug that causes vasodilation in the vessels serving erectile tissues in the male.

Villi Fingerlike projections of the luminal wall of the intestine.

Visceral pleura The portion of the pleural membrane attached to the lung.

Vitamin A A fat-soluble vitamin that is the precursor to rhodopsin.

Vitamin D A vitamin that increases Ca^{2+} uptake from the digestive tract.

VLDL Very-low-density lipoproteins.

VO_2 max The maximum volume of oxygen that an individual can take in and use, measured in liters per minute.

Voltage-gated Na^+ channel A sodium channel that is voltage-sensitive and important in the generation of nerve and muscle action potentials.

White glycolytic fibers Skeletal muscle fibers that rely on glycolysis for most of their ATP production. These fibers generate great power quickly and are used for short-duration burst activity.

Withdrawal reflex Nonvoluntary muscle response to noxious stimuli.

Z-disc A set of membrane-bound proteins that bind myosin and actin holding the sarcomere in register.

Zona pellucida The oocyte membrane.

Further Reading

Now that you've completed this human physiology text, perhaps you would like to continue your study of this subject. Listed below are some texts, monographs, and review papers that will expand your knowledge of physiology. Physiology is a vast discipline, and these readings will carry you further along the road to mastery.

Cellular Physiology

Plopper. *Principles of Cell Biology.* Jones & Bartlett Learning, 2013.

Cassimeris, Lingappa, and Plopper. *Lewin's Cells,* 3rd edition. Jones & Bartlett Learning, 2014.

Blaustein, Kao, and Matteson. *Cellular Physiology and Neurophysiology,* 2nd edition. The Mosby Physiology Monograph Series, 2012.

Keynes and Aidley. *Nerve and Muscle,* 3rd edition. Cambridge University Press, 2001. This is a classic work that goes into much detail on electrical signaling of the action potential.

Einstein. *Investigations on the Theory of the Brownian Movement.* Dover Books, 1956. This may seem like an intimidating addition to this list, but it is short, well written, and the best discussion of diffusion you are likely to find!

Autonomic Nervous System Physiology

Boron and Boulepaep. *Medical Physiology.* Elsevier, 2013. This book covers all of human physiology in great detail, and the section on the autonomic nervous system reveals much of the complexity of this system.

Endocrine Physiology

White and Porterfield. *Endocrine Physiology,* 4th edition. The Mosby Physiology Monograph Series, 2012.

Immune System Physiology

Owen, Punt, and Stanford. *Kuby Immunology,* 7th edition. W. H. Freeman, 2012.

Olefsky and Glass. Macrophages, inflammation and insulin resistance. *Annual Review of Physiology* 72:219–246, 2010.

Somatic Nervous System and Special Senses

Snell. *Clinical Neuroanatomy for Medical Students*, 7th edition. Lippincott Williams & Wilkins, 2009. This is a good reference for neuroanatomy and function. If the somatic nervous system interests you, this is a fine text.

Digestive Physiology

Leonard. *Gastrointestinal Physiology*, 7th edition. The Mosby Physiology Monograph Series, 2007.

Cardiovascular Physiology

Pappano and Wier. *Cardiovascular Physiology*, 10th edition. The Mosby Physiology Monograph Series, 2012.

Opie. *Heart Physiology from Cell to Circulation*, 4th edition. Lippincott Williams & Wilkins, 2004.

Galkin and Ley. Immune and inflammatory mechanisms of atherosclerosis. *Annual Review of Immunology* 27:165–197, 2009.

Respiratory Physiology

West. *Respiratory Physiology: The Essentials*, 9th edition. Lippincott Williams & Wilkins, 2012.

Cloutier. *Respiratory Physiology*. The Mosby Physiology Monograph Series, 2007.

Renal and Acid-Base Physiology

Koeppen and Stanton. *Renal Physiology*, 5th edition. The Mosby Physiology Monograph Series, 2012.

Reproductive Physiology

White and Porterfield. *Endocrine Physiology*, 4th edition. The Mosby Physiology Monograph Series, 2012.

Comprehensive Readings in Physiology

Boron and Boulepaep. *Medical Physiology*. Elsevier, 2013. This is an advanced text, but it does an excellent job of uniting cellular mechanisms with organismal consequences.

Koeppen and Stanton. *Berne & Levy Principles of Physiology*, 6th edition. Elsevier, 2010. This text is less cellular but places more emphasis on physical forces at the organ level. This is a classic approach to human physiology.

Costanzo. *Physiology*, 4th edition. Saunders, 2010. This is a shorter physiology text with some cellular mechanisms and some treatment of physical forces.

Golan, Tashjian, Armstrong, and Armstrong. *Principles of Pharmacology: The Pathophysiologic Basis of Drug Therapy*. Lippincott Williams & Wilkins, 2011. This text is a beautiful example of why a thorough understanding of physiology is essential for the study of pharmacology. It will improve your physiology skills and teach you pharmacology simultaneously.

Index

Note: Page numbers followed by *f* indicate material in figures.

D

E

G

J

K

L

M

N

S

W

Z